Indigenous People and Mobile Technologies

T0320266

In the rich tradition of mobile communication studies and new media, this volume examines how mobile technologies are being embraced by Indigenous people all over the world. As mobile phones have revolutionized society both in developed and developing countries, so Indigenous people are using mobile devices to bring their communities into the twenty-first century. The explosion of mobile devices and applications in Indigenous communities addresses issues of isolation and building an environment for the learning and sharing of knowledge, providing support for cultural and language revitalization, and offering the means for social and economic renewal. This book explores how mobile technologies are overcoming disadvantage and the tyrannies of distance, allowing benefits to flow directly to Indigenous people and bringing wide-ranging changes to their lives. It begins with general issues and theoretical perspectives followed by empirical case studies that include the establishment of Indigenous mobile networks and practices, mobile technologies for social change and, finally, the ways in which mobile technology is being used to sustain Indigenous culture and language.

Laurel Evelyn Dyson is a Senior Lecturer in Information Technology at the University of Technology, Sydney, and President of anzMLearn, the Australian and New Zealand Mobile Learning Group. Dr. Dyson's research interests center on Indigenous people's adoption of mobile technologies, as well as the use of mobile technologies in education.

Stephen Grant is a Lecturer at the University of Technology, Sydney. Since 2002 he has taken a key position in the Indigenous Participation in IT Program, UTS. He is one of a small number of qualified Indigenous IT professionals working in Australia. He researches mobile networks and autonomous systems.

Max Hendriks lectures in Internetworking at the University of Technology, Sydney. He has been an educator for more than 40 years. His research interests are in Internetworking, Indigenous people and their innovative use of technology. Of particular interest to him are security technologies within wireless networks.

Routledge Studies in New Media and Cyberculture

Indigenous People and Mobile Technologies

Edited by Laurel Evelyn Dyson,
Stephen Grant and Max Hendriks

Routledge
Taylor & Francis Group

NEW YORK AND LONDON

First published 2016
by Routledge
605 Third Avenue, New York, NY 10017

and by Routledge
4 Park Square, Milton Park, Abingdon, Oxon OX14 4RN

Routledge is an imprint of the Taylor & Francis Group, an informa business

Library of Congress Cataloging in Publication Data

Indigenous people and mobile technologies / edited by Laurel Evelyn Dyson, Stephen Grant, and Max Hendriks.
    pages cm. — (Routledge studies in new media and cyberculture; 31)
Includes bibliographical references and index.
    1. Indigenous peoples—Communication. 2. Mobile communication systems—Social aspects. 3. Internet and indigenous peoples. 4. Communication and technology.
    5. Information technology—Social aspects. I. Dyson, Laurel Evelyn, 1952- II. Grant, Stephen, 1965- III. Hendriks, Max A. N.
GN380.I348 2015
302.2'2—dc23   2015020158

ISBN: 978-1-138-79331-6 (hbk)
ISBN: 978-1-315-75936-4 (ebk)
ISBN: 978-0-8153-8653-7 (pbk)

Typeset in Sabon
by codeMantra

Contents

List of Figures and Tables

Figures

Tables

1 Framing the Indigenous Mobile Revolution

Laurel Evelyn Dyson

In 2014, the International Telecommunication Union (ITU) recorded that for the first time in history there were almost as many mobile (cell) phone subscriptions as people in the world. As we approach 100% global saturation, Brahima Sanou (ITU 2013, 1), Director of the ITU Telecommunication Development Bureau, describes the rise of mobile phones as the "mobile revolution … this mobile miracle".

Though much has been written about mobile communication and mobile media, one significant group—the Indigenous peoples of the world—has been largely ignored. Yet, Indigenous people are an integral part of the mobile revolution, using a variety of mobile technologies to bring their nations into the twenty-first century. The explosion of mobile devices and applications in Indigenous communities offers the potential to address issues of geographic isolation, build an environment for the learning and sharing of knowledge, provide support for cultural and language revitalization, and furnish a means for social and economic renewal. This book explores how mobile technologies are overcoming disadvantage and the tyrannies of distance, allowing benefits to flow directly to Indigenous people and bringing wide-ranging improvements to their lives.

As we come to the conclusion of the United Nations' Second International Decade of the World's Indigenous People in 2015 (UN 2008), it is opportune to examine the subject of mobile uptake by Indigenous people. For the first time since the computer age began mobile technologies offer a means to overcome the digital divide that has impacted so negatively on many communities. In order to be part of the global Information Society on their own terms and in response to their own needs Indigenous people have asserted their right to access information and communication technologies (ICTs), as well as the financial resources and knowledge to build and maintain their own infrastructure and applications (UNESCO 2003). Where fixed-line technologies failed Indigenous peoples, mobile technologies are appropriate to their geographic locations, their socio-economic circumstances and their cultures.

This book adopts a multidisciplinary approach as the issues go far beyond the purely technical. Chapter authors come from many different fields: from computer science and community informatics, from media studies and journalism, from linguistics and cultural studies, from health, education and ICT for development (ICT4D). Amongst them are a significant number of Indigenous authors, and those who are not Indigenous have many years experience working with communities to improve the lives of Indigenous people and achieve self-determination through mobile and other technologies. Their scholarly contributions cover both people living in remote regions of the planet and those who live in the cities and reach across six continents. The book thus aims to provide a comprehensive overview as well as in-depth accounts of specific issues and solutions.

We begin by examining what "Indigenous" means in the context of the modern world and show that Indigenous peoples, though now often economically, socially and geographically marginalized, are by no means an insignificant group either historically or in terms of their potential to offer sustainable solutions to some major challenges. The role of Indigenous people in the mobile revolution is examined; the technologies they use, the networks that serve them and the applications that fulfill their goals. Finally, we present an overview of the book, its main themes and chapters.

The World's First Nations

Indigenous peoples are, for the most part, those who have a historical connection with their territory that predates colonization and who thus see themselves as different from mainstream society (Martinez Cobo 1987). The United Nations (2009) conservatively estimates that there are more than 370 million Indigenous people living in some 90 countries spread across six continents. They comprise 4% of humanity and speak more than 4,000 of the world's 7,000 languages.

The Americas: In the USA Indigenous populations comprise Native Americans (or American Indians), Alaskan Natives and Hawaiians. More than 5 million people in the USA alone identify as Native or having Native heritage, with more than a third living in the four states of California, Oklahoma, Arizona and Texas (Norris, Vines and Hoeffel 2012). The Cherokee and Navajo are the most populous nations. In Canada, on the other hand, Aboriginal people total about 1.5 million (Statistics Canada 2013). They comprise three groups who are recognized separately by law: First Nations, Inuit and Métis. The latter originated in the 1600s when Aboriginal Canadians and European settlers intermarried, and they are seen today as having a distinct culture. Cree, Inuktitut and Ojibway are the native languages with the most speakers. By contrast, in Latin America, *Pueblos Indigenas* (Native Americans or Amerindians) and *Mestizos* (part Indigenous people) form the majority in a number of countries, for instance, Mexico, Peru, Bolivia, Guatemala, Ecuador and Paraguay (CIA 2014; CONAPO 2005; INE 2012). For this reason, they wield considerable political power in places such as Bolivia and

Ecuador. Moreover, they have been strong in asserting their cultures; for example, the official recognition in Mexico of native languages alongside Spanish and in Bolivia the flourishing of a vibrant Indigenous television, radio and film industry. Some languages, such as Quechua, Aymara and the Mayan languages, have millions of speakers.

Oceania: Beyond the Americas and stretching across the Pacific Ocean are the Polynesian, Melanesian and Micronesian nations. Many countries in the "liquid continent" (Bray-Crawford 1999) are amongst the smallest in the world and the most susceptible to rising sea levels and the threat of climate change (UN 2009). The nation of Tuvalu, for instance, has nine islands totaling a mere 26 square kilometers of land, the highest point of which is no more than 4.5 meters above sea level (IWGIA 2013). Many Oceania nations are in tropical regions of the Pacific and are subject to extreme weather events, such as cyclone Pam that devastated Vanuatu. Although lacking economic development, Pacific cultures are often strong, with some nations like Tonga never colonized and others such as Papua New Guinea recognizing customary land tenure and cultural practices in their constitution (Independent State of PNG 1975). Most Indigenous peoples in the region form the majority in their countries, although the Hawaiians and Maori are notable exceptions.

Australia: On the western shore of the Pacific, Australia's Indigenous population comprises the Aboriginal people of mainland Australia and the Torres Strait Islanders, whose traditional home is the archipelago lying between Cape York and Papua New Guinea to the north. It is estimated that there were about 600 Aboriginal clans in Australia at the time of colonization, speaking 250 distinct languages, each with several dialects (ACME et al. 2008; Walsh 1993). Today, there are more than half a million Indigenous people living in Australia (ABS 2012). Aboriginal Australians are believed to have the oldest living culture in the world, dating back some 60,000 years, but at the same time have been amongst the most active in embracing new technologies, such as television, video, community radio and now mobile technology.

Asia: The continent of Asia has the largest population of Indigenous people. China employs the term ethnic minorities, of which it recognizes 55, comprising 113 million people (IWGIA 2013). Also numerous are the Scheduled Tribes, or Adivasis, of India, who number more than 84 million, with millions of Indigenous people also in Nepal. In fact, there are numerous Indigenous minorities right across the continent, including the Nganasan and Dolgan in Siberia; the Ainu and Okinawans of Japan; many tribal minorities in Tibet, Taiwan and the Philippines; Indigenous fisher folk, hunter-gatherers and subsistence farmers in Vietnam, Thailand and Laos; the Orang Asli of the Malaysian peninsula; many Indigenous people in Borneo, both in Malaysian Sarawak and Indonesian Kalimantan, and a range of Indigenous peoples in other parts of Indonesia. Their socio-political and economic situations vary greatly, but many live in conflict with development interests, and many are not recognized by their national governments.

Africa: As in Asia, the Indigenous minorities of Africa are too numerous to list but include the San (previously known as "Kalahari Bushmen") of southern Africa, the Maasai of Kenya, the Forest Peoples (so-called "Pygmies") of the Congo and the Tuareg camel herders of the Sahel. Although diverse, their traditional pastoralist or hunter-gatherer lifestyles separate them from the sedentary agricultural peoples that now inhabit their territories. In addition, there are African communities that constitute the dominant majority in their countries but, through the era of European colonization, also experienced dispossession of land, deprivation of culture and disempowerment. These are included in this book because they live today with the colonial and postcolonial legacy.

Europe: There are relatively few Indigenous people still practising their customary way of life in Europe, and those who remain are confined to the north, chiefly the Sami, the reindeer herders and fishers of northern Scandinavia and neighboring Russia. Although in the past they suffered suppression of their culture and language, they now often enjoy a standard of living only slightly below their national average.

The Strength of Indigenous Mobile Participation

Despite the ITU's proclamation of the mobile revolution, not all people participate to the same degree. Compared with developed countries, where the subscription rate currently stands at about 121 per 100 inhabitants (including multiple subscriptions held by the same individual), in the developing world the figure is 90% overall and 69% in Africa (ITU 2014). While there are no official statistics available on Indigenous peoples' mobile uptake, research across several continents indicates that they are very much engaged with this technology, even if their rates of ownership are lower than the national average for their respective continents.

Of course, the mobile revolution is not purely a matter of numbers. It also represents a fundamental shift in the nature of people's access to ICTs as a whole. In industrialized countries the growth of mobile technology ownership by the mainstream population has been one of rapid evolution from fixed ICTs with a cabled connection (fixed-line telephones and Internet access via desktop computers) to the acquisition of mobile devices as a complement to existing ICTs (Castells et al. 2007; Katz 2008).

The inherent portability of mobile devices and the convenience of being able to stay in touch from any location is clearly a key to most people's decision to buy a mobile phone when they already live in ICT-rich home and work environments.

By contrast, for many Indigenous people in developing countries, the revolution has often involved the transformation from no ICT access to mobile ICTs (Donner 2008). This is the famous "leapfrog" phenomenon, much talked about in the early years of mobile diffusion, whereby communities on the wrong side of the digital divide skipped a whole generation of fixed technology (Castells et al. 2007). Mobile technology thus forms a substitute for traditional ICTs. A key factor here has been cost. From the Indigenous purchaser's viewpoint advantages include the cheapness of a mobile device compared to paying for a home phone connection or buying a computer, and the superior cost management features of prepaid mobile phones compared to the unpredictable cost of fixed-line bills (Brady and Dyson 2010; Castells et al. 2007). Additionally, for governments, national carriers and new telecommunication operators trying to break into the market, establishing wireless infrastructure is much more cost effective than laying cable in previously unserviced or poorly serviced regions of low population density.

While for many Indigenous people in developing countries mobile telephony is their only access to ICT, by contrast, in industrialized countries it is their only *personally owned* ICT. Their governments, in numerous cases, will have provided public payphones and shared public-access computers to their communities. In these cases, public facilities will still be actively used after the introduction of a mobile network, for example, for making cheaper payphone calls than can be made from their mobile phones or for editing multimedia content on the larger screen of public computers (Brady and Dyson 2009; Kral 2013). The fact that people with often limited incomes are prepared to purchase mobile devices and fund their use, even where there are alternative avenues of ICT access, demonstrates the attraction of this technology in enhancing Indigenous personal autonomy.

The Technologies Powering the Revolution

The key device behind the mobile revolution is obviously the mobile phone. In many parts of the developing world Indigenous people, like other poor people, have employed a variety of cost-effective strategies for acquiring mobile phones. These include used and sometimes damaged phones imported from developed countries and refurbished for cheap sale by former watchmakers and radio repairers, now turned mobile technicians (Hahn and Kibora 2008). Then there are phones presented as gifts by returning migrant workers. Increasingly, manufacturers and mobile providers have targeted the low-end market and imaginatively designed services with simple phones at rock-bottom prices and low-value phone cards. For example, since base stations were constructed along the Omo River in southern Ethiopia in recent years, Indigenous villagers have embraced mobile technology, pooling resources to purchase voice- and text-only phones that sell for under $15 (Max Hendriks personal communication). Many of these people have no electricity, but small portable solar panels prove an inexpensive way of charging their phones.

By contrast, Indigenous people living in wealthier countries often have access to the very latest mobile technology, despite their socio-economic disadvantage compared to the mainstream populace. This applies whether they live in urban, rural or remote areas. For example in Australia, beginning in 2008, a number of Indigenous communities in the Western Australian desert, Arnhem Land, Cape York and the Torres Strait Islands—in short, some of the most isolated parts of the continent—were given access to the then most advanced mobile telephony service when the national telecommunications carrier installed 3G (third generation, Internet-enabled) mobile towers (Auld, Snyder and Henderson 2012; Dyson and Brady 2013; Featherstone 2013). The morphing of mobile phones into Internet-enabled phones and converged devices of multiple functionality (Wilken and Goggin 2012) have provided Indigenous people with voice and video calls, text messaging, social networking, Internet browsing, TV and sports results, music and movies, photography, sound and video recording, contact lists and calendars and the no-cost short-range file-sharing platform Bluetooth. For Indigenous people the multimedia features may sometimes comprise the bulk of their mobile phone time, despite the attraction of calling, texting and social networking: listening to music or taking photos or videos are activities that are cost-free and can be conducted even if there is no remaining credit left on the phone, as well as providing an avenue of entertainment in isolated communities.

Since the advent of touchscreen technology, iPhones, tablet PCs and the touchscreen One Laptop Per Child (OLPC) XO laptops have become the centerpiece for many Indigenous cultural projects and education programs funded externally by governments, NGOs and corporate donors around the world. In the same way an earlier generation of computing devices became more widely accessible through the development of the graphical user interface in the 1980s, the gestural interface has made the new generation of touchscreen devices potentially more accessible to users in the twenty-first century (Burgess 2012). Particularly for Indigenous people, coming as they largely do from pre-literate cultures, direct manipulation of graphical displays on a screen is much easier than learning to type on a keyboard or use a mouse. Furthermore, these technologies offer a larger screen for viewing and creating multimedia content and social networking. In a remarkably short space of time touchscreen phones and tablets have now emerged in Indigenous communities for personal use (Figure 1.1), with cheaper models making them an attractive choice for people living on reduced incomes. Last year I shared a long bus trip with a young Sami girl who was snapping photographs of the Norwegian fjords on her tablet and showing me pictures of her reindeer and herself in national costume.

Figure 1.1 Woman from Poruma (Coconut) Island in the Torres Strait taking photographs with a smartphone. (Photo: Tyler Wellensiek).

Although most studies of the mobile revolution have focused solely on the phone, the MP3 player and iPod should not be forgotten. Music is such a fundamental form of human expression that MP3 players often received high levels of acceptance by Indigenous people before there was any cellular network in place. The affordance of the device for creating a private and mobile world of music wherever one goes, in opposition to the mobile phone's perpetual connectedness (Bull 2007), is valued by Indigenous people as much as by others. For adults on limited incomes or for children, the ownership of an MP3 player avoids the need to sign up for a phone plan and the recurring costs this inevitably brings.

The Networks Enabling Indigenous Mobile Use

The networks that serve Indigenous people vary greatly. For some living in remote regions there might be no access, although even in these cases mobile ownership can be an option, allowing Indigenous people to enjoy the multimedia features of their phones and store photos, music and videos even when out of network range (Featherstone 2013). In addition, a phone is useful for communication when visiting neighboring townships that have connectivity (Vaarzon-Morel 2014). By contrast, city-based individuals can, if they wish, use the latest 4G (fourth generation, fully IP (Internet Protocol) integrated) technologies for high data speeds when social networking and downloading multimedia content from the Web (Casado and Moreno 2008). However, 4G networks are unlikely to become a reality in the foreseeable future in remote regions due their more limited coverage.

In remote regions, it must be said that there is rarely a business case for installing a mobile network because earning a return on investment is unlikely due to low population density and lack of industry. Costs are also higher, with expensive satellite feeds sometimes the only way of getting connected to the outside world, even if the cheaper WiFi mesh technology is deployed to provide "last mile" delivery to schools, businesses and homes. Government subsidies and special programs are usually necessary, realized in the interests of equity and improving Indigenous welfare. For example, VisionOne is a US federally subsidized mobile phone service providing connectivity and low-cost phones to thousands of Native Americans living on tribal lands in Arizona, New Mexico and Utah who otherwise have little access to telecommunications (Castells et al. 2007; VisionOne 2015).

Community-owned wireless networks are a real option but again often installed with some form of subsidy. They enhance Indigenous control and decision-making to maximize social benefit, an instance of "self-determination applied to telecommunications" (O'Donnell et al. 2011, 1). A number of these networks have been established in both developed and developing countries. Often over time the network evolves, beginning with more easily achievable goals, such as party-line communication, where residents share one public channel and can hear everybody's conversations, and move toward standard mobile phone and laptop access (Absolhasan and Boustead 2007; Featherstone 2013). The Tribal Digital Village Network is an example in the USA, operating since 2002, initially supported by funding from the corporate giant Hewlett-Packard and enabling the connection of 18 remote Native American reservations that form the Southern Californian Tribal Community, stretching from the Mexican border to San Diego County (Luna 2002). This Wi-Fi broadband network has promoted expanded opportunities for communication of community news and events, documentation of cultural knowledge and language and education and training through its learning centers. Despite leveraging funding from external sources, control of the network has remained very much under the control of the community (McMahon 2011).

Mobile Applications to Serve Indigenous Cultural and Development Needs

Culture influences the way any technology is taken up, and mobile technology is no exception (Katz 2011). Interesting manifestations of this are the mobile phone covers, lanyards and iPad sleeves that various groups of

Indigenous women around the world have crafted over recent years for their own use or sale to tourists, adapting their traditional techniques of beading, weaving or painting (Figure 1.2).

Figure 1.2 Beaded mobile phone cover made by an Indigenous woman from Sarawak, Malaysia. (Photo: Laurel Evelyn Dyson).

There are many distinctive uses to which Indigenous people are putting mobile technology. As Baron (2008, 131) notes, the practices surrounding mobile phones are determined partly by the devices themselves and partly by "the cultural norms—or pragmatic necessities—of the society in which they are embedded." "Domestication" describes the particular ways in which a cultural group makes a technology its own, adapting the technology to its needs and preferences but also adapting its behavior to the technology. Evidence of Indigenous domestication of the mobile phone has long been documented, as the following examples will show.

The affordance of mobile devices for sound and image recording and viewing fits with the cultural traditions of Indigenous peoples. For those steeped in oral and artistic traditions, and with high rates of illiteracy in many communities, mobile phones, MP3 players and tablet PCs represent a marked advantage over desktop computers with their keyboard and literacy requirements. With many Indigenous cultures and languages eroded by colonization and now threatened further by globalization, the multimedia potential of mobile technologies for cultural and language documentation and revitalization has been recognized in recent years by many researchers. The Wadawurrung people in Victoria, Australia, are using rugged tablets enhanced with a map-based data-capture application for documenting significant sites on their traditional lands: this knowledge is then stored on a web-based cultural management system and can be used in negotiating land rights and resisting unwelcome development (Chester 2013). Other programs include the design of cheap, robust e-book readers for San communities in southern Africa in their own languages, and there is an iStore app and talking dictionary for the Tuvan community in southern Siberia (Traxler 2013).

Furthermore, the multimedia functions allow greater spontaneity and individuality of Indigenous media creation. Building on the historic successes of Indigenous involvement with television and community radio in countries such as Canada, Australia and Bolivia (Alia 2010), Indigenous people are now uploading their own content to popular Web 2.0 platforms such as Facebook or YouTube and sharing their music, photographs and videos widely in their own communities using Bluetooth (Kraal 2013). Where specialist recording skills and sophisticated equipment were once required for TV and radio production and broadcasting, now recording is being placed in the hands of any individual with a mobile device, resulting in informal records of the community's cultural practices and contemporary events. It gives them choices, allowing them to counterbalance the tide of incoming Western media and combine their traditions with the mainstream if they wish, bringing them into the global conversation. For example, in the Sahel in West Africa Tuareg musicians use their phones to make do-it-yourself recordings that mix their traditional guitar music with a drum machine and Bollywood-style autotune; these are disseminated all over Africa through Bluetooth or by swapping SIM cards and now globally on an American record label (Kirkley 2014).

Mobile technology, because of its inherent portability, lends itself to recording in the field, and this is now being used for keeping alive Indigenous people's accumulated knowledge of the environment and their deep understanding of how to live with nature in a sustainable way. Their holistic approach for the management of land and water may well provide lessons for the rest of the world to follow, faced as we are with enormous challenges of environmental degradation and loss of biodiversity (NCIV 2010; UNESCO 2003). Some organizations have been working with communities to build Indigenous environmental knowledge management systems to preserve the wisdom of the elders for future generations. CyberTracker is an environmental system developed in South Africa and deployed internationally to allow Indigenous rangers with low literacy levels to record wildlife sightings and help manage game (Jones and Marsden 2006).

While the preservation and revitalization of Indigenous knowledge, cultures and languages is of great importance, eliminating Indigenous poverty and raising standards of education and health is also vital in order to meet the Millennium Development Goals set by the United Nations. Because of Indigenous peoples' special relationship with their land and their obligations to "care for country", many live where their ancestors have dwelled since time immemorial. These are often remote, underdeveloped parcels of land with small populations, lacking infrastructure and modern amenities. Mobile networks have now penetrated many of these historically marginalized regions and act as enablers of services that were previously unavailable. Africa has often led the way in terms of these applications. In the area of education, South Africa has been at the forefront of mobile learning through innovative mathematics programs like Dr Math, which connects students with volunteer tutors through the mobile instant-messaging platform MXit; and literacy programs like Yoza (previously m4Lit), which supports literacy development by supplying young people with m-novels and conducting writing competitions (Traxler 2013). There are many health programs, including interactive digital stories accessed via mobile phone calls designed to develop better health awareness and SMS reminders to HIV patients to ensure treatment compliance (Bubenzer 2011; Mechael 2008). The contribution of mobile technologies for development (M4D) was recognized early in the adoption of mobile phones when, for example, Senegalese fishermen began phoning in the size of their catch before they disembarked so that the correct number of trucks and quantity of ice could be dispatched and waste avoided (Chéneau-Loquay 2001). Other examples include payment systems like M-PESA in Kenya, village-phone operator businesses such as MTN villagePhone in Uganda, and mobile access to online classifieds for job seekers (Castells et al. 2007; Gitau, Marsden and Donner 2010; Ling and Donner 2009).

Finally, the most obvious application of mobile technology for Indigenous people is communication. While there are those people who choose to remain on their ancestral lands, there are many who have moved away. For some the impetus has been conflict with rival tribes, for others competition for land and forced eviction during the colonial era. In the modern era many have gone to cities to seek the jobs, education and social services missing in their homelands. For example, in Mexico the urban drift comprises one in three Native Mexicans; in Bolivia, Brazil and Chile it is more than half the Indigenous population; more than three-quarters of US Native Americans and Alaskans live outside their tribal reservations, mostly in the cities (UN 2009; US Census Bureau 2012). One effect of the diaspora has been to create an unprecedented demand for cost-effective tools of long-distance communication to link those living in community with those who have moved away. Mobile phones fill this need.

Book Overview

The book comprises nineteen chapters, organized into four sections. The chapters seek to place mobile technologies within a general Indigenous context and deal with specific case studies. Both contribute to broader understandings of mobile adoption by Indigenous people.

Part I: Indigenous Mobile Technology Adoption and Theoretical Perspectives

Following the introductory chapter, my colleague Fiona Brady and I (*Chapter 2*) take up the theme "Why mobile?" What is it about mobile technologies that have made them the technologies of choice for Indigenous people all over the world? Drawing on research conducted since 2007 in four remote Indigenous Australian communities in Cape York and the Torres Strait Islands we note the importance of a rapidly evolving range of mobile technologies in people's lives, despite the fact that at the edge of the continent they often fail to work, particularly when most needed. We come to the conclusion that there is a complex interplay of factors that has led Indigenous people to choose mobile over fixed-line ICT, only some of which are identical to the reasons behind non-Indigenous people adopting mobiles. Socio-economic constraints and cultural preferences are obviously important as are aspects of mobile design, including the portability of the devices, billing structures and cost of hardware. These interact with aspects of place in sometimes unpredictable ways but still offer a decided advantage over previous ICTs.

In the next chapter, Pedro Ferreira and Kristina Höök (*Chapter 3*) present a fascinating study of "plei-plei" from Rah Island in the small Pacific nation of Vanuatu. We read about a wide range of playful behaviors as the inhabitants of Rah encounter mobile phones for the first time—tweaking the settings and appearance of their phones, constantly changing ringtones, playing with information in gossipy text messages on the "coconut wireless," physically "playing" with coverage as they climb palm trees or hike up mountains to find the best signal and use this new technology in their lives. The authors argue persuasively for the importance of play and aspects of play, such as games and entertainment, in the lives of all people, even those from resource-constrained communities. They warn against narrow agendas in ICT4D, which often deny the legitimacy of playful interactions in favor of so-called higher values. Ferreira and Höök emphasize the relevance of play for understanding technology adoption and conclude that, indeed, play may be even more important for people living in hardship as a way of escaping the realities of their lives.

Paul Kim, Karla Alfaro and Leigh Anne Miller (*Chapter 4*) examine the plight of Indigenous children in poverty-stricken regions of Latin America. They note that Latin America has the highest number of undocumented births and children not in school and the largest income inequalities outside of Africa, despite a stronger economy and smaller population. Growing up illiterate and unregistered with no official identity, Indigenous people have no access to even basic education and no voice as citizens of their countries of birth. However, mobile technology promises a way out: it has reached the poorest of the poor more quickly and profoundly than any other in human history. The authors report on the success of a number of mobile-based innovations around the world, including birth registration by text message. Mobile technology may be a tool powerful enough to serve as a catalyst for change, allowing Indigenous people in Latin America and elsewhere to realize their individual human rights to citizenship and access to education and other public services.

Gino Orticio (*Chapter 5*) moves the reader's attention to another part of the world and the iTadian people, who inhabit the mountainous Cordillera region of the Philippines. He invites controversy by critiquing current scholarship on ICT and Indigenous people and the four major metatheoretical approaches that characterize this area of research. He proposes Actor Network Theory (ANT) as a method for overcoming the assumptions that often limit current studies. He retraces the iTadian's current use of mobile phones to the pre-digital era when systems of communication existed and were reliable, even if much slower. ANT allows the author to follow the translations of the communication systems over time and note the role of mobile phones in replacing but also in stabilizing existing systems, bringing improvements but also creating challenges to iTadian ways of life.

Amanda Watson and Lee Duffield (*Chapter 6*) present another study of the introduction of mobile phones into a remote Indigenous community. Contrasting with Orticio's chapter, their research focuses on first contact with the technology. The villagers in Papua New Guinea had little access to modern ICTs, some still using traditional means such as wooden drums for sending messages. The expansion of mobile coverage brought a shift in the communal way of life in the villages from the public communication of the drum to the ability to communicate privately over wide distances. Instead of mobile phones being seen as replacements for traditional channels, the villagers saw the drums as still relevant, embodying as they did powerful spirits and linked to influential leaders in their communities.

Part II: Self-Determination for Indigenous People through Mobile Technologies

Moving to the important topic of network provision, without which Indigenous people would remain excluded from the mobile revolution, Brian Beaton, Terence Burnard, Adi Linden and Susan O'Donnell (*Chapter 7*) present a case study of Indigenous self-determination in action: Keewaytinook Mobile (KM). This is a community built, owned and managed mobile phone service located in the Hudson Bay area of Northern Ontario, home to Canada's most remote First Nations. The authors detail how KM supports partner First Nations communities in the region to develop and operate their own local mobile service, responding to the unique needs and business philosophy of their communities. The case study presents a strong argument for Indigenous control over the technologies that affect their lives and the advantages that can flow from this in terms of addressing critical safety and development requirements facing all remote and rural communities.

On a different theme, Ivo Burum (*Chapter 8*) presents a program in which Aboriginal people from isolated communities in the Northern Territory of Australia are trained as mojos (mobile journalists), recording, editing and publishing news stories directly from their iPhones to the Internet. Mobile journalism is a means of remedying the lack of diversity of opinion about marginalized Indigenous communities and reversing the globalizing trend by which Indigenous people are often swamped with Western media. Mobile journalism practices in isolated Indigenous communities can provide the skills and technologies to empower Indigenous people to tell local stories from their own perspective and create new job opportunities through local enterprise and in partnership with national media. The author outlines a process for introducing mobile journalism education and technologies to Indigenous people living in remote communities and presents a thorough evaluation of a mojo initiative he led.

Scandinavia was one of the world's leaders in introducing cellular networks—Sweden as early as 1955 (Jessop 2006)—and thus it is not surprising to find the Sami embracing mobile technologies to achieve their goals. Coppélie Cocq (*Chapter 9*) questions the notion of "anywhere any time", noting how mobile technologies can more firmly fix Indigenous people's relationship with their land. In an era when most Sami live in urban areas, mobile applications using mapping tools and augmented reality can allow Sami to digitally access the places where their identity is rooted and also increase Sami visibility. This chapter presents InSight Sápmi, a mobile application that recreates a Sami linguistic landscape and demonstrates the potential for Indigenous empowerment in the context of their community.

Part III: Mobiles for Health, Education and Development

The next five chapters deal with the practical issues of improving health, education and socio-economic development, vital concerns for people who often suffer from poor health, low formal educational attainment and poor employment prospects. Stephanie Craig Rushing and her co-researchers (*Chapter 10*), working in the Pacific Northwest of the USA, present a suite of six health applications to promote sexual health and wellbeing amongst American Indian and Alaska Native teens and young adults using mobile, online and multimedia technologies. The programs build on the widespread use of media technologies and mobile health information seeking behavior among this population. The authors demonstrate how technology-based programs can be tailored to the maturity level and interests of the individual, disseminated broadly regardless of geographic location and privately accessed when and where the individual is ready using laptops, mobile phones and tablets, as well as streaming videos. They share lessons learned during the development process and stress the importance of community-based participation to maximize cultural alignment between a program and its target group.

Moving to education, school teacher Lisa Switalla-Byers (*Chapter 11*) presents a mobile learning case study. She describes the use of podcasts to improve the pronunciation of the Maori language as part of an integrated art and storytelling program at her New Zealand primary school. The children practiced hearing, speaking, reading and writing Maori and recorded a speech of greeting and introduction (*mihi*) as a podcast. A strong message from this study is the need to integrate the learning of culture and language: in Maori, *Me ōna tikanaga me ōna reo*—the language of Maori and the culture of Maori are linked. A particular emphasis was on facilitating quality feedback and explicitly building children's ability to reflect: podcasts were the perfect medium to facilitate this because the children could listen to themselves, reflect on what they had heard and improve their pronunciation, and podcasts enabled another child to listen and suggest changes. Furthermore, the podcasts could be shared with family and friends. The chapter makes recommendations on how teachers can enhance students' reflective skills and learning strategies within the classroom, encouraging students to self-monitor, be persistent and take personal responsibility for their learning. The study has important implications for Indigenous language revitalization through mobile learning.

In the second chapter on the use of mobile technologies in education, Maria Augusti and Doreen Mushi (*Chapter 12*) present a project for integrating multimedia into open and distance learning materials at the Open University of Tanzania. Mobile phones have become such a part of the daily life of people in Africa that adopting mobile learning seems an obvious choice. Multimedia learning resources were produced for viewing on mobile phones to enhance access for students located in urban, rural and Indigenous communities. An interesting aspect of the project was the incorporation of sign language into the resources for those with special needs. The authors describe how the m-learning application and its embedded multimedia elements improved the level of access of course materials for students located in areas where ICT infrastructure is poor and Internet connectivity is limited. The chapter also captures the challenges faced during implementation and evaluates the results of the implemented solution.

The concluding chapters in this section focus on development and, like education, mobile technologies are an ideal platform, particularly for Indigenous programs in the developing world. Lorenzo Dalvit (*Chapter 13*) presents the eight-year long experience of the Siyakhula Living Lab, a holistic and multi-disciplinary ICT4D project spearheaded by two South African universities, Rhodes University and the University of Fort Hare, in partnership with the local community. The Siyakhula Living Lab seeks to explore the potential of new technologies for socio-economic empowerment in a group of villages in Dwesa, a rural area representative of much of Africa. The stories of six community members presented in this chapter highlight different dimensions of empowerment through mobile phones. The author concludes that, whereas the overwhelming ICT4D focus has been on providing new e-government services and stimulating income generation, significant improvements in Indigenous people's lives can be affected by more efficient use of resources and the ability to cut costs—strategies mediated by mobile phones.

Sojen Pradhan and Gyanendra Bajracharya (*Chapter 14*) explore the link between socio-economic status and access to mobile phones of people belonging to the four major Indigenous groups of Nepal: the Magar, Tharu, Tamang and Newar. They use data from the Nepalese Population and Housing Census 2011 to investigate the relationship between the adoption of mobile phones and literacy, place of residence and economic activity. The authors found that there was no significant difference in mobile phone adoption rates between Indigenous and non-Indigenous nationalities in Nepal. In fact, Indigenous people's ownership of mobile phones was slightly higher than that of non-Indigenous people due to several social-economic factors such as types of occupation and proximity to the capital. An interesting feature of their study was interviews with Indigenous people that showed mobile phones being used in income-generating activities and to assist them in their employment.

Part IV: Cultural and Language Revitalization through Mobile Technologies

The final section deals with the important issue of Indigenous culture and languages, which lie at the heart of Indigenous identity and to which mobile technologies can contribute greatly. The first of these chapters, by Native American academic Kevin Kemper (*Chapter 15*), traces the history of the Cherokee language over the past 200 years, from word-of-mouth to written, then wired and now wireless forms. Kemper argues that this evolution represents the exercise of "rhetorical sovereignty" by the Cherokee people. He first discusses the origins of the language and the invention by Cherokee scholar Sequoyah of the syllabary for writing down the language, which is used in the bilingual publication *The Cherokee Phoenix*, the first and longest-running Native American newspaper. The insertion of Cherokee syllabary and language into mobile applications, games and websites accessible from mobile phones, shows the cultural resilience of the Cherokee people. Exploring this Cherokee colonization of mobile space, the author notes the cultural hybridity that has kept Cherokee strong and will help preserve the language and culture for centuries to come.

The chapter by Tariq Zaman, Narayanan Kulathuramaiyer and Alvin Yeo Wee (*Chapter 16*) deals with the utility of tablet PCs for cultural revitalization and transmission of knowledge from elders to the younger generation. Specifically, they report on a project in which researchers worked with the Penan community in Sarawak, Malaysia, to co-design and build technology to manage their sacred botanical knowledge through proper recording and documentation. Tablets loaded with specially developed software can be used to collect botanical data for verification at community meetings and then uploaded to the local database for use by the Penan people. A major part of the project involved the development of research methodologies capable of engaging project participants and mobilizing their knowledge resources.

Languages lie at the core of cultural transmission, and support for Indigenous language revitalization can be provided through both Web and mobile technologies. This is the topic of the chapter by Peter Brand, Tracey Herbert and Shaylene Boechler (*Chapter 17*), who present the FirstVoices suite of technologies developed for language documentation and "vitalization" by the First Peoples' Cultural Council in British Columbia, Canada. Communities documenting their languages in the FirstVoices database have the option of repurposing their collections as mobile dictionaries and phrase books for the iPhone, iPod or iPad. Applications include a package delivering language lessons to the iPad and an Indigenous language texting app for Facebook, Chat and Google Talk, with keypads capable of texting in more than 100 Indigenous languages from Canada, the USA, Australia and New Zealand. Since the early days of the mobile revolution, the First Peoples' Cultural Council has pioneered the application of these technologies for the benefit of Indigenous people wishing to document, teach and learn their heritage languages. The authors explore the importance of Indigenous-led innovation in the development of language and cultural resources.

Olga Temple (*Chapter 18*) presents a fascinating analysis of thousands of text messages collected in Papua New Guinea over several years. Her aim was to record changes in the local languages, in particular those spoken in the villages and the more widely understood Creole and English. Her study shows that neologisms are far less frequent in the more localized village languages. Using a dialectical method of linguistic analysis, she discusses the relationship between language and culture and between the individual and society. She questions the conventional concepts of national identity, endangered languages and cultural preservation. She recognizes the value of English language competence for economic development and that more widely understood languages will necessarily be privileged due to the need for a common language of communication in this linguistically diverse country. Her research shows how mobile phones are acting as a catalyst in this trend toward the Global Village.

In our final chapter Traci Morris (*Chapter 19*) examines the uses of technology in Native American language revitalization by both Chickasaw tribal citizens and the Chickasaw Nation government. The Chickasaw tribal citizens, independent of the Chickasaw Nation, are using social media to create content and share access to cultural and linguistic programs. The Official Chickasaw Nation Language Revitalization Program has developed both mobile and desktop technologies in order to perpetuate the Chickasaw Language and to serve those within the tribal lands as well as those who live elsewhere.

To conclude, the *Epilogue* presents the editors' vision for an Indigenous mobile future. We summarize the findings of the book and speculate on the possible directions for mobile technologies in Indigenous communities and provide insights into how these might empower Indigenous people to become once more nations of cultural vigor and social and economic well-being.

References

ABS (Australian Bureau of Statistics). *2075.0—Census of Population and Housing—Counts of Aboriginal and Torres Strait Islander Australians, 2011.* Canberra: Australian Bureau of Statistics, 2012.

Absolhasan, Mehran, and Paul Boustead. "UHF-Based Community Voice Service in Ngannyatjarra Lands of Australia." In *Information Technology and Indigenous People*, edited by Laurel Evelyn Dyson, Max Hendriks and Stephen Grant, 295–297. Hershey, PA: Information Science Publishing (InfoSci), 2007.

ACME et al. *Australian Indigenous Cultural Heritage.* Canberra: Australian Government, 2008. Last updated Jan 7, 2008. http://australia.gov.au/about-australia/australian-story/austn-indigenous-cultural-heritage.

Alia, Valerie. *The New Media Nation: Indigenous Peoples and Global Communication.* New York and Oxford: Berghahn Books, 2010.

Auld, Glenn, Ilana Snyder and Michael Henderson. "Using Mobile Phones as Placed Resources for Literacy Learning in a Remote Indigenous Community in Australia." *Language Learning and Education* 26, no. 4 (2012): 279–296.

Baron, Naomi S. *Always on: Language in an Online and Mobile World.* Oxford: Oxford University Press, 2008.

Brady, Fiona, and Laurel Evelyn Dyson. *Report to Wujal Wujal Aboriginal Shire Council on Mobile Technology in the Bloomfield River Valley.* Wujal Wujal, Qld: Wujal Wujal Aboriginal Shire Council, June, 2009.

Brady, Fiona, and Laurel Evelyn Dyson. "A Comparative Study of Mobile Technology Adoption in Remote Australia." *Proceedings of the Seventh International Conference on Cultural Attitudes towards Technology and Communication (CATaC)*, Vancouver, Canada, June15–18, 2010. Murdoch, WA: Murdoch University, 69–83.

Bray-Crawford, Kekula P. "The Ho'okele Netwarriors in the Liquid Continent." In *Women@Internet: Creating New Cultures in Cyberspace*, edited by Wendy Harcourt, 1999, 162–172. London and New York: Zed Books.

Bubenzer Arnold. "Story-Based Mobile Learning with GSM Phones: Using "Old Tech" to Reach Audiences in Threshold and Developing Countries." *Proceedings of World Conference on Educational Multimedia, Hypermedia and Telecommunications 2011*, 348–350. Chesapeake, VA: AACE.

Bull, Michael. *Sound Moves: iPod Culture and Urban Experience.* London and New York: Routledge, 2007.

Burgess, Jean. "The iPhone Moment, the Apple Brand, and the Creative Consumer: From "Hackability and Usability" to Cultural Generativity." In *Studying Mobile Media: Cultural Technologies, Mobile Communication and the iPhone,* edited by Larissa Hjorth, Jean Burgess & Ingrid Richardson, 28–42. New York: Routledge, 2012.

Casado, Antonio C., and Jose I. Moreno. "4G Systems: Multimedia Content Transmission." In *Encyclopedia of Wireless and Mobile Communications,* vol. 1, edited by Borko Furht, 26–39. Boca Raton and New York: Auerbach, 2008.

Castells, Manuel, Mireia Fernandez-Ardievol, Jack Linchuan Qiu and Araba Sey. *Mobile Communication and Society: A Global Perspective.* Cambridge, MA: MIT Press, 2007.

Chéneau-Loquay, Annie. "Les Territoires de la Téléphonie Mobile en Afrique [Territories of the Mobile Phone in Africa]." *NETCOM* 15, no. 1–2 (2001): 1–11.

Chester, Simon. "Preserving Australia's Cultural Heritage." *Position* 64, April-May (2013): 38–41.

CIA. *The World Factbook.* Last modified June 20, 2014. https://www.cia.gov/library/publications/the-world-factbook/geos/pe.html#People.

CONAPO. *Proyeccionaes de Indígenas de México y de las Entidades Federativas 2000–2010.* Mexico: National Council on Population, 2005.

Donner, Jonathan. "Shrinking Fourth World? Mobiles, Development, and Inclusion." In *Handbook of Mobile Communication Studies,* edited by James E. Katz, 29–42, Cambridge, MA: MIT Press, 2008.

Dyson, Laurel Evelyn, and Fiona Brady. "Mobile Phone Adoption and Use in Lockhart River Aboriginal Community." *Proceedings of the 8th IEEE International Conference on Mobile Business,* Dalian, China, June 27–28, 2009, 170–175.

Dyson, Laurel Evelyn, and Fiona Brady. "A Study of Mobile Technology in a Cape York Community: Its Reality Today and Potential for the Future." In *Information Technology and Indigenous Communities,* edited by Lyndon Ormond-Parker, Aaron Corn, Cressida Fforde, Kazuko Obato and Sandy O'Sullivan, 9–26. Canberra: Australian Institute of Aboriginal and Torres Strait Islander Studies (AIATSIS), 2013.

Featherstone, Daniel. "The Aboriginal Invention of Broadband: How Yarnangu are Using ICTs in the Ngaanyatjarra Lands of Western Australia." In *Information Technology and Indigenous Communities,* edited by Lyndon Ormond-Parker, Aaron Corn, Cressida Fforde, Kazuko Obato and Sandy O'Sullivan, 27–52. Canberra: Australian Institute of Aboriginal and Torres Strait Islander Studies (AIATSIS), 2013.

Gitau, Shikoh, Gary Marsden and Jonathan Donner. "After Access—Challenges facing Mobile-Only Internet Users in the Developing World." *Proceedings of the 28th International Conference on Human Factors in Computing Systems,* Atlanta, Georgia, April 10–15, 2010, 2603–2606.

Hahn, Hans Peter, and Ludovic Kibora. "The Domestication of the Mobile Phone: Oral Society and New ICT in Burkina Faso." *Journal of Modern African Studies* 46, no. 1 (2008): 87–109.

Håkansson, Ann-Kristin, and Kenneth Deer. "Indigenous Peoples and ICTs: Millennium Development Goal 8 and the Information Society." Paper presented at the 1st session of the Global Alliance for ICT and Development, June 2006.

Independent State of Papua New Guinea. *Constitution of the Independent State of Papua New Guinea.* Port Moresby, PNG, 1975.

INE (National Institute of Statistics). *Bolivia: Características de población y vivienda. Censo nacional de población y vivenda 2012,* 2012.

International Telecommunication Union (ITU). *The World in 2013: ICT Facts and Figures.* Geneva, 2013.

International Telecommunication Union (ITU). *The World in 2014: ICT Facts and Figures.* Geneva, 2014.

IWGIA (International Work Group for Indigenous Affairs) *The Indigenous World 2013.* Copenhagen, 2013.

Jessop, Glenn. "A Brief History of Mobile Telephony: The Story of Phones and Cars." *Southern Review: Communication, Politics & Culture* 38, no. 3 (2006): 43–60.

Jones, Matt, and Gary Marsden. *Mobile Interaction Design.* Chichester, UK: John Wiley, 2006.

Katz, James E. (ed.). *Handbook of Mobile Communication Studies.* Cambridge, MA: MIT Press, 2008.

Katz, James E. (ed.). *Mobile Communication: Dimensions of Social Policy.* New Brunswick, USA: Transaction Publishers, 2011.

Kral, Inge. "The Acquisition of Media as Cultural Practice: Remote Indigenous Youth and New Digital Technologies." In Information Technology and Indigenous Communities, edited by Lyndon Ormond-Parker, Aaron Corn, Cressida Fforde, Kazuko Obato and Sandy O'Sullivan, 53–73. Canberra: AIATSIS Research Publications, 2013.

Kirkley, Christopher. "Musique, Téléphone Mobile et Identité Tourègue au Sahel [Music, the Mobile Telephone and Tuareg Identity in the Sahel]." *Téléphone Mobile et Création* [The Mobile Telephone and Creation], edited by Laurence Allard, Laurent Creton and Roger Odin, 2014, 107–116. Paris: Armand Colin.

Ling, Rich, and Jonathan Donner. *Mobile Communication.* Cambridge, UK: Polity Press, 2009.

Luna, Lynnette. "Tribal Gathering." *Telephony* 242, (2002, June 3): 100, 102.

McMahon, Rob. "The Institutional Development of Indigenous Broadband Infrastructure in Canada and the United States: Two Paths to 'Digital Self-Determination.'" *Canadian Journal of Communication* 36, no. 1(2011): 115–140.

Martinez Cobo, José R. *Study of the Problem of Discrimination against Indigenous Populations.* New York: United Nations, 1987.

Mechael, Patricia. "Health Services and Mobiles." In *Handbook of Mobile Communication Studies,* edited by James E. Katz, 91–103. Cambridge, MA: MIT Press, 2008.

NCIV (Netherlands Centre for Indigenous Peoples). "About Indigenous Peoples." Last modified November 1, 2010. http://indigenouspeoples.nl/indigenous-peoples/about-indigenous-peoples.

Norris, Tina, Paula L. Vines and Elizabeth M. Hoeffel. *The American Indian and Alaska Native Population: 2010 Census Briefs.* United States Census Bureau, January 1, 2012. Accessed March 25, 2015. http://www.census.gov/prod/cen2010/briefs/c2010br-10.pdf.

O'Donnell, Susan, George Kakekaspan, Brian Walmark, Raymond Mason and Michael Mak. "Keewaytinook Mobile in Fort Severn First Nation." *Proceedings of the Canadian Communication Association Conference,* Fredericton, Canada, 2011, 1–20.

Statistics Canada. *2011 National Household Survey: Aboriginal Peoples in Canada: First Nations People, Métis and Inuit.* Ottawa, Ontario, 2013.

Traxler, John. Mobile Learning across Developing and Developed Worlds: Tackling Distance, Digital Divides, Disadvantage, Disenfranchisement. In *Handbook of Mobile Learning*, edited by Zane L. Berge and Lin Y. Muilenburg, 129–141. New York & London: Routledge, 2013.

UNESCO. *Indigenous Position Paper for the World Summit on the Information Society (WSIS).* Geneva, 2003.

UNESCO *Atlas of the World's Languages in Danger.* Paris, 2010.

United Nations. *Second International Decade of the World's Indigenous People.* New York, 2008.

United Nations. *State of the World's Indigenous Peoples.* New York: Secretariat of the United Nations Permanent Forum on Indigenous Issues, 2009.

Vaarzon-Morel, Petronella. "Pointing the Phone: Transforming Technologies and Social Relations among Warlpiri." *The Australian Journal of Anthropology* 25 (2014): 239–255.

VisionOne. Accessed March 28, 2015. https://www.cellularoneonline.com/vision-one.

Walsh, Michael. (1993). "Languages and their Status in Aboriginal Australia." In *Language and Culture in Aboriginal Australia*, edited by Michael Walsh and Colin Yallop. Canberra: AIATSIS, 1993.

Wilken, Rowan, and Gerard Goggin. "Mobilizing Place: Conceptual Currents and Controversies." In *Mobile Technology and Place*, edited by Rowan Wilken and Gerard Goggin, 3–25. New York and London: Routledge, 2012.

Part I
Indigenous Mobile Technology Adoption and Theoretical Perspectives

2 Why Mobile? Indigenous People and Mobile Technologies at the Edge

Fiona Brady and Laurel Evelyn Dyson

Invoking the Edge

We borrow the notion of "edge" to reference how differently things work in communities beyond the periphery (Carson et al. 2011). Indigenous communities frequently lie at the edge geographically, pushed to the outermost rim of the "range" of human habitation. Moreover, while at the center of their universe on their traditional lands, they also live at the edge of the dominant culture socially, culturally and economically. Government categorizations of remote, regional or urban mean little to Indigenous people, moving as they often do between locations, their ways of being (culture) always eluding mainstream discourse, power and priorities.

The edge is a place where Western, First World imperatives and explanations have a weaker grip. New concepts are needed. Jon Altman (2001) brought the term "hybrid economy" into use to describe the mix of customary practices (hunting, foraging, arts and crafts), public support (government funding in the form of welfare programs and services, as well as capital and recurrent expenditure on infrastructure) and private economic activity in remote Aboriginal Australian communities. These are in contrast to the prevalence of private enterprise and a relatively small government sector in urban and regional centers. People at the edge navigate life between competing systems, the pull of the new, and resistance from the embedded practices of their place. Being at the edge is precarious.

In this chapter we explore the question "Why mobile?" Mobile technology has been adopted enthusiastically in remote Indigenous communities, whereas previous attempts to introduce other information and communication technologies (ICTs) have been relatively unsuccessful. We also ask whether, compared to major population centers, mobile technology is used differently in remote communities and in remote Indigenous communities in particular. We explore these questions with specific reference to the Australian context, but our discussion will be relevant to many Indigenous people around the world—communities at the edge in Canada, Alaska, Africa, Oceania and elsewhere. An equally valid question is, "Why not fixed-line technologies?" In fact, we used to often ask ourselves this when bringing computer projects to a successful close—carefully negotiated in the initial stages with the Indigenous participants, answering an identified need and providing a good fit with presumed cultural strengths—and yet they died. There was no life beyond the end of the project. Governments, no doubt, have also puzzled over this as they fund just such well-intentioned projects; finance endless training and incentive schemes to encourage Indigenous people to take up landline phones, home computers or satellite Internet connections; or conduct reviews of why these have habitually failed.

While acknowledging the special challenges of providing ICTs for Indigenous people in remote areas, it was only in the opening decade of the twenty-first century that mobile technology provided an alternative. In 2002 it was noted that about 200 Indigenous Australian communities from urban to remote locations enjoyed mobile coverage, in 2008 the Federal Australian Government identified mobile networks as an opportunity to introduce mobile broadband services, and in that same year the Northern Territory Government defined mobile phones as "the product of choice in remote, and particularly, Indigenous communities" (ACMA 2008; DCITA 2002; RTIRC 2008, 75).

In order to answer the question of Indigenous people's relationship with mobile technologies we travel to the Far North of Australia, a region comprising nine discreet Aboriginal communities on the mainland of Cape York, as well as the Torres Strait Islands off the coast with their Melanesian communities. We draw on many years of living and working in the Far North, as well as ICT studies of four Indigenous communities conducted between 2007 and 2015: the island of Dauan in the Torres Strait (population 150) and in Cape York, Lockhart River, Wujal Wujal and Kowanyama (populations 542, 305 and 1,200 respectively) (Figure 2.1). These form an ideal place to

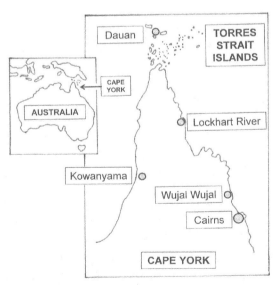

Figure 2.1 Location of the communities in Cape York. (Map: the authors).

explore Indigenous people's motivation for their technology choices because they are all provided with a range of ICTs, including mobile technologies. In addition, a point of comparison is offered by the non-Indigenous people who live in the communities—either permanently or temporarily—most of whom come to work in the businesses and government services. Wujal Wujal, which lies next door to the mainly non-Aboriginal community of Bloomfield, both occupying the same valley, provides a further opportunity to explore whether the reasons Indigenous people have embraced mobile technology, like no other previous ICT, are peculiar to them as a group. All of these communities are indeed at the edge, located in remote or very remote regions, with limited formal economies.

Mobiles at the Edge: "It's the Go!"

To give some idea of the degree to which mobile technologies—mobile phones, tablet PCs, MP3 players and mobile broadband—have enrolled themselves in the lives of Indigenous people, we will consider two examples: one of technology up-take in the early days of mobile phones in the Torres Strait and then a more detailed account of mobile technology recently at Kowanyama. Both demonstrate the high value placed on these new tools, operating from mobile networks installed through government funding as there is no business case for building them in such remote regions.

A CDMA (Code Division Multiple Access) service providing voice calls and text messages was rolled out in the Cape York communities in 2003 (Pearce 2003), but the islands of the Torres Strait had to wait two years more (Brady, Dyson and Asela 2008). Many Torres Strait Islanders have family and friends on the mainland and so were aware of the change before the mobile service was available at home. Some people were using mobile phones on the inner islands where they could pick up the signal to communicate with relatives on the mainland. More remarkably, many people on Dauan (a small outer island) had bought mobile phones even before their service was switched on, even though they were then of no use. Later, within a few short weeks of the implementation of the mobile network in 2005, it was observed that most adults in the community had a phone (Brady, Dyson and Asela 2008). Outer Islanders are mobile people: they move to the mainland for employment or health services, and children must travel away to boarding schools for their secondary education. Communication becomes an imperative for the Torres Strait diaspora to maintain contact and keep their cultural and language links. With only a minority of households in the Torres Strait subscribing to a home phone service, mobiles are the communication tools that allow this to happen. As one young woman on Dauan said, "Everyone I want to talk to has mobiles. Very hard to contact people otherwise."

In Kowanyama, a decade later, the network had evolved (Brady and Dyson 2014). Like all the Cape York and Torres Strait communities, a 3G (third generation, Internet enabled) service had replaced the older CDMA network in 2008. The 3G service had the advantages of providing faster Internet access and enabling mobile broadband. It meant that now all phones came with multimedia functions (camera, video, sound recording) and by 2014 touchscreen functionality (Figure 2.2) Sales of mobile phones were and continue to be phenomenal, not to mention regular sales of MP3 players, laptops and most recently cheaper model tablet PCs: the Post Office reported that it is selling 40 to 50 mobile phones per month, while the Kowanyama Store sells 3 to 4 phones per day. The basic and almost universal 3G multifunction phone of 2008, on sale then for $130, has translated increasingly into the once luxurious flip phones presently at $100 to touchscreen smartphones from $79 for the LG brand to the almost $600 iPhone, not to mention MP3 players, which can be had for as little as $20. The Pendopad Windows tablets are now available from $99 and are apparently in strong demand: an initial four ordered by the Post Office in February 2015 sold out in the first week.

Figure 2.2 Woman receiving training in touchscreen phones and iPads at Kowanyama, 2015. The smartphone with the cover is hers. (Photo: Kelli Boultbee).

By comparing mobile phone sales with the population of Kowanyama it is easy to see that there is a considerable churn rate, with devices damaged, stolen or passed on to family or friends. Even taking this into account, ownership is high. One manager described the dependence of Kowanyama residents on their mobiles: "They're vital. ... Their only form of communication to the outside world is their mobile." Some elderly and middle-aged people may not have phones but receive important messages via a relative who owns a mobile and acts as a contact point. Likewise, young children often do not have their own phones, but instead might be allowed to play with their parents' phone. When children reach high-school age and attend boarding school, 85 to 95% of them are reported to have mobile phones as an essential means of keeping in contact with their parents and family while they are away from country. As people under 18 years of age are not allowed to buy phones, these are obviously bought for them by their families.

What do the statistics say about people and mobile technology? Attempts at quantifying Indigenous Australian ownership of mobile phones have been reasonably consistent across several studies undertaken in remote communities: 55 to 58% (Auld, Snyder and Henderson 2012; Brady and Dyson 2009; Dyson and Brady 2009; Tangentyere Council and Central Land Council 2007). This figure is lower than the national average but conceals the impact of Indigenous norms of reciprocity, which result in much (but not universal) sharing of phones. Auld, Snyder and Henderson (2012) reckoned that access to a mobile phone when people needed one was pretty well ubiquitous. As one of the women we interviewed at Wujal Wujal said, when queried about her ownership, "A mobile phone is the go!"

Mobiles are one of the two technologies that have revolutionized life on the Cape. The first was cars and the roads that act as conduits for people to travel to visit family, to work at Weipa, to attend funerals for "sorry business", or to meet up with old friends on the rodeo circuit or at the agricultural shows in the various regional centers. The coming of four-wheel drive vehicles in the 1960s, particularly Toyota's "Troop Carrier", hefty enough to transport big families, "changed forever the nature and composition of migration by enabling larger groups to travel, by increasing the catchment areas for travel parties and by expanding the spatial reach of migration events" (Taylor et al. 2011, 168). Traveling, people cannot always communicate, but when they at last arrive at their destination and regain signal their mobiles work again, linking them back to the country and the kin they have left. When people cannot travel, are not mobile—when the rains set in and close all roads for six months of the year, as at Kowanyama, when work, family or cultural commitments mean that people stay put on country—then mobile phones again form the conduit of connection.

Mobile Technology: "For Everything!"

During our research it has been obvious that residents put their mobile phones to excellent use, and many have spoken passionately about the importance of mobile phones in their lives. They help parents support their children through the six years of boarding school and help maintain contact when residents marry into other communities or move away for work—or outsiders move to these Indigenous communities to work. One mother from Kowanyama noted that her daughter lived in the Cape York mining town of Weipa, where her husband was employed at the mines, and telephoned her every week: "I miss her. She is my only daughter." Text messages and Facebook are prime facilitators for maintaining these personal networks. In these communities of high unemployment and welfare dependency, automatic reminders of appointments and job vacancy alerts sent to job seekers and welfare recipients are extremely useful: as the service provider who assists residents to find work stated, "Mobiles are definitely important. When the reception goes down, our whole world crumbles." People needing medical treatment receive important messages by mobile phone from the clinic, and some doctors have started texting patients to make appointments: as the daughter of a diabetic patient noted, "It is a real issue that they are contactable." The mobile phone has become the preferred personal organizer, with the contacts function used to record not just phone numbers but other useful information like tax file numbers and bank account details. Residents use their mobiles to receive and pay bills and undertake their banking: for example, one young woman had bought a car and found it convenient to phone in her payments on the loan. In an isolated community, residents enjoy using their mobiles for listening to music, downloading game apps, taking photographs or "for everything!" as one woman noted. Also, we see informal cultural research: community members post old and new photos on a public Facebook site, creating lots of interest and stories. This organic sharing and interest-driven phenomenon could complement the formal Indigenous cultural archive systems developed by researchers and government.

The key role of mobile technology at Kowanyama was illustrated in 2014 by a cyclonic weather event, to which the Far North is very prone. During a total of nine days of non-stop, torrential rain, a place that was safely in the network became cut off, losing connections to roads, air traffic, power, satellite, mobiles and television, and the radio took on an aberrant life of its own: when severed from the regional radio "feed" it defaulted to an endless music loop as the township teetered on the brink of "disaster". The landline phones kept working, but few Aboriginal people own those. The mobile network failure had a major effect on many residents and caused them to focus on how important this technology was in their lives. Because emergencies are most likely to occur in the Wet season the inability to log defects became serious: one resident was anxious about reporting electrical faults and overflowing sewerage at her home but couldn't with her mobile not working.

The cyclone coincided with the beginning of the school year. Two-thirds of the children of high school age were already at boarding school. Parents were unable to let their children know that everything was all right in Kowanyama when the news down south was full of reports of the torrential rain, making the children worried about their families. The start of the school year is always the most intense: the first settling-in period is crucial for students if they are to enjoy academic success. As one Transition Support Officer noted, "It's absolutely vital for kids' wellbeing. The kids get distressed if they can't contact parents. Parents get distressed if they can't contact their kids." Moreover, there was still a plane load of children due to fly down: Transition Support Services had to ring the boarding schools to say that the students would not be coming; eventually a satellite phone was located. Similarly, all the new "schoolies" (school teachers), who had recently arrived in Kowanyama to begin teaching, were cut off from their families down south as they, too, were dependent on their mobiles. Their families were worried because they had heard about the cyclone buzzing around but no one could get through.

Somehow the businesses survived. The cost of redundant systems, developed over years of operating in this extreme environment, paid off. One of the shopkeepers described how he moved all his mobile broadband systems to dial-up when the mobile network ceased working. However, this was extremely time consuming and laborious, taking about one and a half hours to move to dial-up and another one and a half hours to move back to broadband when the network was restored. In addition, transaction speeds slowed and levels of customer service declined, but at least he could continue to operate and his customers could make purchases as long as they were patient, while the town waited for many days for the Telco to come and fix it. If the landlines had gone down—the case during a cyclone two years before—the businesses and the people they service would have suffered.

The desire to communicate and be contactable is universal. As Douglas and Ney (1998, 46) stated many years ago, a "social being has one prime need—to communicate". Communication is not trivial but rather recognized as an important end in itself (Castells et al. 2007). It is not limited to one cultural group, but instead forms a significant motivator in why both Indigenous and non-Indigenous people have taken up mobile phones. As social beings communication is a basic necessity, enabling them to harness the structures of society and familial links to fill other needs. For economies at the edge mobiles offer people the ability to "mobilize" resources and garner small favors from their network of friends and family in order to meet urgent needs (Horst and Miller 2006). Technology supports this by extending people's network beyond those in the community who can be physically contacted, to the diaspora in the world beyond. As such it is an economic strategy. Therefore, mobile technology is not a luxury for such people; it is a highly valued service for which they are prepared to go without food or borrow tiny amounts of credit.

Intersections of Mobility and Place

Talking with people in the Far North leads us to think more critically about the vital element of place: places of failed communications and infrastructure, of extreme weather and of remoteness, which acts as a key socio-economic and technological delimiter. We recognize that we cannot cover the range and nuance of place so we will use it in a general sense, that "place can be understood as all-pervasive in the way that it informs and shapes everyday lived experience, including how it is filtered and experienced via the use of mobile technologies" (Wilken and Goggin 2012, 6).

Stimulated by the cyclone at Kowanyama and the sudden dropping out of the mobile network, stories emerged about mobile technology, stories relating to lack of coverage beyond the town where the mobile footprint has never reached, impacting the Dry season when the roads re-open. Here we compare these accounts at Kowanyama with those gained the year the 3G network was rolled out in Wujal Wujal, one of the smaller communities where the terrain is very different.

A peculiarity of the mobile phone networks installed in Cape York more than a decade ago was that they were built to a minimum standard—one tower each, whether the population of a community was 305 as at Wujal Wujal (ABS 2006) or 1,200 as at Kowanyama (KALNRMO n.d.), whether the homes were concentrated or dispersed. The coming of these "standard" services was not unproblematic. The result of ignoring population differences would be felt a few years down the track when larger communities like Kowanyama were struggling for bandwidth in peak usage times, exacerbated by the increasing take-up of smartphones and social networking and affecting the efficiency of businesses that had switched to mobile broadband. The outcome of having only one tower in each community was felt immediately, with signals often penetrating only a few kilometers beyond the town and inhabitants of outlying camps receiving no coverage whatsoever, even in Kowanyama where the country is dead flat and there are no physical barriers to block the signal. This was made worse by the later introduction of 3G, which residents reported had a lesser reach than the old CDMA service. In Wujal Wujal there was a fight to have the tower placed on the hill overlooking the town, rather than on the flat, so that at least some people further away could be on the network and some of the popular recreation and service areas would be covered (Figure 2.3).

Figure 2.3 ICT ecology of Wujal Wujal: mobile phone tower on the hill, computer access center, shop selling mobile phone cards, offices with computers and fixed-line phones, and public telephone in the foreground. (Photo: Laurel Evelyn Dyson).

In Kowanyama the inadequacy of network coverage affects almost everyone, whether Indigenous or non-Indigenous, as fishing and camping at localities beyond the reach of the network are favorite pastimes in the Dry season: "The signal does need boosting", said a keen fisherman, "If there is an accident, a child breaks a leg, or someone gets bitten by a taipan [snake], you need to be able to make a phone call". In addition, there is an influx of fishing tourists during the Dry, who represent a significant source of income for the community through the issue of camping permits. A culturally significant aspect of Aboriginal life in Kowanyama is the 18 Homelands, which are visited in the Dry by the traditional owners, as one woman noted: "We do go out a lot. If we get stranded we can't contact the community". Lack of mobile coverage raised major concerns amongst interviewees about personal safety in the event of breakdowns, accidents and illness in these areas. As a manager who had been a resident in Kowanyama for several decades stated, "It's a safety issue. This is an unforgiving area".

For Kowanyama the mobile footprint is limited to the community surrounds, with no mobile reception along the major roads in and out of the community. With family members living in other towns and residents needing to shop or transact business in the city, trips by car are common. From very remote Kowanyama this means driving along hundreds of kilometers of unsealed roads with no coverage until one arrives. These roads are also much traveled by service providers, contractors and tourists. Each road has its challenges, but none is good. Breakdowns are particularly common and people get stuck between creeks when waters at the river crossings rise suddenly after torrential rain. One resident told how her vehicle got bogged in a creek at the beginning of the Wet: the only option for her and her husband was to climb onto the roof of the vehicle and wait for help. They clung there for four and a half hours until close to dusk, the swirling waters almost to the roof, afraid to move because a large crocodile was known to dwell in that stretch of the creek. Eventually, a helicopter from a nearby cattle station spotted them and came to their rescue. People's historical memory of particularly bad accidents where suffering could have been averted if only there had been a way of phoning or texting for assistance included someone who broke a limb, a local manager who was found wandering dazed in the wrong direction after rolling his vehicle 20 kilometers out of Kowanyama and the contractor who bled to death by the roadside, unable to get help.

In the smaller community of Wujal Wujal, the coverage problem is exacerbated by the hilly terrain in the Bloomfield River Valley. Thus coverage is patchy. The only residents with a good signal are those actually living in the Wujal Wujal township, which is in direct line of sight with the tower. Those in the nearby village of Ayton (Jajikal) or in the mainly non-Indigenous Bloomfield community receive no signal. As one impassioned Bloomfield resident, who does not own a mobile phone, stated: "I don't want one. They're useless. ... I can't use it where I live." Even people who live at Wujal Wujal may need to contact others who live outside the network: one Wujal Wujal woman stated that, "I can't ring my Mum. She lives at Ayton". For Wujal Wujal residents it chiefly impacts their ability to use their phones when going about their normal daily lives outside the township, fishing at the beach or creeks or shopping at Ayton. Some fishing and camping spots have coverage, but many do not. Once on the roads out of the Valley coverage is again patchy: because this part of the Cape is not as isolated as Kowanyama, with shorter distances between settlements, some stretches of road have a signal, even if it may be weak, while others have none. One man put an aerial on his car to increase signal strength and makes calls from his vehicle even when at home.

So we see somewhat of a mobility paradox. In remote areas like Kowanyama and Wujal Wujal mobile is not an extension of person, enabling communication anywhere, anytime, as in well serviced urban areas. It is much more contingent: an odd, unreliable prosthetic that works partially or fully, depending on where people are, not on when they need it. In a perverse way it usually works best when at home, within walking distance of the public telephone and surrounded by family, and worst when out hunting, camping or traveling, far away from services that might be required. Of course, for those who live at the camps and outstations, or at Ayton or Bloomfield, just around the bend in the river and out of mobile range, it works at neither, and residents relive the past: driving to the signal, just as they once drove to the public phone when no private phone service was available. The need for mobile phones changes depending on location: social, business and service facilitator when in the main community, a desirable safety net when traveling (but rarely, since mobile coverage does not allow it until one is in reach of the better populated regions), and an umbilical cord when away visiting other communities, towns or the city, when one is able to ring back home.

Given that we know that for all people the most important place is home, what does "home" mean for the people of the Far North? For Indigenous people, unquestionably, there is a strong connection to country, but that does not mean that people stay put. In her research of Indigenous Australians Elspeth Young (2002, 84) found "the existence of deeply rooted social networks that linked people in extended families together over long distances, and also linked these groups of people to their traditional 'countries'". She noted the bilocality and multilocality of Indigenous people's residential patterns and, if anything, more movement over greater distances in modern times than in old. In our research we have found Indigenous people move to visit family and friends; to go fishing, foraging or hunting; to go camping; for work or school; for shopping or to shift house. Some are trips of a day while others may last weeks or months. Some are within the community, some to camping and fishing spots outside town, some to other Indigenous communities where family live or to business centers hundreds of kilometers away.

ICT infrastructure was put in place by government as a means to overcome distance and enable people to access services while staying put in their far-flung communities, it is interesting that the only ICT technology into which Indigenous people have bought—mobile technology—is supporting *greater* mobility. Catching up with this trend, what we see now is government making use of this technology to provide the services—and enforce compliance—for a more mobile population. Welfare recipients can lodge their reports online via their mobile phones, rent arrangements can be mediated by mobile phone, and hospitals can contact patients, no matter where they are at the time.

For all its failures in these places at the geographic edge, mobile technology provides a means for staying connected—not always, but often enough—and for allowing peripatetic or relocating people to maintain links with the places to which they belong.

Intersections of Mobiles and Culture

It is easy to fall into cultural stereotypes when considering how people engage with technology. We can see this in comments in the literature, for example, on how mobile phone use by Aboriginal adolescents is "consistent with their collective culture (e.g., communicate with family and friends)" (Johnson 2013, p. 1), as if this were peculiar to Indigenous people.

However, in attempting to answer the question of "why mobile" with specific reference to Indigenous people we inevitably come to a consideration of culture. If not encapsulating the entirety of Indigeneity, we might reasonably assume that cultural norms would express themselves in Indigenous people's technology choices and patterns of use in some way, even if aspects of socio-economics and remoteness, not to mention the peculiarities of the technology itself, also play a role. The problem is that in the Far North it is hard to separate these as they are often inextricably intertwined.

An unavoidable dimension of this question is a reflection on the differences in Indigenous and non-Indigenous people's participation in mobile technologies. Like the Indigenous residents of the Far North, those who are not Indigenous are also people living at the edge, dealing with exactly the same conditions of isolation, dirt roads, intermittent signals, catastrophic weather events and limited work options. Here we draw on our research, first at Kowanyama, where we chose to largely ignore cultural and racial groupings by classifying people based on their technology use (Brady and Dyson 2014) and second in the Bloomfield River Valley, where we focused directly on differences in technology use between the two groups of residents at Wujal Wujal and Bloomfield (Brady and Dyson 2010). Insights from the Lockhart River and Dauan studies are also invoked.

At Kowanyama we could see that communication was of prime importance to everyone who lived in the town, whoever they were. During the cyclonic event and deluge at Kowanyama, many residents wanted to communicate using their mobiles to allay family fears or report faults but couldn't. All felt distress at not being able to communicate.

Similarly, in our study of the Bloomfield River Valley, communication was of great importance to everyone we talked to. All interviewees at Wujal Wujal reported that they made phone calls or sent text messages or did both—to contact family and friends, for work, or to communicate with people while traveling to other locations or for emergencies. Most residents of Bloomfield bought mobile phones purely to be able to use when visiting regional towns and cities and for the increased sense of safety when traveling on roads. As Bloomfield residents could not use their mobiles where they live because of lack of coverage in their part of the valley, they had to use landlines at home.

Yet for Indigenous people, the support for communication while mobile has significant cultural and identity implications. Their enhanced mobility is part of a wider cultural shift as most Indigenous people in Australia, as in the United States and Canada, now live in regional and urban areas and experience their culture at a distance. In the twenty-first century Aboriginal communities are marked by:

> the retention and extension of community by new technologies of transport and communication, the underlying strength of kinship networks, the revitalising of culture through native title claims, the broadening of cultural values at a distance from country.
>
> (Thompson 2009)

For young Indigenous people growing up in the city—physically isolated from their ancestral homeland and their kinship networks dispersed in the wake of the devastating effects of colonization—social networking and other communication via mobile phones form an essential element in their "struggle to reclaim and assert their cultural connections and identity" (Edmonds et al. 2012, 23).

Mobiles, Music and Multimedia

The other exigency of the human spirit that is answered by mobile technology is music. It is not unusual in the communities to see people—particularly young people—walking around wearing earphones with a mobile phone showing from the top of their pockets. MP3 players are used by children and some adults, but the sales of MP3 players declined once 3G phones arrived on the market as they too can be used for playing music. Our Lockhart River study in 2009 was the first to capture the importance of mobiles for music and multimedia for Indigenous Australians. In fact, local managers believed that mobile phones had only become really popular once the 3G phones, with all their multimedia features, arrived. A number of Lockhart residents had novel ways of maximizing the use of their devices. For example, mobile phones provided the music for the Friday night disco that took place at the Church Hall during school holidays: "Some of the kids plug their mobile phone into a speaker to play", reported one teenage boy. A middle-aged woman used her earphones when in public but at home connected her mobile into loud speakers. Footballers sometimes listened to music from the mobiles or MP3 players in their pockets while they played their game.

The pervasiveness of music is supported by our findings from the Bloomfield River Valley, which showed that more than half the people in both the Aboriginal and non-Aboriginal populations owned MP3 players or used their mobile phones for listening to music. Some reasons given for preferring these devices to older music technology, such as CD players, were purely pragmatic—their superior performance when driving on rough dirt roads. Others liked to personalize their music choices: for example, one parent in Wujal Wujal stated that being able to listen to the music she liked via her earphones was important as her musical taste was not the same as her children's.

In addition, playing digital games (cards or PlayStation games) and taking and sharing photographs and videos were reasonably common, again predominantly amongst youth. As young people listed their uses at Lockhart River, often entertainment came up first: "I Bluetooth [music], games, text message, ring", stated one young woman. Likewise in Wujal Wujal multimedia and entertainment uses of mobile phones were prevalent, with interviewees obtaining their multimedia content from various sources, including from people they knew via Bluetooth, downloading content from the Internet using their mobile phones, and downloading from work computers. Now at Kowanyama, touchscreen devices are being used by Land Office rangers to photograph and track weed infestations or native plants of relevance to their job (Figure 2.4).

Figure 2.4 Ranger photographing a tree planted in town. (Photo: Tyler Wellensiek).

The high level of interest in mobile multimedia should not be surprising to anyone who is acquainted with Indigenous Australian cultures. Indeed, there is a close fit between the multimedia features of mobile devices and traditional cultural strengths, namely in oral and audio practices (song, music, storytelling and ceremony) and in pictorial expression (painting, sculpture, carving and weaving). Further, multimedia devices have the advantage that they are always usable, even when the network is down or the owner has run out of credit, a common occurrence in these communities (Auld, Snyder and Henderson 2012).

One significant point of difference that should be noted between the two populations in the Bloomfield River Valley was that in Wujal Wujal mobile phone owners exploited the multimedia and data features of their phones to a much, much greater extent than Bloomfield respondents. Only 2 out of the 20 Bloomfield respondents in our study reported making use of these features. Comparing music listening via mobile devices, Wujal Wujal residents used mobile phones slightly more than MP3 players, while Bloomfield residents nearly always used an MP3 player. Bloomfield people were better equipped at home with landlines, computers, Internet connections and other technology and mainly kept their phones for communication when visiting other locales.

Mobile technology represents the most accessible ICT for the general Indigenous population and, although some may feel that entertainment is a rather hedonistic, time-wasting use of such advanced tools in remote communities with limited amusement options, it can have a significant role in personal wellbeing. All people need to relax, to take "time-out" from the worries of life and to enjoy themselves. Mobile technology has joined TV, radio and CD players in serving that essential part of the human spirit—our endless love of music and storytelling. Playing also provides an opportunity for people to gain familiarity and build confidence with ICT. For some adults mobile devices may be their only exposure to ICT and even entertainment uses as repetitive as some mobile games may be useful: Sey and Ortoleva (2014, 1) acknowledge "playful uses of technology as essential for personal development and adaptation to social and technological change."

While there is concern that downloads from the Internet may promote the Westernization of Indigenous culture, there is evidence of support for Indigenous cultural practices and preferences as well. Sport, and especially football, both a national and an Indigenous obsession, is commonly viewed on mobile phones. The user-generated content that circulates around Indigenous communities powerfully reflects local culture. As Inge Kral (2013, 64) notes, "young people are borrowing or 'styling' ... global cultural and linguistic resources, but also layering these onto traditional modes and speech styles". For example, music from the local band is Bluetoothed around Wujal Wujal. In Lockhart River a teenager, well-known locally as a hunter, shows videos of himself at the end of a successful chase, posing with his kill of turtle, dugong or cow. Text messages are sent in the local language

Kala Kawa Ya on Dauan, reinforcing an already strong oral language tradition and extending it through new technological practice to writing. With fears of cultural and linguistic decline in Indigenous Australian communities, as in many Indigenous communities around the world, this integration of new technology with traditional cultural practices is promising for the future.

The Fit of Mobile Design with Indigenous Ways of Being

Technology as a medium of social practices necessarily has both restricting and enabling implications. Which one dominates depends on multiple factors including the actions and motivations of the designers, the context in which it is embedded and the autonomy and capability of the people who use it. Mobile technology facilitates human action but also constrains it because of the characteristics of its design and capabilities (Latour 1992). In examining the fit of mobile technologies with people's desire for an affordable ICT, which is at the same time usable across locations, we begin to understand the attraction of *this* technology rather than others that presented in the past.

Portability is an attribute of the mobile device that is highly valued by just about everyone in the Far North. For Indigenous people, it allows an escape from the modern-day economic aggregations of population that echo assimilationist missionary times, with their services of housing, health, education and telecommunications designed around a fixed population. Mobile phones, tablet PCs and MP3 players have the "edge" over fixed-line communication tools, desktop computers and boom boxes in assisting non-sedentary people to renew their culture of temporary mobility or "walkabout," viewed in the past as disruptive. More than this, they have been a trigger for structural change in that marginalized and hard-to-reach people have, by choosing mobile technologies (and rejecting other ICTs), changed Federal Government service delivery to mobile-enhanced online platforms. This represents a major change in thinking from a model of pushing out information via text messages to allowing Indigenous people the pull over which services they access from whichever community they choose to reside in at the time.

As portable devices they are also personal devices. Where large extended families living in one house is the norm and overcrowding results, the ability to have a private phone conversation is highly valued: the personal and portable nature of the mobile phone facilitates this by allowing calls to be taken outside. The mobile phone, too, obviates the inconvenience and loss of privacy inherent in using the community's public telephone, where everyone can see you are making a call, or using the bank of public computers, where there is no way of concealing what you are posting to Facebook or which websites you are accessing. MP3 players offer private music listening and allow for the recognition of people's individual tastes. All of these are increasing an individual, personal, private lifestyle, compared to the communal way of life that has been part of the Indigenous stereotype. The technology is thus a facilitator of myriad personal negotiations with family and community, not so much a trigger but a slow fuse for change.

As portable devices they are usually carried at all times. Though sharing of phones is very common, our research also showed that this is not always the case. Inge Kral (2012, 230), in a study of the Yarnangu people of Western Australia, viewed the small size of mobile devices as lending themselves to being tucked away in clothing and kept safe on one's person at night: "In an environment predicated upon demand sharing, these are items of personal ownership that don't have to be shared". This permits Indigenous people to control costs by circumventing culture, that is, by avoiding the Indigenous norm of reciprocity.

This, together with other cost management features, represents a key advantage of mobile phones over landlines for Indigenous people. The fundamental method of managing costs for Indigenous people is pre-paid phones. This is supported by all the shops in Indigenous communities, which only sell these. Despite the fact that pre-paid callers typically pay more per call than those on a monthly plan (Dyson and Brady 2013), it is impossible to accumulate debt. For people on low and irregular incomes, this represents a major advantage. For Indigenous people pre-paid mobiles avoid the problems associated with landline phones: in overcrowded houses where cultural obligations make it difficult for the landline subscriber to refuse phone use to family and friends, large numbers of calls often result and the subscriber may be unable to pay, leading almost inevitably to disconnection of the service (DCITA 2002).

Obviously, as for non-Indigenous people, the text-messaging function of mobile phones is another method of regulating costs. In the case of the Indigenous users—since writing is not part of their traditional culture—this again shows that cultural traditions can be overcome if motivation (to communicate cheaply) is strong. On the other hand, in some communities where the traditional language is alive and well, text messages and phone calls are often conducted in language, showing that the new technology is being both adapted to and adapting traditional practices.

In addition, mobile devices are relatively cheap compared with computers, with prices falling considerably over time. With cheap hardware, malfunctioning phones become disposable items, to be sent south in the Post Office recycling bag when they are no longer of any use. In communities on the edge, there are no repair shops or technicians to fix them. Fixed-line phones and computers require technicians to be flown in long distance if they need repair.

Finally, the people of the Far North recognize the design limitations of mobile technologies and are able to choose when to use them. Whereas the businesses at Kowanyama and the residents of Bloomfield subscribe to fixed ICTs to overcome the limitations they experience in their mobile service,

Indigenous residents will fall back on the publicly provided ICTs, continuing at times to use the public telephone as it provides untimed—and hence much cheaper—local calls than their mobiles or asking a clerk at the Council Office to perform Internet banking for them. During the school holidays the young people will make use of the public access computers to "Facebook" or play digital games for free, while job seekers or welfare recipients utilize the computers set up for their use because it is easier reading job ads or filling in forms on the big screen than on the limited interface of their mobiles. Mobiles are the go, but not always.

Conclusion

We know from theories of technology use that there is no simple answer to why any technology is taken up. Technology choices are influenced by many factors—social, historical, cultural, economic, political and geographic. We should not forget that mobile technologies themselves are not passive players but "transmit effects on their own terms" (Underwood 2008, 1), offering possibilities while discouraging others through their design.

In the Far North, we can see what many other studies of mobile technology have shown in Australia and across the developed and developing world, that there is a huge motivation for owning mobile phones *whoever you are*, whatever your cultural background may be. Their prime (initial) purpose as tools of communication locks squarely into one of the most fundamental human needs, while their translation into the multifunction, multimedia, pocket-size 3G phone provides the ultimate convergence technology, so very valuable to those with limited personal possessions, limited income, high mobility, who may have accessed ICT—publicly or at school or work—but never *owned* ICT before. Whereas tablet PCs and laptops have often been overlooked in studies of Indigenous communities, ICT functions are so highly desired that early indications are that they, too, will play an increasingly significant part in many Indigenous people's lives.

For the Indigenous residents of the four remote communities there is a complex interplay of socio-economic constraints (low incomes and overcrowding in houses), cultural factors (the need for mobility to camp, fish or translocate; the preference for music and multimedia; the difficulty of refusing fixed-lined phone access to family), the portability of the devices (promoting mobility, privacy and possible circumvention of sharing), billing structures (pre-paid, debt-free), cost of hardware (low initial outlay, replaceable) and aspects of place (distance from technicians) that result in a decided advantage of mobile technology over fixed ICTs. They overcome many of the problems associated with fixed ICTs, which in a number of respects reverse the success factors of mobile technologies: a mismatch between billing methods and Indigenous incomes and household situations, prohibitive connection costs and maintenance difficulties in remote regions, lack of multifunctionality (in the case of fixed-line phones) and lack of support for temporary mobility, to mention just the most obvious (DCITA 2002; RTIRC 2008). As a result, compared with most other Australians, for whom mobile technologies represent merely a complement to existing ICTs, for Indigenous people these represent the clear ICT of choice. Ironically, people least able to pay, pay the highest cost (pre-paid) with no choice of provider and get an inferior service, yet value it very highly: for the majority of the Indigenous population mobile is their only personal ICT.

Being at the edge the boundaries falter, systems overlap and we reach for metaphor to give insight to this constantly shifting state. Place and culture are inextricably intertwined in these communities, given Aboriginal and Islander occupation of their lands *ab origine*, from the beginning. Through talking, texting and social networking, through listening to music and sharing photos and videos, mobile technologies provide a sense of constant connection with place and the cultural practices associated with it, whether on country or away from it. The still contingent nature of mobile connectivity in regions such as this (not ubiquitous, not reliable) means that people are still reflective of the medium—grateful yes, but still testy about the limits of service, while avidly trying new technologies. As long as the mobile infrastructure itself is pushed into contradiction—working best domestically in the community where the signal is strong but weaker when people are mobile on the road or living a great part of their lives out bush—Indigenous people will continue to quest for the technology that seamlessly meets their needs. Mobile technologies in their current form are the first to come close.

Acknowledgements

The authors thank the residents of Dauan, Lockhart River, Wujal Wujal and Kowanyama who have so generously given up their time over the years to talk to us about the central place of mobile technology in their lives. We thank the mayors, councilors and managers who facilitated our visits. Finally we thank Dr. Jim Underwood and Max Hendriks for their insightful and valuable comments on our draft chapter.

References

Altman, Jon. "Sustainable Development Options on Aboriginal Land: The Hybrid Economy in the 21st Century." Discussion Paper No. 226/2001. Canberra: Centre for Aboriginal Economic Policy Research, Australia National University, 2001.

Auld, Glenn, Ilana Snyder and Michael Henderson. "Using Mobile Phones as Placed Resources for Literacy Learning in a Remote Indigenous Community in Australia." *Language Learning and Education* 26, no. 4 (2012): 279–296.

Australian Bureau of Statistics (ABS). Canberra: Commonwealth of Australia, 2006.

Australian Communications and Media Authority (ACMA). *Telecommunications in Remote Communities*. Canberra: Commonwealth of Australia, 2008.

Brady, Fiona, and Laurel Evelyn Dyson. "A Comparative Study of Mobile Technology Adoption in Remote Australia." In *Proceedings of the Seventh International*

Conference on Cultural Attitudes towards Technology and Communication *(CATaC)*, Vancouver, Canada, 69–83. Murdoch, Australia: Murdoch University, 2010.

Brady, Fiona, and Laurel Evelyn Dyson. "Enrolling Mobiles at Kowanyama: Upping the Ante in a Remote Aboriginal Community." In *Proceedings of the Ninth International Conference on Culture, Technology, Communication (CATaC)*, Oslo, 179–194. Oslo: University of Oslo, 2014.

Brady, Fiona, and Laurel Evelyn Dyson. *Report to Wujal Wujal Aboriginal Shire Council on Mobile Technology in the Bloomfield River Valley*, June, 2009.

Brady, Fiona, Laurel Evelyn Dyson and Tina Asela. "Indigenous Adoption of Mobile Phones and Oral Culture." In *Proceedings of the Sixth International Conference on Cultural Attitudes towards Technology and Communication (CATaC)*, Nîmes, France, 384–398. Murdoch, Australia: Murdoch University, 2008.

Carson, Dean, Prescott C. Ensign, Rasmus Ole Rasmussen and Andrew Taylor. "Perspectives on 'Demography at the Edge.'" In *Demography at the Edge: Remote Human Populations in Developed Nations*, edited by Dean Carson, Rasmus Ole Rasmussen, Prescott Ensign, Lee Huskey and Andrew Taylor, 3–20. Farnham, UK, and Burlington, USA: Ashgate, 2011.

Castells, Manuel, Mireia Fernandez-Ardievol, Jack Linchuan Qiu and Araba Sey. *Mobile Communication and Society: A Global Perspective*. Cambridge, MA. MIT Press, 2007.

DCITA (Department of Communications, Information Technology and the Arts). *Telecommunications Action Plan for Remote Indigenous Communities: Report on the Strategic Study for Improving Telecommunications in Remote Indigenous Communities (TAPRIC)*. Canberra: Commonwealth of Australia, 2002.

Douglas, Mary, and Steven Ney. *Missing Persons: A Critique of Personhood in the Social Sciences*. Berkeley: University of California Press, 1998.

Dyson, Laurel Evelyn, and Fiona Brady. "A Study of Mobile Technology in a Cape York Community: Its Reality Today and Potential for the Future." In *Information Technology and Indigenous Communities*, edited by Lyndon Ormond-Parker, Aaron Corn, Cressida Fforde, Kazuko Obato and Sandy O'Sullivan, 9–26. Canberra: AIATSIS, 2013.

Dyson, Laurel Evelyn, and Fiona Brady. "Mobile Phone Adoption and Use in Lockhart River Aboriginal Community." In *Proceedings of the 8th IEEE International Conference on Mobile Business*, Dalian, China, 170–175. IEEE, 2009.

Edmonds, Fran, Christel Rachinger, Jenny Waycott, Philip Morrisey, Odette Kelada and Rachel Nordlinger. *Keeping Intouchable: A Community Report on the Use of Mobile Phones and Social Networking by Young Aboriginal People in Victoria*. Melbourne: Institute for a Broadband-Enabled Society, University of Melbourne, 2012.

Horst, Heather, and Daniel Miller. *The Cell Phone: An Anthropology of Communication*. Oxford: Berg, 2006.

Johnson, Genevieve M. Technology Use among Indigenous Adolescents in Remote Regions of Australia. *International Journal of Adolescence and Youth* 2013 (2013): 1–14.

KALNRMO (Kowanyama Aboriginal Land and Natural Resource Management Office). *Kowanyama Aboriginal Community Fishing and Visitor Guide*. Kowanyama: Kowanyama Aboriginal Shire Council, n.d.

Kral, Inge. *Talk, Text and Technology: Literacy and Social Practice in a Remote Indigenous Community*. Bristol: Multilingual Matters, 2012.

Kral, Inge. "The Acquisition of Media as Cultural Practice: Remote Indigenous Youth and New Digital Technologies." In *Information Technology and Indigenous Communities*, edited by Lyndon Ormond-Parker, Aaron Corn, Cressida Fforde, Kazuko Obato and Sandy O'Sullivan, 53–73. Canberra: AIATSIS Research Publications, 2013.

Latour, Bruno. "Where are the Missing Masses? The Sociology of a Few Mundane Artifacts. In *Shaping Technology/Building Society: Studies in Sociotechnical Change*, edited by Wiebe E. Bijker and John Law, 225–258. Cambridge, MA: MIT Press, 1992.

Pearce, James. "UPDATE: Qld Government Muscles Carriers into Better Coverage." *Zdnet*, 7 July, 2003.

RTIRC (Regional Telecommunications Independent Review Committee). *Framework for the Future: Regional Telecommunications Review*. Canberra: Commonwealth of Australia, 2008.

Sey, Araba, and Peppino Ortoleva. "All Work and No Play? Judging the Uses of Mobile Phones in Developing Countries." *Information Technologies & International Development* 10, no. 3 (2014): 1–17.

Tangentyere Council and Central Land Council. *Ingerrekenhe Antirrkweme: Mobile Phone Use among Low Income Aboriginal People, A Central Australian Snapshot*. Alice Springs, Australia: Tangentyere Council Inc. & Central Land Council, 2007.

Taylor, Andrew, Gary Johns, Gregory Williams and Malinda Steenkamp. "The 'Problem' of Indigenous Migration in the Globalised State." In *Demography at the Edge: Remote Human Populations in Developed Nations*, edited by Dean Carson, Rasmus Ole Rasmussen, Prescott Ensign, Lee Huskey and Andrew Taylor, 163–188. Farnham, UK, and Burlington, USA: Ashgate, 2011.

Thompson, David. "Promises and Perils of Transition to Urban Living." Paper presented at the *AIATSIS Conference*, Canberra, October 1, 2009.

Underwood, Jim. "Varieties of Actor-Network Theory in Information Systems Research." In *Proceedings of the European Conference on Research Methods*, London, 1–7, 2008.

Wilken, Rowan and Gerard Goggin. "Mobilizing Place: Conceptual Currents and Controversies." In *Mobile Technology and Place*, edited by Rowan Wilken and Gerard Goggin, 3–25. New York: Routledge, 2012.

Young, Elspeth. "Indigenous Demographic Issues at Australia's Millennium: Population Mobility." In *The Aboriginal Population Revisited: 70 000 Years to the Present*, Aboriginal History Monograph 10, edited by Gordon Briscoe and Len Smith, 81–92, Canberra: Aboriginal History Inc., 2002.

3 The Case for Play in the Developing World

Lessons from Rah Island, Vanuatu

Pedro Ferreira and Kristina Höök

Introduction

Mobile technologies are rapidly entering the lives of many Indigenous people, such as the developing nations of the South Pacific, altering living conditions and the capacity for communication. The field of ICT4D (Information and Communication Technologies for Development) focuses on understanding the relationship between technology and socio-economic development. There has been a great deal of criticism about development and aid, as well as ICT4D more specifically, regarding paternalistic attitudes (Mosse 2001; Escobar 2011) or lack of efficiency more generally (Easterly 2006). More recently some scholars have focused on understanding the motivations that drive technology use beyond strict socio-economic goals such as entertainment and play (Chirumamilla and Pal 2013; Smyth et al. 2010). The topic of this chapter concerns these efforts: the appreciation of everyday playfulness around ICTs.

During our ethnographic fieldwork in Vanuatu, and specifically on Rah Island, we happened to stumble upon the exact moment when mobile connectivity was introduced. What became apparent in our study were all the playful and engaging uses of mobiles—several of which we may remember from our own first encounters with mobiles (such as playing with ringtones). This was not an unproblematic observation for us or for our informants. We had to revisit our own values in order to shed a more honest light on the role of playfulness— beyond instrumental secondary goals, such as education or health. Here, we will provide an account of some such playful behaviors and why they are important and legitimate. We discuss some of the tensions that emerged within these Indigenous communities, ourselves and ideas of development.

We begin with an introduction to the concept of play to get an understanding of how we will use the term throughout this chapter and how it has been understood in the field of ICT4D. We then briefly describe the methodology and follow up with some of the identified playful behaviors around mobiles. We will discuss bodily orientations as a concrete example of how to reframe our thinking around play and technology, following up with a broader discussion on the tensions between needs and desires in ICT4D work. We conclude with a suggestion to reframe the way we approach ICT4D work through changing the status of play.

Understanding Play

Playfulness is an elusive concept. Sociologist Thomas Henricks discusses how different disciplines will approach it in different ways—a property of no single academic field and of all at the same time (Henricks 2006). Prominent play scholar Sutton-Smith (2009) calls for action, emphasizing how compartmentalization is deeply entrenched and that the study of play needs to rise above compartmentalization in order to affirm itself as a wider and more unified discipline.

As Henricks suggests, to escape from each field's narrow framing of play we are encouraged to revisit play at its roots in intellectual thought, namely Dutch historian Johann Huizinga's *Homo Ludens*—the playing man, widely acknowledged as the first comprehensive overview and defense of play as a worthy activity in itself (Huizinga 1955). Huizinga defines influential concepts such as the *magic circle* that players enter when engaged in play. The magic circle is a social contract, an agreement between those playing. Some of the most important aspects of the magic circle are that it: (1) stands outside the seriousness of ordinary life, (2) is an engagement for its own sake and not as a means to some other end, and (3) is bounded by rules and exists in time and space. This is not by any means an extensive analysis of Huizinga's concept but rather some of its most commonly accepted components around which we will define play.

Play is not always a well-defined, bounded, activity. Artist Alan Kaprow (2003) wants to distinguish play from gaming or entertainment. He sees play as possessing less rigid structures and representing a freer mode of exploration. While these and other discussions are important, we would like to make use of a broader view of play, including free play, games as well as entertainment and other leisure activities: an assemblage of activities that resemble each other, not by a strict set of defining characteristics but through family resemblances (Wittgenstein 1953). We take an almost everyday understanding of play, which is autotelic, not necessarily serious and outside of everyday worries, hardships and the seriousness of life.

Methodology

Our study took place in Vanuatu, a Melanesian nation composed of 83 different islands, with approximately 200,000 inhabitants and more than 100 distinct languages, in what has been deemed the highest language density per capita in the world (Crowley 2000). Currently, apart from the inherited French and English languages, deriving from a complex and shared colonial past known as the *Condominium* (Peck and Gregory 2005), possibly the most widely spoken language in Vanuatu is *Bislama*, which shares many elements with the two previously cited languages. Across the different islands and locations, not only languages, but also socio-economic situations and livelihoods vary widely. We focus here on lessons learned during a three-week period of fieldwork on Rah Island, Torba Province, in North Vanuatu. Rah is a small island connected to the larger island of Motalava, often referred to as the mainland. The proximity makes it possible to walk across at low tide, and the lives of people on both islands are strongly interconnected. A little less than 200 people live on Rah Island (Vanuatu National Statistics Office 2009).

We used an ethnographic approach collecting more than 50 pages of field notes, 1300 pictures and four hours of video. We conducted 10 semi-structured interviews and held, or rather were invited to, open assemblies with local communities in order to discuss and explain our project. We relied on an interpretivist approach, reflecting on the ways mobiles were at odds with everyday practices and how this led us to reflect about our own practices with them. This is not, nor does it pretend to be, a thick description of the culture and society on Rah. During our work, we were struck by the playfulness and tensions that mobiles spurred and how these connected to broader discourses. Those playful behaviors are not to be understood as restricted to Rah; some we recognized from our own experiences with mobiles. Some researchers have written how these behaviors can come to be seen as failures of technology adoption within ICT4D while desirable from the communities' point of view (Ratan and Bailur 2007).

We will selectively quote from members of the community on Rah, people such as Fajo, Frelal, Jorege and Father Robert, to illustrate our discussion. These quotes are in no way an exhaustive or full account of the totality of different perceptions and attitudes on Rah.

Playful Findings

Although the mobile phone tower began operating at the time of our arrival, the devices themselves had been around for two months. The tower was installed only after local interest in the technology had been certified through enough people acquiring the devices. Only then did TVL (the government-owned and main telecommunications company in Vanuatu) commit to installing the infrastructure. In practical terms, it is important to take this fact into account while reading the descriptions to come, as it provides some context as to why some people were already fairly technically savvy upon our arrival.

Playing Games

Unsurprisingly perhaps, there was some excitement around the games on the mobile, in particular the version of Tetris, a classic puzzle video game where geometrical shapes are fitted together by a player as they slowly descend from the top to the bottom of the screen. We witnessed numerous instances of people playing the game and some individuals, such as Fajo, complaining about the time children wasted on the activity:

> *The pikinini [Bislama term for children, small ones] spend the whole day playing on the mobiles. … They don't want to work anymore.*

The playing was in fact prevalent during our time there, and it was clearly not restricted to children; many adults played the game as well. There was much more resistance from adults to overtly discuss and admit to playing games, whereas children would usually talk about it. Our observations indicated that a significant amount of play by adults and children was taking place.

This is not the whole story. Jorege explained how the mobile helped him coordinate his businesses on different islands; others like Fajo felt that phones were essentially used for games or joking and thus employed wastefully.

Ringtones

Multimedia is one of the most exciting aspects of modern digital and mobile technologies. The majority of the phones on the island supported nothing more than polyphonic ringtones. These mobiles were nowhere near offering the possibilities of playing music, for instance, or recording and playing voices.

This lack of possibilities was a clear disappointment to many users, who longed for functions such as playing music. While these capabilities were known to be present in modern devices, and most people had understandably hoped to see them embedded in their own, the device was not a complete dead end in terms of playful potential. Perhaps one of the most immediately striking aspects of our stay there was the permanent and ubiquitous ringing of phones throughout the island. This was true even of areas on the island that were not covered by the TVL tower, challenging our initial interpretation that there was a surprisingly large volume of calls being made to mobiles on Rah. In reality, much of this ringing was, for instance, due to people selecting ringtones for their phones, selecting different ringtones for different people, changing them around, and so on. These sounds would echo throughout the island and would last from early morning until late in the evening most days, perfectly audible in the otherwise characteristic silence on Rah.

As interaction designers we were inspired by, and reminded of, the enjoyment of digital interactions, even in these simple and limited mobiles. This was the first digital device many people on Rah owned, and the novelty of the possibilities afforded was likely to be met with some kind of exploratory playfulness and enjoyment.

Settings and Tweaking

The phones' settings were not out of bounds for playful exploration. Everything from wallpapers and color schemes to the previously mentioned ringtones were tinkered with and exploited. Many people spent significant amounts of time personalizing their devices and sharing these outcomes.

While this practice was readily visible around Rah through simple observation, our roles as mobile phone experts, and the frequent requests we would get to fix mobiles, allowed us a deeper understanding of just how often this was occurring. Some of the frequent requests we got for assistance were due to the idea that the phones were somewhat damaged. People would come to us worried that the phones had been physically damaged either through exposure to water or some physical impact. While we were there, however, this was rarely the actual problem. Most often these situations came from what we might call over-tweaking. Through the explorations of the different menus on the phones, people would sometimes switch the language to something other than the more commonly used English, making it extremely challenging to get it back. Only through comparison with a working phone could we navigate the menus and set the language back. Other times, the phone's volume was set to minimum or silent mode, leading some to think that the speakers were somehow damaged. Changing the brightness settings to a state where the screen became unreadable was also common.

As with the ringtones, it is worth emphasizing the novelty effect and the subsequent lack of habituation to using these devices and their settings. Certainly these situations will become less and less frequent and their ability to deal with them will improve, but beyond the awkwardness they created, it's the willingness to interact and explore that should be noted.

Playing with the Surface Aesthetics

The phone that people on Rah were sold, during the promotional period, was a source of disappointment, a concern raised by many purchasers. Apart from usability and functional issues briefly mentioned above, they lacked modern capabilities such as a camera, which many buyers were hoping for, and they were rather uninspiringly designed and ancient looking. Adding to that, almost everyone had the exact same model.

To tackle these issues, several strategies were adopted. Some placed stickers on the phones to make them unique (Figure 3.1). Others adorned them with different types of lanyards. We even witnessed a more extreme case in which someone had made a cigarette burn mark on the mobile's outer casing in order to make it distinguishable. While on a first analysis one could ascribe a strict instrumental value to such decorations, such as telling the phones apart, that is not the whole story. The way lanyards were collected and sought after, how they were tweaked and incremented on, as well as some visible enjoyment in displaying the work that had been put into their mobiles, bore witness to an aesthetic dimension—turning the rather bland-looking phone into something more.

Figure 3.1 Playful adornment of mobile phones with stickers.

Playing with Interaction

We will now look at how connectivity, paramount to the mobile ecosystem, spurred playful engagement. Despite frequent complaints regarding the high costs of mobile calls and claims of a consequential care and moderation in use there were a multitude of playful interactions that required connectivity and sometimes costs.

First, as an example of a surprisingly (to us) and enjoyable activity that did not incur any costs, users would frequently check the credit on their mobile— a service free of charge. By sending a message to a specified service number, a message was returned that contained information regarding the amount of credit left for calls as well as the remaining number of messages one could send for free (a certain number of these came as part of the initial promotional package as well as with every charging). While monitoring the balance was certainly something that served a rather instrumental goal—managing one's resources—the possibility of having a free two-way interaction was also important in itself. The sheer frequency of these episodes reminds us of something that should perhaps have been obvious: even if one does not have credit, it is this feeling of two-way communication that one strives for on a mobile phone. A simple echoing of a command was better than nothing.

Second, there was a service known as *plis kolmi*, which is Bislama for "please call me". This is not a service unique to Vanuatu; it allows for sending a message to another person with the content: "plis kolmi [number of sender]". This service was free of charge and used extensively throughout Rah and Vanuatu. The primary intention was arguably to signal someone the desire to communicate while not having enough credit to initiate the call. It was also used in a variety of more creative ways; for instance, people used this extensively to signal presence, thus extracting social pleasure through the device.

Playing with Information

We also noted a fair amount of information spread. We did not dig deeper into the mechanisms and authenticity behind the information exchange, but it was interesting to observe some of the roles mobiles played in supporting certain practices around gossiping and spreading of rumors. One of the first impacts technologies had was entering common vocabulary and discourse. For instance, terms such as *coconut wireless*, intended as a joke about the unreliable nature of the information that was spread, such as hurricane warnings, which occurred a couple of times during our stay, and which were often started by children's hearsay.

Another situation that brought these dynamics to our attention came in the form of a text message, which many people, including us, received as part of social chains of dissemination:

> *One boy is drinking the blood from a young girl in Freswota Park. Come and share with your friends.*

Here, "Freswota Park" refers to a park situated in the area of Freswota, back in the capital Port Vila. We were asked by many people on Rah (who did not have any acquaintances in Port Vila they could easily contact) to confirm the veracity of the occurrence, and while we tried to make phone calls we had little success, as no one seemed to be able to confirm or deny this report. What was interesting to us was how such stories, this one in particular, quickly became topics of conversation and discussion for days on end in places as far away as Rah. It became a welcome addition to the social life on Rah, fueling social dynamics in a novel way (despite the somewhat gruesome nature of its content in this particular case). Inversely, the hurricane warnings, though discussed, rarely gained this prominence in terms of discussion, perhaps given the frequency and relative uneventfulness of the topic to the people on Rah.

Bodies and Mobiles

An important aspect in our work was the understanding of the relationship between mobiles and bodies. We have published more extensively on this elsewhere (Ferreira and Höök 2011), and here we will only focus on the connection to play. As mobiles are designed in mainly Western contexts and marketed largely as urban devices, where daily crossings on a canoe are perhaps not a common situation, they require some work to fit into different types of environments and livelihoods. There are some ways in which mobiles promote and enforce rigid forms of bodily orientation, such as more stationary behavior, as Father Robert told us: "The mobile phones are making people lazy."

These types of claims, which denoted a sort of generational conflict the reader may be familiar with in other contexts, was echoed throughout:

> *The young ones do not want to help their parents in the garden anymore. All they want to do is make "plei-plei" with their mobile phones.*
>
> (Fajo)

Playing with the devices seemed to encourage a sedentary behavior that some saw as problematic. These may not be novel issues to the reader, but, as we will now see, this was not the whole story.

The Limitations in Coverage

On Rah and Motalava (and The Banks group of islands at large) coverage was far from reaching every location. In fact, one would get radically different connectivity behind a geological obstacle such as a mountain, for instance, or during certain weather conditions:

> *Sometimes, if it's cloudy, we don't get reception at all, for days.*
>
> (Frelal)

It was very common for people to walk around looking for reception in the most varied ways. One such example happened when Jorege got a call. He was playing cards at the time with some friends and stood up while reaching for his phone. He then proceeded to pick up the call, but given the poor quality in reception, he started walking about, looking for a better signal, while shouting:

> *Hello ...? Hello ...? [...] Yes! Kaspa [His brother in Port Vila]? Hello ...?*

Jorege and Kaspa eventually proceeded with their call, with frequent interruptions. As Jorege eventually hung up, he walked toward me explaining that it was his brother calling. Jorege then complained about the poor quality of the reception:

> *You see? I call and "Hello! Hello!" then nothing [...] I call again, and again ... sometimes I call ten times just to say "how are you", but then I spend ten calls.*

On the neighboring island of Vanua Lava, which managed, in strategic places, to benefit from the tower installed in Motalava, Jimmy, a local of Vanua Lava, claimed that the reception was not such a big issue:

> *Here we are very lucky, we get reception not far from here [...] In some places people do not get reception at all.*

As we got no network anywhere, even near his village, we inquired as to what he meant:

> *Well, on top of the hill [pointing to the direction] we can get very good reception. We just walk there and make our calls from up there [...] We go there almost every day [...] just about 2–3 hours' walk.*
>
> (Jimmy)

Some of our informants from Rah who had accompanied us laughed at this and made light of this somewhat backward situation. Other physically demanding tasks executed to gain some coverage were for instance climbing a coconut tree:

> *"They climb the coconut tree just to talk on the phone", said Jorege. "When they are up there they can talk, but they cannot drop the phone", he joked.*

The physically available landscape differs from person to person (i.e., climbing up trees or walking uphill for three hours is not something everyone can easily achieve). As mobile telephony settles in, it is not necessarily the case that people stop moving or become constrained, but rather that there is a reorientation of bodily practices to these new superimposed landscapes that now have new meanings. While better coverage might address some of these issues, it will not change the fact that there is no one-way mapping between mobile telephony and movement, and we find rather complex interplays between the two.

Playful Coverage

Another aspect connected to this theme of physicality is how the network connectivity became a playful activity in itself. Intermittent coverage was a reality throughout Rah and Motalava, and exploring the landscape looking for reception became an everyday reality.

One interesting encounter took place in Vanua Lava, as we were being led up to the Mount Suretamate volcano near the locality of Lalntak. As we got farther away from the TVL tower in Motalava and deeper into the bush, the signal became increasingly unreliable. Rob, one of our informants from Rah who was leading us, had brought his mobile phone along. As we progressed, he would climb trees and rocks, holding his mobile in the air, looking for a signal. When he managed some connectivity, he announced it enthusiastically and others, including us, laughed out of surprise and joked about it.

This was not the only instance of playfulness around coverage. In other places we witnessed similar behaviors of looking for coverage in unexpected places, in a game-like way. The pleasure in this type of interaction was present even in everyday activities with a constant monitoring of connectivity:

> *Look, now I have no network again [...] Here I have one stick.*
>
> (Jorege)

The "stick" here refers to the bars displayed on the top right corner of the phone's screen to display the connectivity status. As with the somewhat more extreme example of the hike up Mount Suretamate, where constant coverage-seeking became part of the journey, we witnessed the emergence of a kind of pervasive gaming dimension associated with the simple aspect of the mobile phone's connectivity, an otherwise purely functional aspect of the device.

The seams in network coverage have been discussed extensively in the Human Computer Interaction (HCI) literature. Some research has turned these apparent limitations (i.e., irregular network reception and GPS coverage) into useful resources for design. This practice has been called seamful design (Chalmers and MacColl 2003) and argues for ways to expose those seams in technology to allow for the exploration of novel, interesting and even playful interactions. This phenomenon was, in a way, already taking place on Rah.

Play was evidently highly desired on Rah. The financial resources people were willing to invest in these devices and the excitement around them was visible throughout. Despite the apparent limitations discussed above, people were willing to put in the necessary work to derive playful value from the devices they had acquired, such as playing with coverage. This is a strong argument to take play seriously, as it defies such constraints and will occur regardless of whether it is explicitly intended or not. We have argued that we should, as interaction designers, help bridge that gap by engaging with design within these socio-economic restrictions (Ferreira and Höök 2012). Only by acknowledging, legitimizing and promoting play are we able to honor people's desires, achieve a more honest, less condescending view of technology usage and avoid a strict emphasis on needs rather than desires.

Somaesthetics and Play

To provide a concrete example in which thinking about technologies as something more than productivity-related devices, we will extend the discussion on bodies and technology in light of Shusterman's Somaesthetics (2008). Somaesthetics is a field that deals with improving our bodily and somatic agency. Through increased awareness of our somatic experiences, we train our sensory-aesthetic appreciations, approaching these as we would any other subject we aim at exceling in. Shusterman discusses how certain movements bring about critically aware somatic training, movements that are good for us.

We see in this study how people have to perform new kinds of bodily work, when canoeing or looking for coverage for instance. Some invoked rigid bodily postures in order to protect the mobile device from the elements, such as the water; others introduced different ways of moving around the environment to find coverage and thus maintain the functionality of the device.

We are encouraged to think of how these devices impact not just people on Rah, but anyone who adopts this new technology into his/her everyday life. We all undergo the task of bodily re-orientation, which often goes unchecked. The rather inflexible physical design of such technologies forces us to adapt our bodies in a machine-like manner, which may harm our bodies in the long run and needs to be addressed in the design processes of these devices, in particular within the field of ergonomics.

Ergonomics, like most disciplines that study technology, focuses on a functionalistic approach—how to engage in certain tasks without harming our bodies and achieve those tasks. Bringing values of playfulness to designing bodily interactions encourages us to think beyond the non-harming of our bodies and design for aesthetic bodily experiences. Some efforts have been made to translate somaesthetic insights into technology-design practices, for instance by Schiphorst (2009): "In the context of inter-action, somaesthetics offers a bridging strategy between embodied practices based in somatics, and the design of an aesthetics of interaction for HCI".

By exposing the bodily practices these mobile devices promote and encourage, we see how some of them will hardly become, even after long periods of adaption, pleasant, or more broadly, somaesthetical experiences. We see this fact within more mature bodily experiences regarding mobile devices in the Western world, where the design focus is still mainly anchored in rigid occupational modes of thinking and not in engaging directly with users' pleasant, playful, physical experiences. Perhaps we should, as Shusterman suggests, instead of being solely devoted to traditional ergonomics goals, draw on other mature practices such as Alexander technique or Feldenkrais, to arrive at more aesthetically pleasing, corporeal practices around these devices.

Tensions Around Play

The tension between work and play that we see on Rah, present in other places as well, echoes in academia, for instance in our discussion of ergonomics above. In the field of ICT4D, those tensions can become particularly acute given the framing of the recipient communities around needs and a tendency to forget about other dimensions, such as pleasure or entertainment.

There is a growing discussion within the field of ICT4D concerning the issue of play (Chirumamilla and Pal 2013; Ratan and Bailur 2007). Perhaps one of the earliest manifestations in this field is given to us by Raitan and Bailur (2007), in which the authors examine different ICT4D projects, considered failures despite the benefits they may have brought to the recipients through sheer access to the technology. The authors describe how the dynamics between funding development NGOs and end users' dynamics were at odds with users' feeling shame for engaging with the technology in ways other than the prescribed goals for which it was deployed, for instance, entertainment. The need to comply led to initial reports of the project's success, which did not hold up under later scrutiny. This lack of validity is of course problematic as it undermines ICT4D efforts, but more fundamentally it does not give users a voice over what they desire and instead places that decision in the hands of development experts and donors, who may be more quickly seduced by visions around needs and urgencies, downplaying these other fundamental facets of life.

We see how the field struggles with understanding motivations behind acquiring technology. For instance, a study on mobile phone ownership in Tanzania stated fewer than 15% of mobile owners believed the benefits of owning mobiles justified the costs (Mpogole, Usanga and Tedre 2008). Prominent ICT4D scholar Richard Heeks (2008) suggests that a richer story may exist: "Um ... so if you believe that guys, why on earth do you own a mobile?"

We should not be quick to judge or ascribe intentions to how socio-economically constrained communities allocate their resources. In certain places the amount spent on mobiles exceeds 50% of disposable income (Song 2009). That would be missing the point of academic research, which is first and foremost to understand the phenomena. Heeks (2008) summarizes:

> The significant amounts being spent by the poor on mobiles indicate that phones have a significant value to the poor [...] We [the ICT4D community] have long known [...] that "poverty" is not just about money and, hence, that poverty interventions and tools can usefully target more than just financial benefits.

To be fair, these issues are not just in ICT4D as a discipline, but within broader discourses in technology studies as Genevieve Bell pointed out in her 2010 keynote at *CHI*, a leading Human Computer Interaction conference. Bell points out how we have avoided discussing activities such as sports, sex or religion within the field, despite the fact that these represent a significant portion of our digital interactions (Bell 2010). When we do discuss these, it is under the light of other societal values such as social isolation or health, rarely discussing them as legitimately pursued activities in and of themselves.

In ICT4D, and whenever studying resource-constrained communities, those tensions become exacerbated, as Smyth and colleagues note:

> Are needs really more urgent than desires? Who defines which is which? Do researchers exaggerate the urgency of "needs" due to their own biases and preconceptions?
>
> (Smyth et al. 2010)

Conclusion and Repurposing Play

Play and aspects of play, such as leisure, games or entertainment, are the main factor making modern technologies so desirable. We saw, as others have (Smyth et al. 2010), that people will work hard to derive such value from their technologies even when these are not so readily available. On Rah, this phenomenon was manifest through the effort spent looking for coverage, decorating their phones or playing with their somewhat limited interaction possibilities. We need to appreciate these efforts rather than dismissing them as something else. As researchers, we need to acknowledge the legitimacy of play rather than downplaying it in favor of other higher-held values. Of course, this is easier said than done as dynamics within the communities generate similar tensions echoing throughout ICT4D and academia at large. While, as interaction design researchers, we urge others to take designing for play in resource-constrained contexts seriously, these are not tensions we can address through design alone. They require a deeper reexamination of needs and priorities, as well as who gets to decide these criteria.

How can we shift visions of Indigenous people living in the developing world, construed from models based on needs and precariousness, that push aside wholesome activities, such as playing? On the other hand, how can we do this without underappreciating some of those needs and without undermining efforts aimed at addressing them and providing often essential relief and support? The need to play exists and has been appreciated under the most severe circumstances, such as George Eisen's account of how children, as well as adults, played during the Holocaust (Eisen 1990). Playing was fundamental to escape from the harshness of the situations they were facing, as if giving a meaning to life itself. We need to come to terms with the fact that play is an extremely strong driver in people's lives, regardless of external hardships, and Indigenous populations are no exception. Play and hardship are not mutually exclusive but to some extent mutually constitutive.

References

Bell, Genevieve. "Messy Futures: Culture, Technology and Research." Keynote Address presented at the *CHI' 10*, Atlanta, GA, 2010.

Chalmers, Matthew, and Ian MacColl. "Seamful and Seamless Design in Ubiquitous Computing." In *Proceedings of the Workshop at the Crossroads: The Interaction of HCI and Systems Issues* in *UbiComp*, 2003.

Chirumamilla, Padma, and Joyojeet Pal. "Play and Power: A Ludic Design Proposal for ICT4D." In *Proceedings of the Sixth International Conference on Information and Communication Technologies and Development: Full Papers-Volume 1*, 25–33, 2013. ACM.

Crowley, Terry. "The Language Situation in Vanuatu." *Current Issues in Language Planning* 1, no. 1 (2000): 47–132.

Easterly, William Russell. *The White Man's Burden: Why the West's Efforts to Aid the Rest Have Done So Much Ill and so Little Good*. New York: Penguin Press, 2006.

Eisen, George. *Children and Play in the Holocaust: Games among the Shadows*. Amherst, MA: University of Massachusetts Press, 1990.

Escobar, Arturo. *Encountering Development: The Making and Unmaking of the Third World*. Princeton, NJ: Princeton University Press, 2011.

Ferreira, Pedro, and Kristina Höök. "Bodily Orientations around Mobiles: Lessons Learnt in Vanuatu." In *Proceedings of the 2011 Annual Conference on Human*

Factors in Computing Systems, 277–86. CHI '11. New York, NY, USA: ACM, 2011.

Ferreira, Pedro, and Kristina Höök. "Appreciating Plei-Plei around Mobiles: Playfulness on Rah Island." In *Proceedings of the 2012 Annual Conference on Human Factors in Computing Systems*. New York, NY, USA: ACM, 2012.

Heeks, Richard. "Mobiles for Impoverishment?" *ICTs for Development*. December 27, 2008. Accessed March 22, 2015. http://ict4dblog.wordpress.com/2008/12/27/mobiles-for-impoverishment/.

Henricks, Thomas S. *Play Reconsidered: Sociological Perspectives on Human Expression*. Champaign, IL: University of Illinois Press, 2006.

Huizinga, Johann. *Homo Ludens: A Study of the Play-Element in Culture*. International Library of Sociology. Milton Park, UK: Routledge, 1955.

Kaprow, Allan. *Essays on the Blurring of Art and Life*. Oakland, CA: University of California Press, 2003.

Mosse, David. "'People's Knowledge', Participation and Patronage: Operations and Representations in Rural Development." *Participation: The New Tyranny*, 16–35, 2001.

Mpogole, Hosea, Hidaya Usanga, and Matti Tedre. "Mobile Phones and Poverty Alleviation: A Survey Study in Rural Tanzania." In *Proceedings of 1st International Conference on M4D Mobile Communication Technology for Development*, 69–79. Karlstad University, Sweden, 2008.

Peck, John G., and Robert J. Gregory. "A Brief Overview of the Old New Hebrides." *Anthropologist 7*, no. 4 (2005): 269–282.

Ratan, Aishwarya Lakshmi, and Savita Bailur. "Welfare, Agency and 'ICT for Development.'" In *Proceedings of the International Conference on Information and Communication Technologies and Development*. Bangalore, 2007.

Schiphorst, Thecla. "Soft(n): Toward a Somaesthetics of Touch." In *CHI '09: Proceedings of the 27th International Conference Extended Abstracts on Human Factors in Computing Systems*, 2427–38. New York, NY, USA: ACM, 2009.

Shusterman, Richard. *Body Consciousness: A Philosophy of Mindfulness and Somaesthetics*. Cambridge, UK: Cambridge University Press, 2008.

Smyth, Thomas N., Satish Kumar, Indrani Medhi, and Kentaro Toyama. "Where There's a Will There's a Way: Mobile Media Sharing in Urban India." In *Proceedings of the 28th International Conference on Human Factors in Computing Systems*, 753–62. CHI '10. New York, NY, USA: ACM, 2010.

Song, Steve. "Nathan and the Mobile Operators." *Many Possibilities*, March 20, 2009. Accessed March 22, 2015. http://manypossibilities.net/2009/03/nathan-and-the-mobile-operators/.

Sutton-Smith, Brian. *The Ambiguity of Play*. Cambridge, MA: Harvard University Press, 2009.

Vanuatu National Statistics Office. *2009 National Census of Population and Housing*. Port Vila, Vanuatu: Vanuatu National Statistics Office, 2009.

Wittgenstein, Ludwig. *Philosophical Investigations*. Hoboken, NJ: John Wiley & Sons, 1953.

4 Ecosystemic Innovation for Indigenous People in Latin America

Paul Kim, Karla Alfaro and Leigh Anne Miller

Introduction

In Latin America, there are approximately fifty million underserved Indigenous people residing mostly in Bolivia, Ecuador, Guatemala, Mexico and Peru (UNDP 2004; Banco Interamericano de Desarrollo 2012). Many are still denied their right to an equitable primary education and find themselves unable to break away from the vicious cycle of extreme poverty. Wherever they reside, many Indigenous people are among the poorest of the poor in their nation-state system (Psacharopoulos and Patrinos 1994; Tomei 2005; Hall and Patrinos 2005). The results are significant differences in literacy rates and access to equitable education for non-Indigenous and Indigenous populations (UNESCO–OREALC 2004). For example, in Ecuador, a mere 1% of Indigenous people receive full-time education (Gradstein and Schiff 2006). In Bolivia, Indigenous children receive, on average, four years less schooling than other children (Hall and Patrinos 2005). In Mexico, the majority of Indigenous children live in communities so small that there is no school. Another half-million Mexican children migrate with their parents every year, never staying in one place long enough to be enrolled. The objective of this paper is to examine the potential of mobile phones and existing relevant initiatives around the world while deepening our understanding of the challenges our Indigenous children are facing today. Without an innovative intervention, the challenges will only increase and magnify, further excluding the uneducated, unemployed, and undocumented extremely poor Indigenous people from society and leaving them without the necessary skills to secure their well-being.

Many contributing factors circle and fortify the life-long barriers that prevent Indigenous children from breaking the poverty cycle, a cycle often inherited from their parents. Many Indigenous children do not have birth certificates. Most live in isolated, remote areas without access to government services. Others frequently migrate for work or because of the environmental effects of climate change. With the poverty cycle in perpetual motion, it is clear that the efforts and resources allocated by their responsible governments have not been significant enough to remove the barriers and improve their conditions.

Interestingly, there has been a rapidly substantiating global phenomenon in recent years—personal mobile phones. Haapkylä (2012) summarized the phenomenon as follows:

> *Mobile phones are one of the only devices reaching nearly all consumers at the Base of the Pyramid. This group of people is the largest, but the poorest socio-economic group in the world. These approximately 4 billion people have been ignored by the multinational companies until very recently. ... The spread has been fast and mobile phones have made a bigger difference to the lives of more people and quicker than any previous technology. The world has crossed the 5 billion in mobile connections mark. At the end of 1990 there were just 11 million mobile subscribers. Mobiles provide a unique way of reaching the masses of people, which have been hard or impossible to reach otherwise, especially in the remote places of the world. ... The wide reach and sheer magnitude of mobile communications change the functioning of societies.*

Mobile technologies coupled with increasing affordability, portability, shareability, computability, storability, versatility and connectivity seem to create innumerable opportunities for wide and deep societal changes. One might argue that the consideration of mere mobile phones in the context of human rights may lead to a narrow-sighted doctrine, but today's mobile phones are highly integrated digital computation and wireless network devices that fit in the palm of your hand; the consideration of the highly probable and multifaceted impact of such technology seems worthwhile.

Unlike any other discrete or standalone intervention that might cause a change in one level, mobile technology has the potential to catalyze changes at multiple levels or even transform an entire ecosystem, thus causing a change at an ecosystematic level.

Overall Inequalities in Latin America

The complex puzzle of Latin America is that the region has the largest income inequality outside of some states in sub-Saharan Africa (Birdsall, Lustig and McLeod 2012). Latin America also has the highest number of undocumented births and children not in school outside of Africa. Considering the differences in economies and populations, this should not be the case: Latin America has three times as large an economy and half the people of Africa (OECD 2011a; OECD 2011b). Latin Americans should certainly be able to register births, exercise their rights, get children into school and enjoy greater equity of income among its citizens. Unfortunately, most of the wealth is extremely concentrated at the top of the pyramid, which has profoundly negative implications for long-term stability, equity and quality of government services in these countries (Gasparini and Lustig 2011).

Poor Indigenous Populations

According to the Economic Commission for Latin America and the Caribbean (CEPAL 2010), the poverty of Indigenous people is generally attributed to the fact that they live in rural areas. Nonetheless, the ethnic origin is found to be the principal factor either in urban or rural areas linked with poverty and its consequences, such as the deprivation of children's rights. Studies also show that the ethnic origin is a significant factor of discrimination and therefore of social and economic exclusion (Hopenhayn, Bello and Miranda 2006). As a result, many Indigenous people remain in extreme poverty, generally having little or no formal education, few productive resources, few work skills applicable in the market economy and limited political voice (Kronik and Verner 2010). In addition, due to the overall low education-attainment level within communities of Indigenous populations, they end up working in low-income, low-skill jobs without social security, which ensures that they remain in a poor economic and education ecosystem.

Hidden Voices

The human rights requirement to register births parallels the economic need: children have a right to citizenship and government services, like schooling, which lead to their ability to pursue healthy, productive lives (Mackenzie 2008). Children who do not attend school due to an undocumented status contribute to the continued income inequality in Latin America (Mackenzie 2008). The insurmountable disadvantages of being an Indigenous person and being poor are often rooted in their inability to attain a national ID. Whereas the causes are diverse, this inability has serious repercussions on the quality of life for Indigenous people. The government may give the ID for free, but the costs and processes involved in obtaining an ID are often beyond poor families' incomes and capabilities. With an income of only a few dollars a day per family, even basic necessities such as food are difficult for poor families to obtain (World Hunger 2012). This difficult condition continues even if they try to change their situation. Poor and undocumented Indigenous people of Latin America moving from one place to another to seek a better life (e.g., from their rural village areas to a city) face enormous difficulties in proving who they are, let alone proving what skills they might have.

It is obvious that Indigenous people need to be educated about their rights and how to demand and exercise them as lawfully documented citizens. In other words, the poverty of Indigenous populations needs to be tackled quickly and systematically by securing a reliable and equitable national ID system that protects every citizen regardless of ethnic or economic background. It is also necessary to help Indigenous people become literate, particularly about the importance of national ID, through a sustainable quality education and skill development program.

Status of National IDs in Latin America

Access to a national ID allows someone to legally prove his or her identity and become recognized by the State. Rights that may be violated by the lack of a national ID include: equality before the law; voting; freedom to move to seek a better life and access to public services such as education, health and special aids. In developing regions where there are potentially high levels of human trafficking, governmental corruption, discrimination based on socio-ethnic backgrounds or human rights violation, national IDs can play an important role. Although not all national ID systems are the same, each nation in Latin America has a system in place. The needs for national IDs may vary and some examples are as follows:

Mexico

Cédula de Identidad Ciudadana: In 1990, it was declared by the General Population Law that every Mexican should have the *Cédula de Identidad Ciudadana* (National ID). In 2009, President Felipe Calderón recognized the urgency of having identity cards with biometrical information, instead of using the electoral card (Montalvo 2011). The new ID is supposed to be dispatched through schools to 25.7 million children between 4 and 17 years of age. The cards provide a shield with 8 security features to protect children from impersonation and child trafficking. However, because it is a new program, numerous challenges remain.

Argentina

Documento Nacional de Identidad: In September 2011, Fernández de Kirchner announced that starting January 2012 all babies born in hospitals will be identified and registered with a biometrical ID system. She expects that by 2014 all 40 million Argentineans will be registered in the system (Clarín 2011). Public debates on issues around invasion of privacy and mass surveillance abound and still need to be addressed.

Brazil

Registro General: In 2010, Brazil intended the creation of a biometrical identity document (Biometría 2010). With the issuance of the new ID cards, theoretically, each Brazilian will be registered with a unique national number, which will avoid registration of individuals in more than one state. The card will come with a smart chip containing the person's fingerprints, Internal Revenue Service number and voter card number, as well as information on gender, nationality, date of birth and filiation, photograph, signature and issuing registrar. The new ID card has 17 security features designed to deter fraud and is expected to completely replace the old cards by 2019 (Marques 2010).

Risks of National IDs in Latin America

A national ID program is often a controversial topic, and issuing one to everyone can be an insurmountable challenge. Most nation-states focus on issuing an ID for their legal-aged citizens to participate in elections, monitor and deter criminal activities and trace taxable transactions, but only a few are concerned with the protection of human rights and security of their people from the time they are born. At the same time, there are several reasons registering everyone with a national ID is difficult. Heap and Cody (2008) assert that:

> *The reasons why parents do not register their children include a lack of awareness of the importance of registration; the costs in both time and money of registering a new birth; the distance to a registry office; uncertainty that the child will survive; political turmoil; legal, social or cultural barriers; and the fear of persecution by the authorities.*

Whatever the reason, the consequences of not participating in registration and identification (especially in regions where human rights are often violated) are that marginalized populations in most countries are more vulnerable to victimization. In addition, governments often do not make a sufficient attempt to protect the populations. In Guatemala, an Indigenous person was denied registration because the government administration could not spell Mayan names in the system; in many other countries having unregistered individuals often opens up more opportunities for corrupted officials (Eulich 2011).

Universal IDs would decrease the incidences of corruption and various types of exploitations, yet it faces fierce opposition (The Economist 2012). Without such robust identification systems, poor people of all ages continue to be victimized and exposed to numerous adversities including modern slavery (caused by lack of lawful protection and minimum age or wage restrictions) in their hometowns or elsewhere.

Resolving Risks

UNICEF and the UNHCR organized the Organization of American States (OAS) and the InterAmerican Development Bank (IDB) into a partnership to ensure all two million children born each year in Latin America are registered by 2015 (Mackenzie 2008). Protecting children from trafficking or exploitation is a goal of utmost importance, but registering children with information regarding access to school and public health services, life-long progress from education to employment, and scheduled vaccinations must be considered as equally important objectives. In this regard, a national intervention must be integrative and comprehensive, addressing issues at all levels in an ecosystem, while avoiding potential threats such as fraud and political and economic exploitation. For this and various other reasons, a few measures have been examined; one of them is biometric identification.

International Examples of Universal Biometric ID to Resolve Risks

The concept of establishing a biometric-based national ID database system has quickly been globalized (Lyon 2010). The World Economic Forum links such cards with basic human capital questions such as financial inclusion (WEF 2009, 2011). Although no one has yet studied the full scale impact of the biometric ID and access to financial services, the Center for Global Development estimates that it could increase the ability of foreign aid to directly reach intended recipients by as much as 40% (Gelb and Decker 2011). With such intervention—combined with technology much closer to the people at the base of the pyramid—numerous possibilities can become realities.

One strong example is in India, which has one of the most ambitious biometric identification programs, a system using a remote, wireless biometric scanner that takes information from citizens to create their unique IDs. These permanent, nontransferable IDs are critical for allowing people to start bank accounts or collect deeds to lands for the first time (Government of India 2012). The objective of the program is to accelerate "the emphasis on de-licensing, entrepreneurship, the use of technology and decentralization of governance to the state and local level" (Government of India 2012).

Migration of Indigenous Populations

The India example is of particular importance with respect to their ability to serve large migrant populations. Hundreds of millions of Indians live in slums, on construction sites or even on streets; many have no documents such as residential addresses, passports or income tax certificates to prove who they are. As a result they find it difficult to open bank accounts, register their children for school, get insurance policies or mobile phone numbers or to exercise their right to be protected by the law (Pasricha 2011). According to the International Organization for Migration (IOM 2008), like the economic refugees in India, Indigenous peoples in Latin America often migrate for economic reasons and face similar challenges, with one important additional pressure.

Impact of Climate Change

Although there are many, one of the recently emerging reasons for Indigenous populations in Latin America to migrate is climate change, including desertification and aridity, along with increasing pests and diseases. Recently, many governments have begun to approach climate change as an economic problem, not a mere ecological phenomenon. Most unfortunately, many Indigenous communities end up blaming themselves for the environmental changes they observe in nature (Kronik, Jakob and Verner 2010).

Need for Systemic Intervention

To date, the human rights implications of climate change have not been thoroughly studied. Nonetheless, the increasing and evident effects of climate change threaten a wide range of internationally accepted human rights (e.g., the rights to life, to food and to a place to live and work). Consequently, we must acknowledge that

> *Climate change has a big impact on indigenous peoples around the world. It impacts them in a unique way, due to the deep engagement they have with the land. ... There are communities around the world already being displaced by climate change. While some migration policies have been introduced, to date there has been no coordinated response from the international community to address the needs of "climate change refugees."*
>
> (AHRC 2012, 94, 97)

In short, governments must consider climate change and the migration phenomenon of Indigenous populations in their human rights policy making and support programs.

Possibilities for Systemic Intervention with Mobile Technology

The mass proliferation of mobile phones is improving the quality of life for impoverished populations, including Indigenous peoples, in multiple ways. Aside from the obvious advantage of facilitating communication, access to this technology is increasing income, improving health and safety, making life more convenient and government services more reachable and increasing the GDP of entire countries. More importantly, mobile technology is also helping governments register and bring more people out of the shadows and reducing discrimination and restoring human dignity and rights. In some sense, this simple yet powerful technology is delivering the fruit of the digital revolution to the poorest people in the world (Deninson 2008).

Mobile Innovations around the World

Encouragingly, there have been relevant projects gaining well-deserved global interest and support. However, there are no substantive examples of impact of programs in Latin America among Indigenous populations. Correlative proof of success is found in projects across Africa and Asia among similarly marginalized societies. Some examples with short descriptions are below.

In many regions in Africa, deaf females face a triple stigma: gender, poverty and disability. Often times, these girls are violated because they cannot speak out. To make a difference in their lives, "Cambridge to Africa" employed mobile phones to design special education programs to help deaf children in Uganda (Banks 2012). Self-empowerment, social cohesion and improved literacy skills were all key outcomes of the project.

"Seeds of Empowerment" has implemented several global education projects involving mobile technologies to promote children's literacy and enhance formal and informal education programs in numerous countries including Rwanda, Tanzania, Indonesia, Mexico, Argentina, India and Palestine. Some of their projects involve mobile learning solutions for extremely underserved areas where there is no school or even electricity (Seeds of Empowerment 2011).

Another example of the implementation of mobile technologies for the benefit of the underserved is vouchers for Iraqi refugees in Syria (Irin News 2010). The project is designed to distribute food vouchers to Iraqi refugees in Syria. Families receive one voucher per person every two months (worth $22). They do not have to travel long distances to a distribution center, and they can make their own food choices. The project is for 32,500 refugees across Syria.

Governance out of the Box (GooB) is a birth registration project that has shown success in Liberia and Kenya (Toivanen and Kotipelto 2011). GooB, in partnership with Nokia since 2006, has been funded by the Ministry for Foreign Affairs of Finland and the Alfred Kordelin Foundation with in-country partnership from Nokia Oyj. Basically, GooB utilizes mobile text messaging. A family sends a text message containing the family name and baby name to a designated number, and a healthcare worker provides the family with an actual paper birth certificate and a "new baby" health kit.

iCount is a similar project in the Pacific island nation of Vanuatu. In 2009, Nokia and UNICEF collaborated to deliver mobile birth registration via mobile e-mail using basic Nokia handsets. Health care workers or parents send the family name and baby name to a specific phone number, advertised throughout the island. The pilot project resulted in widespread adoption by health care workers on the main islands.

mBirth, an SMS-based app developed by Tigo Tanzania in collaboration with UNICEF, has early results for the region around the southern city of Mbeya. Mobile registration volunteers now relay birth registrations to central hospitals. This has increased registrations from 9% to 40% in the last six months of 2013.

VaxTrac, a Gates Foundation grantee organization, is working to use basic mobile phones to track vaccinations from birth, which includes conducting birth registrations. They use Android phones for a mobile-based registry with small fingerprint scanners attached via the headphone jack. The system is in place in 120 health facilities across two health zones in the coastal region of Benin (VaxTrac 2010).

A project called Medic Mobile, from FrontlineSMS, led by an NGO called "Plan" (formerly Plan International) is coordinating efforts to document the 51 million children born every year, unregistered, around the globe (Medic Mobile 2012; Plan International 2012). The ownership scale of basic mobile phones is the compelling reason Plan and the partner organizations are employing them.

Mobile Innovations for All for Life

Although the statistics show variations in the patterns of ICT penetration (Cecchini 2005) in different groups (e.g., gender, ethnicity, income, etc.), we cannot deny the penetration of mobile phones into the Indigenous populations in Latin America. Obvious outcomes of the penetration of mobile phones into Indigenous communities are communication, small business transactions such as market price search, relaying daily labor opportunities, mobile minute transfer as means of payment, and social and political awareness.

Spurred by the development of the core wireless technology needed for adoption and use on a massive level, mobile innovations can close the gap between registered and unregistered (Gates and Tolle 2010; Drygajlo 2011; World Economic Forum 2011) and increase educational access and digital literacy through mobile technology (Kim, Miranda and Olaciregui 2008; Kim 2009). As a powerful computation, communication, entertainment, business, government and education tool, mobile technology will only continue to advance in its rapid advancement clock-speed and affect all facets of human life. What is needed is a concerted effort to make sure such technology is designed and implemented for social inclusion and the empowerment of all (i.e., in a way that every unique individual in a society is precisely counted through a national ID program and equally supported to reach her or his full potential) because everyone matters.

According to Fulton (2012) "mobile devices and fingerprint biometrics will come together to enable more widespread civil ID programs that can deliver a multitude of services to even the most rural locations". There are two trends in the implementation of ID programs: 1) acceptance of biometric technology and 2) use of mobile technology (small ID terminals). Developers of this technology are working on the improvement of algorithms for accuracy, efficiency and easier integration.

Mobile cameras, such as those used for finger and iris scanning on slightly "smarter phones", coupled with voice and facial recognition software, will soon be ordinary features. For illiterate elders, smart-strips with preregistered codes can help them make choices for various transactions (Drygajlo 2011). In short, mobile innovations providing support from "right to access" to "birth to education, health, and employment" require a life-long support system for people of all origins and ages.

Conclusion

The end goal is to improve national economies for all and, more importantly, realize individual human rights to citizenship and public services including education. Despite the costs and scale of affordable mobile technology, there is incomplete institutional capacity in Latin America and most of the developing world, which inhibits using them to their full potential (Wilson, Best and Klein 2005). These challenges are varied and substantial, from the limited range of mobile networks to basic corruption that threatens new birth registrations and the achievement of legal status. Nonetheless, Latin America has the greatest opportunity to use mobile innovations for bridging the gap in economic opportunities and fulfilling their legal obligations for children of all socio-ethnic backgrounds.

We hope this paper helped widen the understanding of numerous challenges that poor Indigenous people face today and the positive impact that mobile innovations may have at all levels of the ecosystem locally and globally. Only mobile technology has reached poor people very quickly and profoundly. It has even managed to penetrate the Indigenous world and is an advantage that must be leveraged. The use of mobile technology by and for Indigenous people in Latin America has to be studied on a deeper level, and best practices must be shared in order to cause a paradigm shift at an ecosystematic level.

References

Australian Human Rights Commission. *Climate Change and Human Rights—International and Domestic.* Sydney: Australian Human Rights Commission, 2012. Accessed March 17, 2015. http://www.humanrights.gov.au/sites/default/files/content/social_justice/nt_report/ntreport08/pdf/chap4.pdf.

Banco Interamericano de Desarrollo. "El BID y Los Pueblos Indígenas." 2012. Accessed March 17, 2015. http://www.iadb.org/es/temas/genero-y-diversidad/pueblos-indigenas,2605.html.

Banks, Ken. "Talking with Texts. How Cell Phones Empower Deaf Children in Uganda." 2012. Accessed March 17, 2015. http://voices.nationalgeographic.com/2012/02/07/talking-with-texts-how-a-cellphone-empowers-deaf-children-in-uganda.

Biometría. "Brasil Lanza Documento Biométrico de Identidad." 2010. Accessed March 17, 2015. http://www.biometria.gov.ar/noticias/2010/09/18/brasil-lanza-documento-biometrico-de-identidad-.aspx.

Birdsall, Nancy, Nora Lustig and Darryl McLeod. "Declining Inequality in Latin America: Some Economics, Some Politics." Working Paper 251: Center for Global Development: Publications, 2012. Accessed March 17, 2015. http://www.cgdev.org/content/publications/detail/1425092?utm_&&&%27.

Cecchini, Simone. "Oportunidades Digitales, Equidad y Pobreza en América Latina: ¿Qué Podemos Aprender de la Evidencia Empírica?" 2005. Accessed March 17, 2015. http://www.eclac.org/publicaciones/xml/7/24287/lcl2459e.pdf.

CEPAL. *Pobreza Infantil en América Latina y el Caribe.* United Nations, 2010. Accessed March 17, 2015. http://www.unicef.org/lac/Libro-pobreza-infantil-America-Latina-2010%281%29.pdf.

Clarín. "Lanzan un Sistema Más Amplio de Identificación de Personas." 2011 Accessed March 17, 2015. http://www.clarin.com/sociedad/Lanzan-sistema-amplio-identificacion-personas_0_587341354.html.

Deninson, N. *10 Ways Cell Phones Help People Living in Poverty.* 2008. Accessed March 17, 2015. https://www.newsday.co.zw/2013/02/14/10-ways-cellphones-help-people-living-in-poverty.

Drygajlo, Andrzej. "Biometrics for Identity Documents and Smart Cards: Lessons Learned." *Lecture Notes in Computer Science* 6583 (2011): 1–12.

Eulich, Whitney. *Mayan Guatemalans Disenfranchised because their Government Can't Spell?* 2011. Accessed March 17, 2015. http://www.csmonitor.com/World/Americas/2011/1104/Mayan-Guatemalans-disenfranchisedbecause-their-government-can-t-spell.

Fulton, Jim. "Biometrics, Mobile ID Tech Drive Expansion of Civil ID Programs." 2012. Accessed March 17, 2015. http://www.secureidnews.com/news-item/biometrics-mobile-id-tech-drive-expansion-of-civil-id-programs.

Gasparini, Leonardo, and Nora Lustig. *The Rise and Fall of Income Inequality in Latin America* (Working Papers). Buenos Aires: CEDLAS, Universidad Nacional de La Plata, 2011. Accessed March 17, 2015. http://econ.tulane.edu/RePEc/pdf/tul1110.pdf.

Gates, Bill, and Kristin Tolle. "2010 mHealth Summit: Keynote Luncheon: William (Bill) H. Gates III and Dr. Kristin Tolle." 2010. Accessed March 17, 2015. http://vimeo.com/17647061.

Gelb, Alan, and Caroline Decker. "Cash at Your Fingertips: Biometric Technology for Transfers in Developing and Resource-Rich Countries." *SSRN eLibrary*, Working Paper no. 253, 2011. Accessed March 17, 2015. http://papers.ssrn.com/sol3/papers.cfm?abstract_id=1888376.

Government of India. *Aadhaar Usage: Banking*. New Delhi, India, 2012.

Gradstein, Mark, and Maurice Schiff. "The Political Economy of Social Exclusion, with Implications for Immigration Policy." *Journal of Population Economics* 19, no. 2 (2006): 327–344.

Haapkylä, Heli. "Review of Benefits of Mobility in the Base of the Pyramid (bop) Markets." Master's thesis, 2012.

Hall, Gillete, and Harry A. Patrinos. "Latin America's Indigenous Peoples." *Finance & Development* 42, no. 4 (2005). Accessed March 17, 2015. http://www.imf.org/external/pubs/ft/fandd/2005/12/hall.htm.

Heap, Simon, and Claire Cody. "The Universal Birth Registration Campaign." *Forced Migration Review* 32 (2008): 20–22.

Hopenhayn, Martín, Álvaro Bello and Francisca Miranda. "Los Pueblos Indígenas y Afrodescendientes ante el Nuevo Milenio." Comisión Económica para América Latina y el Caribe (CEPAL), United Nations: Santiago de Chile, 2006.

International Organization for Migration. *Indigenous Routes: A Framework for Understanding Indigenous Migration*. Geneva: IOM, 2008.

Irin News. "SYRIA: WFP Rolls out SMS Food Vouchers for Refugees." 2010. Accessed March 17, 2015. http://www.irinnews.org/Report/90560/SYRIA-WFP-rolls-out-SMS-food-vouchers-for-refugees.

Kim, Paul. "Action Research Approach on Mobile Learning Design for the Underserved." *Educational Technology Research and Development* 57, no. 3 (2009):415–435.

Kim, Paul, Talia Miranda and Claudia Olaciregui. "Pocket School: Exploring Mobile Technology as a Sustainable Literacy Education Option for Underserved Indigenous Children in Latin America." *International Journal of Educational Development* 28, no. 4 (2008): 435–445.

Kronik, Jakob, and Dorte Verner. *Indigenous People and Climate Change in Latin America and the Caribbean*. Washington, DC: World Bank, 2010.

Lyon, David. "IDs in a Global World: Surveillance, Security, and Citizenship." *Case Western Reserve Journal of International Law* 42, no. 3 (2010): 607–623.

Mackenzie, Yamilée. "The Campaign for Universal Birth Registration in Latin America: Ensuring All Latin American Children's Inherent Right to Life and

Survival by First Guaranteeing Their Right to a Legal Identity." *Georgia Journal of International and Comparative Law* 37 (2008): 519–554.

Marques, Luciana. "Governo Lança Novo Modelo de Identidade Nesta Quinta." 2010. Accessed March 17, 2015. http://veja.abril.com.br/noticia/brasil/novo-registro-de-identidade-e-lancado-nestaquinta- e-passa-a- valer-em-2011.

Medic Mobile | Right Tools. Real Impact. 2012. Accessed March 17, 2015. http://medicmobile.org.

Montalvo, Tania L. "La Primera Cédula de Identidad en México Avanza en Medio de la Polémica." CNN México, 2011. Accessed March 17, 2015. http://mexico.cnn.com/nacional/2011/01/14/la-primera-cedula-de-identidad-en-mexico-avanza-en-medio-de-una-polemica.

OECD. *Latin American Economic Outlook 2012.* OECD, 2011a.

OECD. *African Economic Outlook.* OECD, 2011b.

Pasricha, A. "India Requires Citizens to Register for Biometric Identity Number." 2011. Accessed March 17, 2015. http://www.voanews.com/content/india-requires-citizens-to-register-for-biometric-identity-number-130774278/145882.html.

Plan International. "Birth Registration Highlighted at UN Event." 2012. Accessed March 17, 2015. https://plan-international.org/where-we-work/geneva/news/birth-registration-highlighted-as-protection-tool-at-un-event.

Psacharopoulos, George, and Harry A. Patrinos. *Indigenous People and Poverty in Latin America.* World Bank, 1994.

Seeds of Empowerment. "1001 Stories." 2011. Accessed March 17, 2015. http://seedsofempowerment.org/whatwe- do/1001-stories.html.

The Economist. *India's Identity Scheme.* 2012. Accessed March 17, 2015. http://www.economist.com/node/21542763.

Tigo Tanzania. "Mobile Birth Registration." Accessed March 17, 2015. http://www.tigo.co.tz/mobile-birth-registration.

Toivanen, H., & H. Kotipelto. "Birth Registration Using Mobile Phones Advances Civil Rights in Developing Countries." *Crisis Management Initiative and VTT Technical Research Centre of Finland.* 2011. Accessed March 17, 2015. http://phys.org/news/2010–10-birth-registration-mobile-advances-civil.html.

Tomei, Manuela. *Indigenous and Tribal Peoples: An Ethnic Audit of Selected Poverty Reduction Strategy Papers.* Geneva: International Labour Organization, 2005.

UNDP. *Human Development Report 2004: Cultural Liberty in Today's World.* New York: UNDP, 2004.

UNESCO-OREALC. *Education for All in Latin America: A Goal within Our Reach.* Regional EFA Monitoring Report 2003. Santiago: UNESCO/OREALC, 2004.

VaxTrac/Bill & Melinda Gates Foundation. 2010. Accessed March 17, 2015. http://www.gatesfoundation.org/How-We-Work/Quick-Links/Grants-Database/Grants/2010/11/OPP1025341.

Wilson III, Ernest J., Michael L. Best and Dorothea Kleine. "Moving Beyond 'The Real Digital Divide.'" *Information Technologies & International Development* 2, no. 3 (2005), iii–v.

World Economic Forum. *ICT for Economic Growth.* Davos, Switzerland, 2009.

World Economic Forum. *The Global Information Technology Report 2010–2011.* Davos, Switzerland, 2011.

World Hunger. "2012 World Hunger and Poverty Facts and Statistics." 2012. Accessed March 17, 2015. http://www.worldhunger.org/articles/Learn/world%20hunger%20facts%202002.htm.

5 The Indigenous Digital Collectif

The Translation of Mobile Phones among the iTadian

Gino Orticio

Introduction

This chapter reports on an adaptation of Callon and Law's (1995) *hybrid collectif* to research conducted on the usage of mobile phones and Internet technologies among the iTadian people. The term iTadian is an ethnonym of the Indigenous people living in Tadian, Mountain Province, in the Cordillera region of the northern Philippines.

The term *collectif* is derived from Callon and Law (1995, 485) and describes an emergent network of heterogeneous elements, or actants. Using actor-network-theory (ANT) as a methodological approach, the results expose the fluid and heterogeneous character of the Indigenous hinterland, replete with actor-networks of human and non-human actants, or participants. By adopting this approach we seek to understand complex and changing situations, looked at without bias or preconceptions. The major modes of communication across recent history in the study area are presented, along with their shifts and diversity of character. Associations of mobile technologies to other actants are retraced. The resulting heterogeneity reveals multiple realities where mobile technologies dynamically compete for existence, indispensability, stability and even irreversibility. Along the way an *Indigenous/digital collectif* coalesces, resulting in an emergent assemblage translated by Indigenous enactments and digital technologies.

The chapter starts with a review of academic literature on the field of information technologies (IT) and Indigenous communities, identifying specific theses that emerge within the texts. Through these emerging theses, I invite controversy on the current scholarship, its paucity and a need for an alternative approach.

Literature Review

Research scholarship on information technologies and Indigenous peoples tends to gravitate toward four major metatheoretical presuppositions, namely: *instrumentality, conditionality, annexation* and *human agency*. This section will examine each one, together with its proponents and their premises.

The Thesis of Instrumentality

This first thesis shows the efficacy of information and communication technologies (ICTs) in promoting Indigenous peoples' welfare and development. A study of Canada's First Nations' Aborigines saw the potential of the Internet as a medium to "speak in their own voices" and form alliances with other like-minded organizations (Alexander 2001, 295). Then, in one cross-country comparison of Latin American Indigenous communities, Internet cafés were significant in promoting solidarity (Delgado, 2002).

By using the public sphere (Habermas, 1989) as an analytical framework for the Mapuche movement, Internet technology facilitated the constitution of a "counter-discursive" movement, as opposed to the depiction of an informed and free public sphere by the Chilean state authorities (Salazar 2003). Another study among Indigenous youth in the Sydney suburb of Redfern, Australia, highlights the educational approach of problem-posing propounded by Freire (1970) in the adoption of web-based technology, by using it as a tool to promote critical thinking and expression (Sengara 2005). Further, Leclair and Warren (2007) argue that Indigenous communities can increase their general capabilities by using the Internet for electronic commerce, e-learning and community web portals (Leclair and Warren 2007, 2–11). In another case, mobile phones are seen as potently "placed resources" (Blommaert 2010 in Auld, Snyder and Henderson 2012) where stored individual and collective media can enhance literacy among the children and youth of the Maningrida community in Arnhem Land (Auld, Snyder and Henderson 2012).

The thesis of instrumentality is predicated on an inherent capacity of digital technologies to effect development gains among Indigenous people. These include increasing knowledge and literacy (Auld, Snyder and Henderson 2012; Leclair and Warren 2007; Sengara 2005), affecting material gains (Leclair and Warren 2007; Longboan 2011), building coalitions across boundaries (Alexander 2001; Soriano 2012), promoting community solidarity (Delgado 2002) and/or enhancing communicative competence (Longboan 2009; Salazar 2003; Sengara 2005), critical thinking (Sengara 2005) and the narratives of celebrating difference (Longboan 2009, 2011). The premise is that information technologies are instrumental in affecting these changes. This thesis portrays Indigenous peoples as reactive to information technologies although not necessarily passive. There is a subtle yet persistent insinuation that Indigenous peoples need development, which could be delivered through information technology.

The Thesis of Conditionality

The second thesis focuses on preconditions for information technologies to achieve developmental or other gains. Mizrach (1999) identified the importance of technical knowledge among the Cheyenne River Sioux. Practical knowledge, in terms of skills in operating a personal computer, navigating the web and creating web pages, was regarded as a necessary condition for Indigenous peoples to utilize IT as a tool for cultural resistance (Mizrach 1999, 379). Another study of Internet usage by the Maori runs parallel to the above findings, again with an emphasis on building the capacity of Indigenous people to gain Internet access and the technical skills to enhance their expression of Indigenous identity and heritage (Gorre 2007). A study of web-based media on the Nishwabe Aski Nation in Canada noted the important role of local leaders in shaping digital technologies as a means of facilitating inter-community communication and personal skills development (Budka 2009). Further, another article on the pilot implementation of Internet access among the Pirlangimpi people of Northern Australia saw the importance of technical skills to operate the Internet, together with the decisive role of community-based councils, in utilizing technological access (Morrison 2000). In a similar vein, Vaughan (2011) points to community participation as a contributing factor for the success of government-initiated IT programs within a number of remote Indigenous communities in Australia.

The foregoing literature tends to foster the idea that technical skills (Gorre 2007; Mizrach 1999), community leadership (Budka 2009; Vaughan 2011) or a combination of both (Morrison 2000) relate to stabilized social conditions that affect cultural resistance, self-identification and community welfare—outcomes considered to be developmental to Indigenous peoples. This thesis of conditionality simply states: existing information and communication technologies will help Indigenous peoples, provided that the people possess (or strive to possess) the distinct personal or organizational qualities necessary for their operation. This thesis shows causation determined by the premise of the inherent capacity of information technologies to evoke gains *provided* Indigenous people *configure* themselves according to its technical workings. It also suggests an image of Indigenous communities as local, configurable and yet dynamic units of analyses amidst a global information infrastructure. Conditionality presupposes Indigenous people as active, knowledgeable subjects lacking the necessary IT skills for self-development.

The Thesis of Annexation

Another emergent theme, that of annexation, tends to point toward attributing Indigenous features to information technologies. Glowczewski (2005) notes the importance of *"reticular"* or networked thinking. By arguing that reticular thinking is an "ancient Indigenous practice", Glowczewski finds its congruence to Internet hyperlinks, allowing Indigenous peoples to explore new meanings, encounters and creations (Glowczewski 2005, 34). Reticular thinking has been associated with ancient and Indigenous cultures, in contrast to the epistemologies of conventional mainstream thinking. A further instance is the emergence of the creative use of IT and old technologies by the international Indigenous peoples' movement and Indigenous peoples themselves (Alia 2009). Based on these two cases, the thesis of annexation appropriates emergent elements of the Indigenous (or what authors regard as Indigenous), which then ascribes power through Indigenous exceptionalism. It depicts Indigenous peoples and their world views as an exceptional, coherent and homogenous portrait. The annexation suggests that Indigenous culture and societies are inherently exceptional. It further creates an image of cultural integrity that is inert across space and time. On the other hand, annexation sees indigeneity as a condition that is hidden in the first instance. It waits to be discovered, and information technology—as a product of modernity—has the propensity to unlock and bring out its potential to engage in the broader conversation of cultures.

The Thesis of Agency

The last thesis focuses on Indigenous peoples as reflexive, sophisticated participants in the use of IT in accordance with determined goals and projects. For example, Aikenhead (1997) emphasizes "autonomous appropriation"—reflexively appropriating aspects of Western knowledge systems without complete assimilation (Aikenhead 1997, 551). A comparable idea is international Indigenous peoples' organizations utilizing Internet technology to (re)construct memory and selfhood, claim historical continuity and assert collective attachment to Indigenous language, land and socio-political systems while consciously hiding imperfect acts from public scrutiny (Niezen 2005). Another dramaturgical variant presented by Belton (2010) and Soriano (2012) reveals how members of Indigenous peoples' organizations negotiate and appropriate online spaces in order to reclaim, create alliances and advocate their rights, while building up credibility and administrative control in offline spaces. Whether it is autonomous appropriation or the practice of dramaturgical strategies, the foregoing literature portrays Indigenous groups as reflexive entities. This thesis of human agency highlights Indigenous people as competent, reflexive, and sophisticated participants in harnessing information technology. In contrast with annexation, human agency accentuates the idealized image of Indigenous as proactive while recognizing and attenuating the mundane and ignoble parts of social action. Further, there is a subtle tendency to invoke an anthropocentric argument that human beings are intelligible subjects having the utmost systematic competencies for structuring modes of action, enabling them to exert domination and power over technology.

Oligoptica

It is argued that the metatheoretical presuppositions of instrumentality and conditionality are related as they adhere to ontological stability within hierarchical levels of structure. The narrative of structuration is evinced through the inertness of technological systems through which resources flow from global structures into human experience and interaction. Tangible and intangible elements of high technology are seen flowing from centers normally contiguous to global centers of capital accumulation. These elements trickle down to local and micro levels of structuration where Indigenous communities are seen to be located. Conversely, Indigenous peoples and communities are depicted as local, configurable and dynamic yet fragmented amidst these technological flows. The scaling is palpable. Through this structuring, the analytical power is contingent on how these elements hierarchically flow and trickle down arbitrarily nested levels of scale. There is no recursive causation and reflexivity. Latour (2005, 181–183) describes this kind of ontology of a totalizing analogue of a well-connected homogenous whole as an *oligoptica*. An oligoptic perspective depicts few, narrowly defined, albeit strongly connected, elements of social reality with the intent of portraying a stable, coherent system; but it only remains as long as its structures hold and inner betrayals and messy extraneous elements are avoided in the narrative (Latour 2005). As such, oligopticas are illusory because considerable elements are rendered absent, ignored or effaced in the process when considering social reality (Latour 2005, 49).

Panorama

Conversely, the metatheoretical presuppositions of annexation and human agency depict the Indigenous as one reflexive, homogenous formation, with banal and ignoble forces effectively obfuscated or even made absent within the narrative. The Indigenous is further employed as a coherent category, with clear distinctions in the tropes of exceptionalism, struggle, historicity, culture and reflexivity. The notion of information technologies is reduced to inert artifacts, with rational and reflexive Indigenous peoples utilizing them in their day-to-day lives. In contrast to oligoptica, Latour (2005) describes this ontological perspective as a *panorama* (2005, 187–188). A panoramic ontology fundamentally depicts social reality as a "well-ordered zoom", which attempts to assemble a fully coherent vision (Latour 2005). However, like any panoramic vision, such portrayal is simply located within the confines of the zoom itself and nothing more, because its frontiers are skillfully closed from scrutiny. It is illusory because the depiction it renders is so excessively coherent (Latour 2005, 188). In the case of IT and Indigenous people, a panoramic depiction romanticizes Indigenous peoples and shows the passivity of technological artifacts.

The apparent fascination of panoramic and oligoptic ontologies in contemporary studies on IT and Indigenous people orientates research toward creating recurrent themes of order, structure, romanticism and Indigenous exceptionalism. These recurrent themes call for re-examination of current epistemic and methodological strategies in the field, a re-examination that acknowledges the existence of heterogeneities (Law 2004, 2009), the agency of human and non-human actants (Latour 1992, 1993, 2005), multiple realities (Haraway 1985; Mol 1999, 2002) and betrayals or *translations* (Callon and Latour, 1981; Latour, 2005, 106–108) within a "system".

Critical to methodological and theoretical development in the current study is the investigation into how technologies associate themselves across other sites within the Indigenous hinterland in order for these technologies to exist, stabilize and maintain irreversibility. The sociology of translation starts with the premise that there are no strong and stable social ties in existence but only associations traceable through things that have equivalent or ambivalent meanings (Latour 2005, 106–108). These meanings allow at least two actants to co-exist. The basis of translation is not simply determined by association through direct causation or a parsing of equivalent meanings of stable elements, but is more about taking into account how human and non-human actants transiently stabilize other actants apparently belonging to competing ontologies (Latour 2005, 6–12).

Data collection in this research involved focus group and semi-structured interviews from different periods from December 2009 to January 2011 involving a number of field visits to communities within the municipality of Tadian. Ethnographic techniques were conducted to describe enacted practices and cultural definitions of digital technology usage in the community.

Background: The Study Area

The municipality of Tadian, Mountain Province, is located in the Cordillera mountain range in the northern part of Luzon Island, the Philippines. The area is up to 1,350 meters above sea level and is characterized by a steep and rugged terrain. The town center of Tadian can be reached by public transportation from the Philippine capital of Manila in about 12 hours. From Baguio City, its nearest metropolitan area, Tadian can be reached in about five to seven hours via hired van or bus. Travel time increases and gets more difficult during the monsoon season (June to October) due to regularly occurring landslides.

The ethnonym *iTadian* (pronounced *ee-tah-jan*) is the term for a native person of Tadian, or as a collective noun, the "people of Tadian". The iTadian is a sub-group of the larger Indigenous population in the Cordillera collectively described as the Igorots. It has been established that iTadian communities were living in their ancestral region before Spanish colonization. Before Spain took claim to some parts of the Philippine archipelago in 1521, the landscape of the area now called Tadian could be characterized as decentralized clusters of communities called *ili*, each consisting of an average of 2,000 persons administered by a council of "petty plutocrats" called *kadangyan* (Scott 1982, 135). The *kadangyan* were pre-eminent men who had gained prestige and influence through the maintenance of conditional lineality, ownership of land and domestic animals, as well as ostentatious (and competitive) ritual feasting (Scott 1982). As part of tradition, *kadangyan* status was not directly passed down through generations but rather by showing one's capability to increase personal and community resources substantial enough to be passed on to the succeeding generation. Display of ownership is done by the regular upkeep of the farm and through delegation among close kin groups. Domesticated beasts, such as water buffaloes, pigs and chickens, are also accumulated by the *kadangyan* to display wealth, prestige and power. These beasts are regularly used for ostentatious feasts but are also sacrificed during times of crises like low harvests, epidemics and natural calamities. Aside from the butchering of beasts and the distribution of meat among the villagers, customary feasting includes the performance of traditional music and dances within the vicinity of the *kadangyan*'s household. This display of ostentation can last for two or more days, and its performance reinforces the stature of *kadangyan* as a figure of prominence not only within the community but to other communities as well (Scott 1982, 135–136). Decisions concerning the *ili* are enacted by an assembly of elder fathers called the *lallakay* or *amam-a,* particularly on matters concerning interpersonal disputes about resource use, management of resources, such as irrigation, intertribal conflicts and crimes, and days of planting, harvesting, recesses and feasts (Scott 1982, 135–139).

The above sociopolitical institutions have persisted through different waves of colonizers and more recently a centralized government bureaucracy but have adapted to contemporary social conditions. The *lallakay* system of community decision-making is still practiced on top of government institutions, although it has now been replaced with a more egalitarian council of elders that includes females. The *kadangyan* elite still exists within the community but has lost its strong influence in community affairs.

Most iTadian today are multilingual as a result of historical and contemporary dynamics. The mother tongue of the iTadian is Kankanaey, but they also use the Ilocano language, predominantly spoken in northern Luzon as a means of communicating with people from the nearby lowland regions. English is an American colonial legacy taught from primary school and now through mass and social media. In fact, there are some accounts of English as the preferred choice of discourse among the elders when communicating with strangers and acquaintances. It is also the preferred choice of formal correspondence within the state bureaucracy and schools. The mainstream Tagalog, the officialized national language, is taught at schools and transmitted by the Manila-based mass media and digital technologies like mobile phones and Internet. It can be safely stated that Indigenous people in the Philippines are polyglots; it is a rarity for an iTadian to be monolingual because of the various exposures to cultural and linguistic enactments.

The iTadian engage in a variety of livelihood strategies or "portfolios" (Rovillos, Orticio and Bangaan 2004). The most common is a mix of subsistence and crop farming on ancestral lands. Most subsistence farms are planted with rice, vegetables and root crops. Together with raising livestock and poultry in small numbers, subsistence farming acts as a buffer against food shortages. The planting of commercial crops such as beans and leafy vegetables is one of the primary livelihood strategies of which households avail themselves in order to access cash. A few households run convenience stores called *sari-sari* (Tagalog for "variety" or "assortment") stores catering to community needs. Other cash-based strategies are dependent on the individual's skill and qualifications which range from daily wage labor to contractual and permanent employment based on technical and higher-level skills. There is a shortage of available jobs in Tadian, enough reason for a large proportion of its population to leave the place in search of greener pastures. Most emigrants settle in adjacent regions and cities while several seek contractual employment and sometimes permanently settle abroad.

Results

Pre-Digital

Before the emergence of second-generation, general service mobile technology (GSM) mobile phones in 1999, the iTadian mostly availed themselves of a traditional system of transmitting letters and packages, a public terrestrial phone system and a municipal telegraph office.

The Paw-It

Prior to the advent of digital technologies, the iTadian's conventional mode of communicating across their geospatial boundaries was the *paw-it* system (pronounced *"pow-eet"*, Ilocano "to send, relay"), still in use. Mail and packages are sent by the few buses shuttling between different communities each or every other day. The system starts with the preparation of the package. Clearly written on each item is the recipient's name, address and/or the address of the nearest drop-off point. These are normally the nearest *sari-sari* store in the community. The sender then takes the item to the nearest bus transit point. The sender waits for the bus and coordinates with the bus conductor. After both parties reach a common understanding of the whereabouts of the addressee store, the conductor charges the sender a fee based on the distance covered by the bus. The item is received by the conductor who then places it in a secure spot in the bus together with other packages. As the bus arrives at the store, the conductor temporarily disembarks and promptly hands over the package to the storekeeper. The storekeeper writes down the name of the recipient on a list, which is then displayed on a wall outside the store. The item is eventually collected, usually within a few days. In spite of being the de facto post office in the community, there are no paid arrangements made to the delegated store or to the storekeeper as an intermediary to the system.

For some iTadian parents whose children study in Baguio City or nearby regions, the *paw-it* system means a lot in sustaining their children's daily cost of living. It is particularly onerous for those living in far-flung communities who must travel for two to four hours, crossing rivers and mountain passes to reach the nearest bus transit point. They must schedule their time of departure from home in order to catch one of the few buses plying the route.

There was one case of a parent in the small community of Mabalité who had to walk from home at four o'clock in the morning in order to reach the bus terminal at Sumadel before the bus left at about seven o'clock the same morning. He then had to walk back home after the transaction was made.

The *paw-it* system is circumscribed by a network of roads, stores, trails and existing bus routes. A *paw-it* package relies on the performance of a concatenation of human and non-human actants in order to reach its recipient. These include the bus, the road system, the memory of the bus conductor, the existence of the nearby stores as drop-off points, the climactic conditions and notices. Letters and packages travel by various means from the sender until they materialize in the hands of the recipient. The *paw-it* is contingent on its intermediaries, such that if one fails (through mechanical breakdown, landslides, road closures, monsoons, frailty of memory and so on), the reliability of the system is in question. However, the *paw-it* system holds up as one of the most durable means of long-distance communication in Tadian. This is because of the translation of mobile phones, which will be discussed in detail further.

Terrestrial Telephone and Public Calling Station

In 1965, a terrestrial telephone service was planned by the national government to interconnect the offices of constituent communities in Tadian. However, its construction was abruptly terminated when a strong typhoon destroyed its main infrastructure and the project was quickly abandoned due to paucity of funds. What replaced it in the early 1970s was the installation of a telephone system for local government offices as well as a public calling station located at the Tadian town center. Not many people patronized the station because using telephones was not seen as a necessity except during times of urgency. When locals availed themselves of the service, most calls were well planned and abridged because of the expensive calling fees. Toward the end of the 1990s the landline calling station was succeeded by calling stations operated by private enterprises using first generation, analogue mobile phones. These stations were fitted with phones connected to an aerial antenna in order to enhance transmission. The antennas were oriented to far-flung base stations. When compared to their landline counterpart, mobile phone calling stations were easier to access but still very costly for providers to operate. Nevertheless, a few of these enterprises had sprouted in several communities where electricity was available together with a steady stream of radio waves from base stations. By 2000 these stations gave way to privately owned, second generation mobile phones.

The failure of public calling stations can be traced to a number of factors. The natural topography and climactic conditions besetting the region made it difficult to install a working network of phones. This was compounded by low government support, which affected its installation completion. The lack of interest from local people due to high transaction costs and the low number of contacts who could be reached by phone was also a main contributor.

Official Post

The official postal service is funded and operated by the national government through the local post office. Official post is mostly patronized by government offices for transmitting official correspondence. Postal deliveries are only for those inhabiting areas adjacent to the town center where the post office is situated. On occasions when letters are addressed to far-flung communities, the official post is eventually sent through the traditional *paw-it* system. The post office is also reliant on community-based networks to send letters on its behalf. These could be municipal staff, from whom post office personnel can seek favors, living within the same village as the addressee and/or affiliated through kinship. Proxies can also be people who happen to live close by and are known to be associated with the addressee. Proxies are requested to send the letter directly or via the *sari-sari* store near the address indicated on the postal item.

The formal state bureaucracy, together with informal networks of kin and *kaîlian* ("fellow community member"), sustain the survival of the official post. The post office still holds relevance through the sustained budgetary and administrative allocation from the national government, as well as the existing need for paper-based correspondence as a standard system of communication among government agencies. In addition, customary networks based on kinship and community assume the task of sending letters to remote villages. Hence, the official postal system could not operate effectively without the help of enduring enactments of kinship and domicile-identification.

Municipal Telegram

Like the postal systems, the municipal telegraphic office is operated by the national government. Community members avail themselves of telegraphic services for two reasons: to send ordinary and social telegrams and to remit money. The process requires customers to fill out a form with their names and addresses as well as those of their recipients. A maximum of 20 words is the standard for a telegraphic message, and any extra words entail an excess fee. Once the appropriate amount is paid, the message and contact details on the form are transcribed into Morse Code by the radio operator using a telegraph. The operator selects the assigned frequency of the receiving station by turning a dial on the radio transceiver. He or she then presses a series of long and short audio pulses on the telegraph paddle. The first series of tones denote the sender's and receiver's call signs. The operator pauses for a few seconds, waiting for a response from the other end. Once a response is heard and acknowledged, another series of transmissions in Morse Code is made until the sender's message is transmitted and acknowledged by the receiver.

Receiving ordinary telegrams entails the same process. Using a manual typewriter, received messages are typed directly onto a *pro forma* sheet of paper loaded on specially built rollers. The sheet contains markings, indicating an official government document along with serial numbers and perforated lines to separate each form. Once the message has been typed, the form is cut from the roll, folded and enclosed in a plastic envelope. It is then sent to the receiver through official post, or by *paw-it* as a last resort. Special social telegrams are typed on a decorative card marking an occasion (for example, Valentine's Day, Christmas, birthdays and graduations). These are then enclosed in a colored envelope and sent to the recipient. A higher fee is charged for these.

People also remit money by means of telegraphic transfer. The sender hands the money to the radio operator who then writes an official receipt of the transaction. Then, just like a telegram, the radio operator sends information of the transfer to the receiving office and receives the proper acknowledgement. The recipient is notified about the money via telegram and proceeds to the telegraph office to redeem the money. The telegraph officer then checks the serial number of the notification telegram with the records. If both numbers match, the recipient is required to sign a receipt and is handed the exact amount remitted by the sender.

According to the municipal telegraph officer patronage of the telegraph system was regular until the advent of mobile phones in 2000. She recalled that event as a sharp drop in clients from about 20 telegrams a day to only a few weekly transactions, mostly coming from inter-agency correspondence from government offices. She related that it would be a very good day if the office transmitted a total of five messages. In early 2012, the telegraphic equipment of her office just gave off a crackling sound characteristic of radio static.

The sustainability of the municipal telegraphic office is heavily reliant on the conventions of state bureaucracy, particularly in formal inter-agency communication. If not for its continued subsidy by the national government, the telegraph would likely have been decommissioned sooner. The abrupt loss of patronage by the iTadian happened when they decided to subscribe instead to the more convenient, quicker and easier to use short message service (SMS) available on mobile phones. Betrayal was quick as the telegraph teetered toward obscurity.

Mobile Phones

The emergence of mobile phones contributed to the evanescence of analogue-based communication systems in Tadian, namely the public telephone station and telegraph office. Most subscribers of analogue technologies have shifted loyalties to the mobile phone not only because of its portability, affordability and convenience, but also its personification of an emergence of a new personal and intimate means of communication.

Mobile phones were introduced at around 1999 by a few iTadian who had relatives overseas working as contract workers. The first few phones were given as gifts (or in a few cases long-term loan) by these relatives as a means of communicating with them while abroad, mostly by SMS texting, which is considerably cheaper and more convenient. However, kinship-based mobile communication eventually gave way to more texting in general as the number of contacts increased, particularly kin and friends based in metropolitan areas—who were already mobile users, having access to the technology since 1997. The use and access of mobile phones increased gradually over time, especially after the construction of two base stations at the highest altitude providing more individual access.

Ten years after it was first introduced among a few iTadian, the mobile phone was already a fixture within the community. Mobile phone usage rose dramatically in 2006, when two cell sites were installed near the municipality. There are claims that about nine out of ten adults and youth have their own mobile phones and that most households have at least one prepaid mobile phone unit. Most phones are second-hand units, either bought second-hand or handed down. Most units used are earlier models, having a small, low-resolution, LCD displays and keypads. There is a preference for phones that have built-in LED torches (flashlights) that are important for traveling during the night. These phones are the cheap, rugged and compact ones that are not easily destroyed and whose users do not mind having them scratched or appearing worn out, as long as they continue to transmit calls and text messages.

Youth are the heaviest mobile phone users and are the ones generally bold enough to explore their phones' system and inner workings. Students are the more frequent texters, sending almost 80 to 200 text messages daily, especially on days when they can avail themselves of unlimited texting. Calling someone through a mobile phone is considered a last resort, because it costs more and uses more battery power. The iTadian youth are more adept with the intricacies of mobile use. They are more dexterous and up to date with the latest models. They know when texting is free to subscribers. They are also most likely to respond with belligerence when losing a mobile phone. Asked whether they could live a day without a mobile phone, they reported that they would become uneasy, immediately borrow from a friend when needed, get angry/cry/curse when the phone got lost and gradually experience a feeling of distance from their family and friends if this occurred.

Adults would use mobile phones more sparingly than their younger counterparts, but they also share the thought of making sure that these are within reach most of the time. Some adults take their cell phones with them to their mountainside farms as they tend to their crops. In areas where cell site signals are intermittent, the farmers make sure that they place their phones in an area where they can access reliable streams of signal during the day. This can be anything from a rock protrusion to a tree or a sturdy branch that can be used as a pole to which they can secure their phones with string or rubber bands.

Responses vary among communities when asked how dependent they are on mobile phones for their daily activities. Some iTadian living in areas where mobile signals are intermittent relate that they may feel initial discomfort when they lose their phones but would not hesitate to use *paw-it*, official post or face-to-face communication. On the other hand, participants living in communities where mobile signals are stable and reliable admit that they would be more concerned about how they can be contacted without their mobile phones. A few people express concerns that some contacts might forget them, saying *"Awan samet makalaglagip kanyak"* ("Nobody might remember me"). They further claim that they would resort to borrowing other people's phones if necessary.

Most community elders report that mobile phones are not an essential part of their lives and that they usually seek help in using them (Figure 5.1). They rarely send texts and mostly rely on children or grandchildren to read aloud the text messages. Their difficulties arise not only in reading small characters on the display unit, but also in the task of sending messages by entering words and navigating through menus with the keypad. Another difficulty is decoding text messages. Messaging commonly involves the use of abbreviated words and a phrasing structure that the elders find difficult to understand. Aside from the difficulty of decoding, the elderly recipient may get the impression that the sender is rude or angry. One elderly participant saw it as a "deterioration of the English language". This concern is particularly poignant for the elderly because they are especially known to be good English speakers, having been mostly taught by the American colonial school system. Another interesting point by several elderly respondents was the issue of privacy. One centenarian pointed out that "there are no more secrets to hide" because texting exposes the message to anyone who can access the phone and read its messages.

Figure 5.1 A sample of a mobile phone used by an elder. The mobile phone numbers are taped and in plain sight for the user's convenience. Note that a few numbers were intentionally blurred by the author for privacy.

Traditional respect for iTadian elders, generational seniority and strong kin-based relations make it easier for them to seek help in mobile phone use. The older generations are customarily living with an adult child and are taken care of by younger members of the kin group as a reciprocal practice of deference where elders are still regarded as a living heritage and embodiment of old wisdoms and oral traditions.

Stable Black-boxes and Highly Competent Mediators

State bureaucracy and kinship networks based on kin and community are among the black-boxed realities operating within Tadian. The state bureaucracy asserts itself as the singular, effective entity providing the warranted services to its constituents. It recognizes other realities within the hinterland but asserts its ascendancy over them across space and time. The state bureaucracy partially exists on the basis of a continuous flow of government-controlled resources, expert systems and information that regulate its survival. On the other hand, customary kinship and community networks operate translocally by invoking familial associations.

There are a number of highly competent mediators across the iTadian hinterland. These fully connected actants oscillate between enrolment and betrayal across various competing realities. They can easily transform themselves within the exigencies of the hinterland. These highly competent mediators could be the *sari-sari* stores that transform themselves into post offices through the *paw-it* system, or they could be kinsfolk shifting their role as dispatchers of letters or spokespersons to adopters of mobile phones as a preferred means of communication. They could also be the road system being used by a plethora of enactments. These mediators exercise their indispensability within the hinterland, either by asserting their materiality (in the form of road systems), simplicity (for example kinship) or deferral (for example *sari-sari* stores).

Translations of Communication Systems

All past and present modes of communication in Tadian involve the process of stabilizing enrolment in order to sustain the interest of the human and non-human actors. Communication relies on a concatenation of human and non-human elements to fully function and sustain itself. However, dissidence always poses a threat to the intermediaries' survival, sometimes resulting in its full termination, as with the community-based telephone system.

Shifts

Ten years after the emergence of the mobile phone, there is an indubitable shift in the way the iTadian communicate. Mobile phones have eventually emerged as stabilized enactments among the Tadian hinterland, their stability relying on the operation of important intermediaries like the on-grid national electric system and the base station. The base station, known by its common term "cell site", is a black-boxed actor composed of different intermediaries whose main function is to send and receive messages to their mobile subscribers through digitized electromagnetic waves that transport information and communication to human mobile phone users. Mobile phones also enrolled mediators in the mountainous terrain of Tadian and deployed close kin and the youths as allies.

Mobile phones made the iTadian do things they had never done before. Mobile phones carved up new cartographies and resultant courses of action. New cartographies occur when the landscapes are redefined as a result of enrolment in mobile phone networks. New places within the home (for example, the windowsill) and community (for example, trees or mountain tops) quickly become matters of concern as a requisite for mobile communication. With these new cartographies emerged new courses of action, including taking mobile phones to work, placing them on window-sills and fastening mobile phones to poles and branches.

Mobile phones introduced new standards among the iTadian. New communication codes were introduced in the form of texting. Texting uses a combination of common and abbreviated words that could accommodate the character limit provided by mobile phones and telecommunications companies. Texting relies on the mutual understanding of local languages, as well as the acceptance of brevity by the mobile phone users. The use of text abbreviations that suit local pronunciations represented a response among human actants. By subverting official lexical structures, users are able to conserve keystrokes, thereby saving transmission costs.

To address the complexities of the configuration of the mobile phone, elders taped their own phone numbers on the outside of their mobile phones. They thus dealt with one less difficult process, remembering their phone numbers or navigating the phone's menu to locate them for reloading credit.

New forms of kin-based social arrangements emerged through mobile phone use. One is the willing subscription of younger members in teaching their parents the fundamentals of mobile phone usage. Two points are evident here. The first is that young members are the most effectively enrolled in this actor-network. The second is that respect for elderly members is reinforced through enactments derived from mobile phone usage. Both observations point out that by effectively enrolling youth and kin-based enactments, the mobile phone formed a strong alliance in sustaining its relevance among the iTadian.

With these new standards come new forms of actions and anxieties. As mobile phones have enrolled human actors, life and presence are now constituted in terms of their competence in replying to texts as an indicator of life. The technology of the short message service, or SMS text, and not the mobile phone *per se,* has become associated with one's life and well-being. The cold formalism of mobile phones immediately dissipates when translated as things of affection. This is the case among a number of elderly iTadian women who view mobile phones as appendages of their life partners. The phone becomes multivalent, an allegorical device for which various enactments are rendered present or absent. Its hidden heterogeneities expose multiple realities struggling for survival. In this case, the mobile phone has not only become strictly an object of communication, but also a surrogate "life partner", a repository of endearing messages, a purveyor of presence and a mediator of kinship. It can be seen that iTadian fluently oscillate from heterogeneous enactments through mobile use.

Liberation and Inayan *Ethics*

A sense of liberation (De Leon 2007) is also evoked through mobile phones. Mobile phones allow human actants to express affection and even anonymity without the gaze and judgment of traditions. However this does not go unnoticed. Indigenous enactments translate acceptable use of mobile phones through the iTadian concept of *inayan*, a shared concept among Kankanaey-speaking Igorot communities, the iTadian included. It is invoked to dissuade a person from acting on something that may cause harm to others, with the strong admonition that the action may cause retribution and adverse consequences to the actor (Delima 2006; Scott 1957). The Indigenous ontology where *inayan* is practiced recognizes an unknown force operating beyond individual action powerful enough to incur adverse consequences against socially unacceptable behavior.

The application of the *inayan* can be seen in one case where mobile phones were used to take photos and upload digital images of iTadian who perished in a tragic landslide in the Kayan community. The photos showed the corpses partially covered in blankets with faces showing various grimaces of death. The people who took and posted the photos were the subject of widespread consternation among the iTadian community. It was met with dismay for they were warned that taking photos of the victims was *inayan* and *abig* (taboo) in accordance to the *ngilin* (solemn observance of the dead). The dead were exploited like a freak show, and the "perpetrators" had to face the elders in shame.

The translation of the mobile technologies and Indigenous culture can be seen in the above situation where *inayan* is applied to the acceptable use of mobile. However, it is also interesting to note that the "perpetrators" decided to take photos regardless of *inayan*. This case shows that not only does Indigenous tradition exert coercion and regulation within the exercise of mobile technologies, but other factors also guide individual action.

New Anxieties

New anxieties have also emerged, particularly when a mobile phone user cannot send text messages or a mobile phone is lost or taken away.

These anxieties arose when one user could not inform others of her presence. Conversely, this is also manifested through positive feelings when someone is able to send a text message or even a sense of affirmation or presence—of being remembered—as in the case of the elderly female textmate. This lethophobia derived from mobile phone usage translates life or existence into a digital version of presence, manifested through digitized text. A momentary therapy for lethophobia is to get hold of a phone and assure oneself of being remembered. A new encumbrance is formed, that of anxiety of being effaced from memory by kith and kin.

The Mobile Paw-It System

The *paw-it* system has been an acceptable method of transporting letters and packages in Tadian. It has outlasted other methods of communication but only through the translation of mobile technologies into its system. Mobile phones are used by the bus conductor not only to coordinate actions of his trade but also as a means of assuring the parties involved of the status of the package.

The inclusion of mobile phones did not make the *paw-it* system more efficient since respondents claim there were no significant improvements in delivery times brought about by mobile phones. In fact, today more work is done by the conductor who now has the added responsibility of answering and coordinating deliveries through text messages.

Mobile phones operate to allay doubt, insecurity and distrust among its human participants when betrayals come into the *paw-it* system. Betrayals may be in the form of breakdowns, landslides, road closures during monsoons or accidents, among other things. What has significantly changed is the additional knowledge and capacity of both sender and recipient to coordinate their activities while the item is in the system. First is the capacity to be informed that a package is on its way whereas before it was a matter of regularly checking local drop-off points, such as *sari-sari* stores for *paw-it* arrivals. Second is the knowledge that the package is available for collection, which allows the sender and recipient to coordinate their actions through text messaging. Third is the capacity to learn about the status of the package itself, as an assurance that it was not lost or broken while in transit.

Both *paw-it* and mobile phones are well stabilized modes of communication. Potential betrayals and anti-programs (Akrich and Latour in Bijker and Law 1992, 259) present themselves as uncertainties and are placed out in the open. The *paw-it* system and mobile phones each have something to offer that the other has not. However, and more important, both systems expose the uncertainties of the other. Mobile phones cannot transport bulky messages and large packages, as the *paw-it* system does, hence it cannot produce material outcomes to the recipient. On the other hand, mobile phones stabilize the system by assuring both sender and receiver of the status of packages. As with all systems of transport and communication, the *paw-it* system is replete with risks that may jeopardize its operation. Inclement weather, road closures, mechanical breakdowns, vehicular accidents and criminal activities among others threaten the trustworthiness of the system. Mobile phones act as a mediator to the parties, assuring them that the package will be received. Hence, the portability and speed of mobile communication assures the materiality of the parcel, which consequentially translates into stabilization of the *paw-it* system and ensures its survival against other means of transporting mail and packages.

Discussion

The above results reveal a fluid Indigenous hinterland replete with competing enactments, among them Indigenous customs and digital technologies. The existence and stabilization of mobile technologies require a dynamic enrolment and counter-enrolment of other realities within Tadian. In contrast with the theses of instrumentality and conditionality, mobile technologies do not stand on their own, hierarchically transforming the Indigenous hinterland. They instead enroll other human and non-human actants and render themselves interesting enough by creating new forms of meaning and equivalences. They also have to navigate and compete with other technologies and their enactments, such as the case of the mobile *paw-it* system. Furthermore, while Indigenous socio-political practices, such as the *inayan*, may have been transiently annexed into the ethical practice of mobiles, this has not resulted in total and stabilized submission; this reveals ontological politics (Mol 1999) among competing fractional realities. Therefore, Indigenous communities must be regarded not as inert objects fixed through time and space, but as living within an "arena of multicultural strife" (Pieterse 2001, 68).

It is here that I address the theoretical, epistemic and methodological paucity of panoramic and oligoptic research as it deals with the pitfalls of associating topological formalisms (Fuller and Goffey 2012), linear/causal analyses (Marres 2012) and hierarchical or nested structurations of global, macro, meso, micro, local configurations (Hand and Sandywell 2002) in the analysis of socio-technical relations. Methodologically, tracing the actants and their associations ensures that research results are not overtaken by ideological interests (Latour 2005). More importantly, the panoramic and oligoptic depictions of such interactions are avoided and the messiness (and translations) of the *Indigenous/digital* are brought out in the open.

The Indigenous/Digital Collectif

Therefore the Indigenous/digital collectif is an emergent assemblage of a heterogeneous and fractional reticulation of human and non-human actants translated within the realities of Indigenous institutions and digital technologies. Non-human actants such as (but not limited to) mobile phones translate the Indigenous hinterland by enrolling existing traditional institutions and stabilizing its importance through time and space as it struggles for relevance and irreversibility. The use of a typographic slash ("Indigenous/digital") instead of a hyphen ("Indigenous-digital") is deliberate, as the former highlights the fractional character of these two realities, instead of an oscillation between two inert states. Hence, the *Indigenous/digital collectif* is stabilized by the dynamic enrolment and counter-enrolment of digital and Indigenous realities along with other pre-existing and emergent enactments. New modes of action, meanings, cartographies and anxieties have emerged within the *Indigenous/digital collectif* as a consequence of stabilization. However, it must be emphasized that a *collectif* is neither stable nor permanent. Like any actor-network, it continuously struggles for irreversibility by creating alliances and checking against betrayals that may jeopardize its relevance. The *Indigenous/digital collectif* of the iTadian successfully rendered the obsolescence of a few modes of communication while it co-opted others like the *paw-it* system. It operated along the rules of Indigenous institutions while it warded away the risks and uncertainties of the hinterland.

The *Indigenous/digital collectif* among the iTadian is an emergent concatenation of heterogeneous and partially connected networks of human (i.e., Indigenous elders, youth, adults, distant kin and spouses) and non-human (i.e., schools, shops, diaspora, traditions, farms, markets, base stations, mountains, terrain, money, etc.) using digital technologies as a reference point. Heterogeneity and ontological politics operate through the translations of a variety of other networks such as kinship relations, traditional exchange of goods, traditional knowledge and beliefs, and even academic and bureaucratic requisites.

References

Aikenhead, Glen S. "Toward a First Nations Cross-Cultural Science and Technology Curriculum." *Science Education* 81 (1997): 217–238.

Alexander, Cynthia J. "Wiring the Nation! Including First Nations? Aboriginal Canadians and Federal e-Government Initiatives." *Journal of Canadian Studies* 35, no. 4 (2001): 277–296.

Alia, Valerie. "Outlaws and Citizens: Indigenous People and the 'New Media Nation.'" *International Journal of Media and Cultural Politics* 5, no. 1 and 2 (2009): 39–54.

Auld, Glenn, Ilana Snyder and Michael Henderson. "Using Mobile Phones as Placed Resources for Literacy Learning in a Remote Indigenous Community in Australia." *Language and Education* 26, no. 4 (2012): 279–296. doi: http://dx.doi.org/1 0.1080/09500782.2012.691512.

Belton, Kristy A. "From Cyberspace to Offline Communities: Indigenous and Global Connectivity." *Alternatives* 35, no. 3 (2010): 193–215.

Bijker, Wiebe E., and John Law. *Shaping Technology/Building Society : Studies in Sociotechnical Change.* Cambridge, MA: MIT Press, 1992.

Blommaert, Jan. *The Sociolinguistics of Globalization.* Cambridge, UK and New York: Cambridge University Press, 2010.

Budka, Philipp. "MyKnet.org: How Northern Ontario's First Nations Communities Made Themselves at Home on The World Wide Web." *The Journal of Community Informatics* 5, no. 2 (2009).

Callon, Michel, and Bruno Latour. "Unscrewing the Big Leviathan: How Actors Macro-Structure Reality and How Sociologists Help Them Do So." In *Advances in Social Theory and Methodology : Toward an Integration of Micro- and Macro-Sociologies*, edited by K. Knorr-Cetina and A. V. Cicourel, 277–303. Boston: Routledge & Kegan Paul, 1981.

Callon, Michel, and John Law. "Agency and the Hybrid Collectif." *South Atlantic Quarterly* 94, no. 2 (1995): 481–507.

De Leon, Kristinne J. L. "A Study of Internet Cafes: Identity, Freedom and Communicative Extension." *Philippine Sociological Review* 55 (2007): 37–49.

Delgado, Guillermo. "Solidarity in Cyberspace: Indigenous People Online." *NACLA Report on the Americas* 35, no. 5 (2002): 49–53.

Delima, Purificacion G. "The Polysemy of 'Inayan' across Tribal Groups in Mountain Province: Exploring Evidence of Culture-Specific Ethical Concepts in Language." Paper presented at the ninth Philippine Linguistics Congress, University of the Philippines, January 25–27, 2006.

Freire, Paolo. *Pedagogy of the Oppressed*. New York: Herder and Herder, 1970.

Fuller, Matthew, and Andrew Goffey. "Digital Infrastructures and the Machinery of Topological Abstraction." *Theory, Culture and Society* 29, no. 4–5 (2012): 311–333.

Glowczewski, Barbara. "Lines and Criss-Crossings: Hyperlinks in Australian Indigenous Narratives." *Media International Australia Incorporating Culture and Policy* 116 (2005): 24–35.

Gorre, Ligeia D. "Expression of Identity: Maori *Ta Moko* and the Utilization of the Internet." Masters thesis, University of Southern California, 2007.

Habermas, Jurgen. *The Structural Transformation of the Public Sphere: An Inquiry into a Category of Bourgeois Society*. Cambridge, MA.: MIT Press, 1989.

Hand, Martin, and Barry Sandywell. "E-topia as Cosmopolis or Citadel: On the Democratizing and De-Democratizing Logics of the Internet, or, toward a Critique of the New Technological Fetishism." *Theory, Culture and Society* 19, no. 1–2 (2002): 197–225.

Haraway, Donna. "A Manifesto for Cyborgs: Science, Technology and Socialist Feminism in the 1980s." In *Reading Digital Culture*, edited by D. Trend. Oxford, UK: Blackwell, 1985.

Latour, Bruno. *Reassembling the Social: An Introduction to Actor-Network-Theory*. Oxford and New York: Oxford University Press, 2005.

Latour, Bruno. *We Have Never Been Modern*. Cambridge, MA.: Harvard University Press, 1993.

Latour, Bruno. "Where Are the Missing Masses? The Sociology of a Few Mundane Artifacts." In *Shaping Technology/Building Society: Studies in Sociotechnical Change*, edited by W. Wiebe E. Bijker and John Law, 225–258. Cambridge: MIT Press, 1992.

Law, John. "Actor Network Theory and Material Semiotics." In *The New Blackwell Companion to Social Theory*, edited by Bryan S. Turner. (140–158). Chichester, U.K. and Malden, MA: Wiley-Blackwell, 2009.

Law, John. *After Method: Mess in Social Science Research*. 1st ed. London and New York: Routledge, 2004.

Leclair, Carol, and Sandy Warren. "Portals and Potlatch." In *Information Technology and Indigenous People*, edited by Laurel Evelyn Dyson, Max Hendriks and Stephen Grant, 1–13. Hershey, PA: Information Science Publishing, 2007.

Longboan, Liezel C. "E-gorots: Exploring Indigenous Identity in Translocal Spaces." *South East Asia Research* 19, no. 2 (2011): 319–341.

Longboan, Liezel C. "Igorots in the Blogosphere: Claiming Spaces, Re-constructing Identities." *Networking Knowledge: Journal of the MeCCSA Postgraduate Network* 1, no. 2 (2009): 1–16.

Marres, Noortje. "On Some Uses and Abuses of Topology in the Social Analysis of Technology (Or the Problem with Smart Meters)." *Theory, Culture and Society* 29, no. 4–5 (2012): 288–310.

Mizrach, Steven. "Natives on the Electronic Frontier: Technology and Cultural Change on the Cheyenne River Sioux Reservation." PhD diss., University of Florida, 1999.

Mol, Annemarie. "Ontological Politics. A Word and Some Questions." In *Actor Network Theory and After*, edited by John Law and John Hassard, 74–89. Oxford, UK, and Malden, MA: Blackwell/Sociological Review, 1999.

Mol, Annemarie. *The Body Multiple : Ontology in Medical Practice*. Durham: Duke University Press, 2002.

Morrison, Perry. "A Pilot Implementation of Internet Access for Remote Aboriginal Communities in the 'Top End' of Australia." *Urban Studies* 37, no. 10 (2000): 1781–1792.

Niezen, Ronald. "Digital Identity: The Construction of Virtual Selfhood in the Indigenous Peoples' Movement." *Comparative Studies in Society and History* 47, no. 3 (2005): 532–551.

Pieterse, Jan N. *Development Theory : Deconstructions/Reconstructions*. London and Thousand Oaks, CA: SAGE Publications, 2001.

Rovillos, Raymundo, Gino C. Orticio and Clint Bangaan. "The State of Food Insecurity of Indigenous Peoples: An Integrative Report on Indigenous Peoples' Communities across the Philippines." In *Our Harvest in Peril: A Sourcebook on Indigenous Peoples' Food Security*, edited by Ann Loreo Tamayo and Maurice Malanes, 13–56. Quezon City, Philippines: EED Philippine Partners' Task Force on Indigenous Peoples' Rights, 2004.

Salazar, Juan Francisco. "Articulating an Activist Imaginary: Internet as Counter Public Sphere in the Mapuche Movement, 1997/2002." *Media International Australia, Incorporating Culture and Policy* 107, May (2003): 19–30.

Scott, William H. *A Vocabulary of the Sagada Igorot Dialect*. Chicago: Philippine Studies Program, Dept. of Anthropology, University of Chicago, 1957.

Scott, William H. *Cracks in the Parchment Curtain and Other Essays in Philippine History*. 1994 ed. Quezon City: New Day Publishers, 1982.

Sengara, Ryan. *Redfern Kids Connect: Technology and Empowerment*. Master's thesis, University of Western Sydney, 2005.

Soriano, Cheryll R. "The Arts of Indigenous Online Dissent: Negotiating Technology, Indigeneity, and Activism in the Cordillera." *Telematics and Informatics* 29, no. 1 (2012): 33–44.

Vaughan, Donna. "The Importance of Capabilities in the Sustainability of Information and Communication Technology Programs: The Case of Remote Indigenous Australian Communities." *Ethics and Information Technology* 13, no. 2 (2011): 131–150. doi: 10.1007/s10676-011-9269-3.

6 Private Mobile Phones and Public Communication Drums in Rural Papua New Guinea

Amanda H. A. Watson and Lee R. Duffield

Introduction

The voice of a traditional communication drum can be heard over great distances. Yet now in Papua New Guinea (PNG) it is hearing, by phone, the voice of a loved one who has moved far away from home for work, marriage or studies that brings the greatest delight. As recently as 2007, most areas of this Pacific island nation had no form of telephony available. Apart from radio, modern communication forms have been restricted predominantly to the urban areas where only a small percentage of the people reside. Landline telephones, television, Internet, facsimile machines and so on have never reached the majority of the inhabited areas.

Commercial ownership and a mixed system of market competition, with public and private providers, were introduced to the mobile telecommunication sector on 1 July 2007, and this resulted in a rapid extension of mobile network coverage to rural areas. This chapter examines the role of mobile telephony in rural communities in PNG with particular emphasis on the attitudes expressed by rural villagers. It reports on a threshold study, with research conducted during the earliest stages of mobile phone adoption in ten such villages in Madang Province.

PNG is situated to the East of Indonesia and North of Australia. Although culturally rich (Reilly 2004, 480), PNG is a developing country even while the economy is growing, largely due to resource extraction projects (Gouy et al. 2010). There is a population of approximately 6.5 million people (CIA World Factbook 2014), most of whom reside in rural areas (Watson 2011b, 170; Watson 2012, 44), where there is often very limited access to services, such as transport, health, education, electricity, landline telephones, postal service and banking. Until 2007, communities in such regions fell outside of mobile phone coverage areas.

Developing nations have experienced significant growth in mobile telecommunication markets, particularly since 2000 (ITU 2010), but the Pacific has been one of the last regions of the world to experience widespread mobile phone access and uptake. At the same time smartphones were taking off in Japan and Korea, most people in PNG were only just starting to gain access to telephony for the first time. When this field research was conducted (2009), nearly half of the adults surveyed owned a mobile phone. In most cases, these were cheap, basic handsets. Thus this early adoption context is very different from the contemporaneous situation in Asian countries and elsewhere. As one of the last countries to achieve widespread mobile phone coverage, PNG is unique, important and worth understanding.

This research sheds light on the uptake of mobile phone technology among an "early adopter" community, and all the findings are conditioned by that fact. The uses and impacts of mobile phones identified in the study are those of a community making first contact with a technology that is already in advanced form—digital with several capabilities and functionalities. Mobile phone penetration is already so high in most other countries that comparable opportunities to understand fresh import of the technology have been lost. This research project provides a rather unique insight into the impact of the introduction of mobile phone technology on communication patterns within a specific society. Further, the research contributes to academic debates around the interplay between communication and technology and presents wider conceptual ramifications.

Theoretical Basis for Study

In establishing context for this study, the authors referred to the concept of communicative ecology, defined by Tacchi, Slater and Hearn as the range of communications in a given setting (2003, 15). It includes "the complete range of communication media and information flows within a community" (Horst and Miller 2006, 12), such as mass media and traditional communication techniques. Studying the mobile phone in isolation would not adequately convey the "repertoire of communications skills and resources" (Tacchi, Slater and Hearn 2003, 15) that exists in the field sites in this study.

Traditional communication methods vary across PNG. In Madang Province, a key method involves drumming on a large, wooden drum or slit gong (Leach 2002) known locally as a *garamut*. This instrument is made from a "hollowed tree trunk" (Blades 1975, 44; also Herzog 1964, 313; Leach 2002, 713) and has, when struck with a wooden stick, "a resonant sound with considerable carrying power" (Blades 1975, 44), reaching many kilometers in favorable conditions (Leach 2002, 718). Such devices are or were usually "used to communicate over distance utilizing a series or code of beats" (Leach 2002, 715). For hundreds of years this traditional communication method has "enabled limited messages to travel very rapidly over great distances" (Unwin 2009a, 17). The continued use of the *garamut* in some parts of Madang Province is important as it provides some perspective regarding the introduction of the mobile phone. Through using the *garamut*, people in many parts of Madang Province "have never actually needed to move physically to be able to communicate with one another" (Unwin 2009a, 18), which means that "very rapid communication over considerable distances is not a particularly new concept" (Unwin 2009a, 18).

An important part of the communicative ecology in many places is now mass media, although "for most people even this is not available since they live in rural poverty without electricity supply, movie theatres or transmitters" (Hamelink 1995, 2). This disparity in access to media and communication technologies is referred to as the "digital divide" between those who do and those who do not have access to modern communication technologies (van Dijk 2005, 1). While the digital divide is often discussed with reference to differences between countries, inequities are also evident within countries, commonly between rural and urban areas.

People residing in rural parts of PNG and other developing nations have missed out on many stages of technological development (Harvey 2000, 62; Unwin 2009a, 19) that have been experienced in developed countries and some urban centers in the developing world. In cases where advanced technologies (such as mobile phones) are available in a place that has never had access to its predecessors (such as the telegraph and landline telephones), this is a phenomenon referred to as "leapfrogging" (Sanzogni and Arthur-Gray 2006, 671).

Many mobile phone studies in various countries have found that users identify social uses of mobile telephony as amongst the most important (Aakhus 2003, 30; Bakke 2010, 365; Bell 2005, 71–72; Chib 2009, 3–4; Donner 2008, 150; Heeks 2008; Johnsen 2003, 163–166; Law and Peng 2008, 55; LIRNEasia 2007; Mpogole, Usanga and Tedre 2008; Pelckmans 2009, 30; Souter et al. 2005; Tabinas and Guzman 2010; Walsh, White and Young 2007, 126), more important than business uses or other such "functional" practices. This finding contradicts expectations in some literature that mobile telephony will aid in business success, job-seeking and income generating activities (for example, Belt 2008; Beschorner 2007, 2–3; Brinkman, de Bruijn and Bilal 2009, 74, 87; Jensen 2007; Mariscal and Bonina 2008, 76; Tabinas and Guzman 2010; The Economist 2005, 2005b, 2007). Recent, insightful papers on mobile phones in PNG make up a useful literature in this field (Andersen 2013; GSMA 2014; Jorgensen 2014; Lipset 2013; Logan 2012; Singh and Nadarajah 2011; Sullivan 2010; Suwamaru 2014; Telban and Vavrova 2014; Temple 2011; Yamo 2013).

Research Design

This research project set out to determine the roles of mobile phones in the communicative ecologies of rural villages in PNG. The research intended to address the attitudes, behaviors and experiences of people and communities during the earliest phase of adoption. It also sought to explain changes in social relationships or economic activities. The research questions, research methodology and ethical considerations have been well documented elsewhere (Watson 2011a, 53–78; Watson 2013, 159–162). Suffice it to say that the main methods employed were semi-structured interviews with 17 key informants and a survey questionnaire in ten villages (n=748). The research was conducted during the early days of mobile phone adoption, with the length of time that the villages had been within mobile network coverage ranging from three to 24 months. The map in Figure 6.1 shows the locations of the ten villages included in the study, in Madang Province. The town, also named Madang, is indicated on the map, and this is where services such as public phones and banking are available.

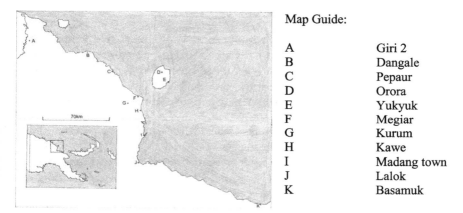

Map Guide:

A	Giri 2
B	Dangale
C	Pepaur
D	Orora
E	Yukyuk
F	Megiar
G	Kurum
H	Kawe
I	Madang town
J	Lalok
K	Basamuk

Figure 6.1 Map of Madang Province field sites. (created by Amanda Watson).

Results

The research findings have been documented in detail elsewhere (Watson 2010, 112–118; Watson 2011a, 79–236; Watson 2011b, 173–176; Watson 2013, 162–167), and a summary of key findings is provided here. In the case of all ten villages, there has never been any landline telephone infrastructure available. Therefore, this research on mobile phones documents the first ever access to any kind of phone in these communities. About half of the respondents owned a mobile phone, but there was low usage overall. Men were more likely to own mobile phones than women, and young people were more likely to own mobile phones than older people. Respondents who were not mobile phone owners did report some rare use of phones, on occasion borrowing a mobile phone belonging to a relative to make a specific phone call. Mobile phone services offered by financial institutions had not been adopted by rural villagers at the time of the field research, although people frequently requested, sent and received mobile phone credit.

The research indicates that the introduction of mobile telecommunications has generally been viewed positively, although several negative concerns have been strongly felt. The main benefit reported related to enhanced communication with relatives and friends living away from their home villages. The social value of the mobile device far outweighed the perception of other benefits, such as use in emergencies (68 out of 748, i.e., 9.1%) and business transactions (31 out of 748, i.e., 4.1%). People in rural areas were particularly pleased to be able to contact relatives and friends residing in other parts of PNG, especially in the towns and cities. This social contact was highly valued and was seen as the main benefit of the new technology in people's lives.

The cost of owning and operating a handset has proven to be a challenge. Recharging handset batteries has also proven to be problematic, particularly in areas with no main electricity supply. Perceived damaging effects of mobile phone access, as described by village respondents, related to disruptive sexual liaisons, crime and pornography. Commonly held beliefs were that the use of these devices was causing an increase in adultery and sexual promiscuity. There were frequently expressed concerns about the possibility of criminals using mobile phones to plan illegal activities such as roadside hold-ups and thefts. A small number of people mentioned concerns about children and young people accessing or being exposed to pornographic images on mobile phone handsets. Others were concerned about mobile devices distracting students from their schoolwork.

The survey established that villagers had limited access to mass media and extremely limited access to computers, with little understanding of computer-based communication technologies such as Internet or e-mail. At the time of the field research, traditional communication techniques remained an important part of the communication landscape in Orora and were used on an almost daily basis but had died out in Megiar (and their use was not explored in the other eight villages). During field research in Orora, the primary researcher frequently heard rhythms being beaten on *garamuts*, indicating regular, almost daily use of *garamuts* to convey a range of messages. Specific rhythms indicated particular, differing messages. Local leaders and interview participants likened the *garamut* to the mobile phone.

> *The garamut is the mobile belonging to the people of Papua New Guinea. And the mobile which has come in now, it comes from you people. Now it's come in to Papua New Guinea. So now if you're in Australia, I will be able to contact you easily. ... The garamut is ours, it is our mobile phone. If we want to ask everyone to come to work, attend a meeting or a gathering and so on. If we don't have the garamut, how will everyone come? We've got a garamut and if we drum it, everyone will come and gather.*
>
> (Interview respondent, 2009; this text is an English translation of a quote in Tok Pisin)

The *garamut* is no longer used for communication within Megiar or between Megiar and surrounding villages. When there is a traditional song and dance performed, the *garamut* can still be used to provide the beat for the music, but it is not used for sending and receiving messages as it was in earlier times. The reasons for the discontinuation of *garamut* communication are unclear. The change pre-dated the introduction of mobile phone reception. One interviewee in Megiar suggested that the *garamut* had not been used as a communication tool since about 1970, and that no *garamuts* had been created since that time. He was unsure of the reasons behind the change but wondered if the construction of a road network may have caused the *garamut* to fade from daily life. In the remaining eight villages where research was conducted, data was not collected on traditional communication techniques.

Discussion

Since the completion of this field research in 2009, several other studies have been published, which in the main support its findings. Regarding positive perceptions, communication with relatives far away is mentioned in several papers (see below) and emergency use of mobile phones is cited (GSMA 2014, 23; Sullivan 2010, 9; Yamo 2013, 83). The finding regarding women's lower ownership of mobile phones is supported by GSMA (2014, 5), and meanwhile two papers suggest that increased access to mobile telephony could empower women (GSMA 2014; Logan 2012, 8). It is suggested that increased access to mobile phones, the Internet and social media could increase political engagement (Logan 2012). In another study conducted in 2013, it was found that there was potential for mobile phone service to be used in development efforts (Watson 2014), and this is supported by Suwamaru (2014) in the health and education sectors and Yamo (2013) in health.

Regarding negative concerns about mobile phones, three papers discuss the high cost of mobile telephony for people in PNG (GSMA 2014, 5, 40; Singh and Nadarajah 2011, 6; Sullivan 2010, 9), three papers mention concerns raised about criminals utilizing mobile phones (GSMA 2014, 39; Lipset 2013, 344, 350; Sullivan 2010, 10), and several papers discuss concerns about marital discord (Andersen 2013, 323, 329; GSMA 2014, 5, 30–39; Jorgensen 2014, 9–10; Lipset 2013, 344–345; Sullivan 2010, 9, 10). Other negative concerns that have been raised in the literature include students being distracted at school (Sullivan 2010, 11) and lack of electricity for recharging mobile phone handset batteries (GSMA 2014, 30–31, 38–40; Sullivan 2010, 10; Yamo 2013, 96–97).

This study adopts a theoretical framework that utilizes the communicative ecology concept in order to understand the place of mobile telephony within the broader communication context of each setting. It is a premise highly useful to the researchers, that understanding the context, or the communicative ecology, will help to understand mobile phone use. For example, kerosene lamps are commonly used in the evenings for lighting, but their glow does not spread widely and they are awkward to carry; therefore, people appreciate being able to use torches (or flashlights) in their mobile phone handsets when walking about at night (also Telban and Vavrova 2014, 226). In the villages studied, modern communication forms were not found to be widely accessible or commonly used. In keeping with other research findings, radio was the most popular medium (InterMedia et al. 2012, 10; Rooney, Papoutsaki and Pamba 2004, 8), while newspapers were not delivered to rural areas (InterMedia et al. 2012, 10; Rooney, Papoutsaki and Pamba 2004, 7).

Examining the communicative ecology of these ten villages in Madang Province, where previously there had been no access to any kind of phones, showed impacts of the very beginning of modern, instantaneous, two-way, distance communication. As the research showed, mobile phones were entering spaces where people did not have access to other modern communication and media forms. The introduction of mobile telephony was the first major shift in these communicative ecologies since radio services began broadcasting in PNG some decades earlier (see Issimel 2011, 148 and Nash 1996).

PNG rural areas had been situated on the far side of the digital divide, with almost no access to mass media and modern communication devices. The introduction of mobile network coverage meant that these places were leapfrogging the telegram, the pager, public phones and landline telephone infrastructure, and instead people were becoming able to use portable, digital communication devices in their rural home settings. The immediate outcome of change could be observed as something of a feat of human adaptation. Village people would adopt the new technology and apply it cautiously to subsume it into a pattern of life. It is the observation of this study that the new form of contact with outside modernity would not be permitted to overturn lives in any catastrophic way, although problems with it were noted from the beginning.

This study found very limited evidence of positive impact upon income generation and business uses of mobile phones, while later studies have repeated the assertion that mobile phones can aid in these two ways (GSMA 2014, 22; Sullivan 2010, 21; Suwamaru 2014, 1). Contrary to expectations generated by a body of literature that presents a dominant view that communication technologies can and will lead to economic improvement (for more information, see Unwin 2009b, 1; von Braun and Torero 2006, 4–5; Weigel 2004, 16), poverty reduction was not a perceived benefit mentioned by villagers, at least not in the early days of mobile phone uptake.

Social uses of mobile phones, not strictly functional for business or service purposes, have been repeatedly referred to by users as the main benefits of mobile phone access. Thus, the research adds weight to mobile phone studies suggesting that the primary advantages of mobile phones in many settings are for social interaction among loved ones, particularly those residing far away (in PNG, see Andersen 2013, 319; GSMA 2014, 22; Jorgensen 2014, 3; Lipset 2013, 341; Singh and Nadarajah 2011, 6). In rural PNG, the mobile phone addresses the needs of marginalized communities by providing a much-appreciated ability to communicate. It is communication itself that is so important for respondents, particularly due to the lack of modern communication tools in rural areas of PNG up until 2007 (the digital divide).

In rural PNG, as in other developing countries, a key way the mobile phone serves the needs of marginalized communities is by providing a much-appreciated new means of social communication. In very many instances across the country, the voice of a close family member has been heard by siblings or parents for the first time in years. Community members have given evidence that in such ways the new technology is helping to fulfill a felt need for richer human communication. The communication chances of the people concerned, up until now largely left out of media and communication flows around the globe, have been much enhanced. In this research, communication was mentioned as one of the main benefits of mobile phone technology. Communication itself was valued by rural villagers—not as a means for increasing income, gaining material goods or addressing other, more pragmatic 'needs', but as an intrinsic benefit. To suggest that the roll-out of mobile phone reception into rural areas has resulted in better, more effective and more efficient communication appears to be stating the obvious. However, for these rural people who have lived in a veritable vacuum of formal and mass communication, the advent of extensive mobile network coverage in rural areas has had major outcomes. It has both filled a gap in practical service, as in sending messages to town, and it is felt as restorative in helping to sustain the social and emotional reality of communication itself. Miller has argued that communication itself should be viewed as an essential human need, and therefore studies should include "the evaluation of communication in its own right" (2006, 41).

Apprehension about mobile telephony negatively impinging upon marital stability was found to be substantial in the village survey, while similar concerns were expressed in interviews. There were similar concerns about telephony facilitating extra-marital affairs in late 1800s United States, with the advent of landline telephones (Marvin 1988, 69), and within the last decade regarding mobile phones in Jamaica (Horst and Miller 2006, 169–170) and Sudan (Brinkman, de Bruijn and Bilal 2009, 80). Mobile phones allow for more private communication than may have been possible previously (particularly in the communal lifestyle of rural villages in PNG), and thus the mobile phone "provides its users a site to explore their desires" (Ellwood-Clayton 2006, 125). However, mobile phones can also be used to uncover illicit relationships (Lipset 2013, 344–345), particularly if suspicious partners check text messages and call logs in their partners' handsets. The text-based communication of SMS exchange "provides a main means by which infidels are exposed" (Ellwood-Clayton 2006, 140) in various countries (Horst and Miller 2006, 170; Mustafa, Siarap and Suan 2009, 223). In the Madang area, monogamy is common (Morauta 1974, 12), but there were concerns raised regarding the use of mobile phones in adulterous behavior. In fieldwork in Orora and Megiar, there was evidence that mobile phones may have aided in the conduct of extra-marital affairs and also that mobile phones had assisted in their detection in some instances.

The common practice of people dialing random mobile phone numbers in the hope of making new acquaintances was mentioned by some respondents in this research and has been explored in two key papers (Andersen 2013; Jorgensen 2014) and mentioned elsewhere (GSMA 2014, 31; Lipset 2013, 350; Sullivan 2010, 10, 23). The capacity for private communication creates an opportunity to dissemble and present a persona that may not be authentic, particularly when interacting with new acquaintances: "the lure of the phone friend operates by disguising social differences as well as through the contraction of physical distance" (Andersen 2013, 331). Thus, it is not only the capacity to communicate in private, but also the ability to do so over great distances that is enabled through access to mobile telephony. Due to difficulties and costs associated with travel, "the mobile phone's capacity to facilitate intimate contact across geographical and social distance is perhaps particularly exciting in PNG" (Andersen 2013, 318–319). Privacy itself "is viewed as an important advantage of mobile phone communication" (Jorgensen 2014, 3), particularly for young people in rural, communal villages who enjoy being able to form new friendships without the oversight of village leaders and older relatives (Jorgensen 2014, 11; Watson 2011b, 173).

In this research project, the first use of a metaphor comparing the mobile phone and the traditional communication drum known in the Madang area as the *garamut* was by an informant during fieldwork in Orora (Watson 2010, 117–118). In fieldwork by Sullivan in the Sepik area, an informant wondered whether *garamuts* would continue to be relevant in the wake of spreading mobile phone reception (2010, 9). But Telban and Vavrova argue, also with reference to the Sepik, that "it would be a mistake to see mobile phones simply as 'modern' replacements" (2014, 226) for *garamuts*. In their field site, *garamuts* are viewed as "powerful spirits, with their own names" (2014, 226), owned by senior male leaders of specific clans. Further research would need to be undertaken in various PNG village contexts to determine whether the presence of mobile phone reception is decreasing the use of traditional communication methods.

The mobile phone and the *garamut* differ in some key respects. Importantly, messages transmitted using *garamut* beats are typically public whereas the very nature of private communication enabled by mobile phones is at the root of some of the primary negative concerns raised by respondents (such as the coordination of extramarital sexual liaisons). There are "attributes intrinsic to the new technology" (Lipset 2013, 336) that enable new ways of interacting with others. In particular, mobile phones allow for private communication, which has not previously been possible in many PNG village communities. Both devices have played (or now play) an important role in connecting people to family, extended family and community networks. In modern PNG, most families are split between the home village and the nearest town (Jorgensen 2014, 3; Watson 2011a, 239). The mobile phone has helped to link family members in these two locations, and this key aspect of communication was prominent in the research data.

On the evidence available now it might be surmised that in the future increasing use of mobile phones may invoke a change in society to further adjust to modernity. If not the potentially shattering impact of "first contact" with Europeans and their possessions, it would see an involuntary shift in social relations. The instance is given of the change from *garamut* to phone as a manifestation of a shift from public to private social interactions. Further, the matter of sexual relations would form part of the same dynamic. In both situations (phones instead of drums, and disruptive sexual liaisons facilitated by phones), persons once subject to approbation or sanction, and regulated in their behavior within the community, might now act privately and individualistically, which may contribute to changing patterns of social relationships and lead to conflict.

Despite differences between mobile phones and communication drums, both are mediums that help to fulfil an enduring human need for communication. The *garamut* met communication needs in an effective manner during earlier times as its voice was heard over a wide area and it conveyed key messages quickly. In a previous era, the *garamut's* voice reached far enough to encompass all the members of the extended family group. In contemporary PNG, as families find themselves dispersed across the nation, the mobile phone fulfills a felt need to be able to keep in touch and check up on one another. But like the drum, the mobile phone is imperfect. The drum did not afford people a means of private communication, and it has been overtaken by modernity in many places. The mobile phone enables communication over greater distances and fits in with the literacy and numeracy taught in schools, but it does not operate without the availability of some form of electric power. Nor—not even with the advent of "social media" in the offing (Logan 2012)—can it be expected to directly broadcast community-wide messages. Clearly the mobile phone is not the same as the *garamut*. However, the mobile phone has not entered a void; instead it has joined communities where people have been using their voices (and the voices of their communication drums) for centuries.

Conclusion

This research was undertaken during the very early days of mobile phone network expansion in PNG. It provides a rare opportunity to study early adopter attitudes and behaviors. As most parts of the globe are now covered by mobile phone networks, it may be that this was one of the last chances to discuss first impressions with people who had never previously had access to any kind of phone or any other modern communication device. During fieldwork in Orora in 2009, the researcher was told by the elderly matriarch of the host family that she remembered when the first metal axes came to the village. In her lifetime, she has witnessed Orora change from being, as it were, literally in the "Stone Age", to being within the coverage area of handheld devices that enable conversations with people at vast distances.

As this research was conducted in the earliest stages of mobile phone adoption, it would be most valuable for follow-up research to re-visit the same villages and ascertain whether attitudes toward and uses of mobile phones are evolving. Even so, the current findings provide substantial information about the uptake and use of mobile phones in broader communication settings. For example, community leaders in Orora were still using communication drums to convey messages on an almost daily basis in 2009. The chapter provides insights about the diffusion of mobile phones in rural areas of PNG in the context of local communicative ecologies, including people's access to and use of mass media and traditional communication techniques. During the early adoption period, villagers were adjusting to a change from public communication forms to the capability to communicate privately, which caused some unsettling of social mores. It might further be surmised that essential changes may follow, for example, from public, consensus-based resolution of social differences toward private conflict, worked out through argument over fealty. The research may be regarded as offering insights into the phenomenon of early adopters who are in the process of leapfrogging the digital divide.

References

Aakhus, Mark. "Understanding Information and Communication Technology and Infrastructure in Everyday Life: Struggling with Communication-at-a-Distance." In *Machines That Become Us: The Social Context of Personal Communication Technology*, edited by James E. Katz, 27–42. New Brunswick, NJ: Transaction Publishers, 2003.

Andersen, Barbara. "Tricks, Lies, and Mobile Phones: 'Phone Friend' Stories in Papua New Guinea." *Culture, Theory and Critique* 54, no. 3 (2013): 318–334.

Bakke, Emil. "A Model and Measure of Mobile Communication Competence." *Human Communication Research* 36, no. 3 (2010): 348–371.

Bell, Genevieve. "The Age of the Thumb: A Cultural Reading of Mobile Technologies from Asia." In *Thumb Culture: The Meaning of Mobile Phones for Society*, edited by Peter Glotz, Stefan Bertschi and Chris Locke, 67–87. New Brunswick, NJ: Transaction Publishers, 2005.

Belt, Don. "Fast Lane to the Future." *National Geographic* 214, no. 4 (2008): 72–99.

Beschorner, N. "Financing Rural Communications Projects: Some Approaches and Experiences." November 28: The World Bank Global Information and Communication Technologies Department, 2007.

Blades, James. *Percussion Instruments and their History*. 2nd ed. London: Faber and Faber, 1975.

Brinkman, Inge, Mirjam de Bruijn and Hisham Bilal. "The Mobile Phone, 'Modernity' and Change in Khartoum, Sudan." In *Mobile Phones: The New Talking Drums of Everyday Africa*, edited by Mirjam de Bruijn, Francis Nyamnjoh and Inge Brinkman, 69–91. Leiden: Langaa and African Studies Centre, 2009.

Chib, Arul. "The Aceh Besar Midwives with Mobile Phones Program: Design and Evaluation Perspectives Using the Information and Communication Technologies for Healthcare Model." Paper presented at Mobile 2.0: Beyond Voice?: Pre-conference workshop at the International Communication Association (ICA) Conference, Chicago, May 20–21, 2009.

CIA World Factbook. "Papua New Guinea." Last modified June 20, 2014. https://www.cia.gov/library/publications/the-world-factbook/geos/print/country/countrypdf_pp.pdf.

Donner, Jonathan. "Research Approaches to Mobile Use in the Developing World: A Review of the Literature." *The Information Society* 24, no. 3 (2008): 140–159.

Ellwood-Clayton, Bella. "Unfaithful: Reflections of Enchantment, Disenchantment ... and the Mobile Phone." In *Mobile Communication in Everyday Life: Ethnographic Views, Observations and Reflections*, edited by Joachim R. Höflich and Maren Hartmann, 123–144. Berlin: Frank & Timme, 2006.

Gouy, Jonathan, Joe Kapa, Alfred Mokae and Theodore Levantis. "Parting with the Past: Is Papua New Guinea Poised to Begin a New Chapter Towards Development?" *Pacific Economic Bulletin* 25, no. 1 (2010): 1–23.

GSMA. *Striving and Surviving in Papua New Guinea: Exploring the Lives of Women at the Base of the Pyramid*. GSMA, 2014.

Hamelink, Cees J. *World Communication: Disempowerment & Self-Empowerment*. London: Zed Books, 1995.

Harvey, David. *Spaces of Hope*. Berkeley: University of California Press, 2000.

Heeks, Richard. 2008. "Mobiles for Impoverishment?" Accessed March 21, 2015. http://ict4dblog.wordpress.com/2008/12/27/mobiles-for-impoverishment/.

Herzog, George. "Drum-Signaling in a West African Tribe." In *Language in Culture and Society*, edited by Dell Hymes, 312–329. New York: Harper & Row, 1964.

Horst, Heather A., and Daniel Miller. *The Cell Phone: An Anthropology of Communication*. Oxford: Berg, 2006.

InterMedia, NBC, ABC International Development and AusAID. *Citizen Access to Information in Papua New Guinea: Citizen Survey*, June 2012.

Issimel, Anisah. "Radio Madang: Tuned in for Development?" In *Communication, Culture & Society in Papua New Guinea: Yu Tok Wanem?*, edited by Evangelia Papoutsaki, Michael McManus and Patrick Matbob, 147–156. Madang, Papua New Guinea: DWU Press, 2011.

ITU. "World Telecommunication/ICT Indicators Database: Mobile Telephony." International Telecommunications Union, 2010. Accessed March 21, 2015. http://www.itu.int/ITU-D/ict/statistics/index.html.

Jensen, Robert. "The Digital Provide: Information (Technology), Market Performance, and Welfare in the South Indian Fisheries Sector." *The Quarterly Journal of Economics* 122, no. 3 (2007): 879–924.

Johnsen, Truls Erik. "The Social Context of the Mobile Phone Use of Norwegian Teens." In *Machines That Become Us: The Social Context of Personal Communication Technology*, edited by James E. Katz, 161–169. New Brunswick, NJ: Transaction Publishers, 2003.

Jorgensen, Dan. "Gesfaia: Mobile Phones, Phone Friends, and Anonymous Intimacy in Contemporary Papua New Guinea." Paper presented at the meeting of the Canadian Anthropological Society, Toronto, April 30, 2014.

Law, Pui-lam, and Yinni Peng. "Mobile Networks: Migrant Workers in Southern China." In *Handbook of Mobile Communication Studies*, edited by James E. Katz, 55–64. Cambridge, MA: The MIT Press, 2008.

Leach, James. "Drum and Voice: Aesthetics and Social Process on the Rai Coast of Papua New Guinea." *The Journal of the Royal Anthropological Institute* 8, no. 4 (2002): 713–734.

Lipset, David. "Mobail: Moral Ambivalence and the Domestication of Mobile Telephones in Peri-Urban Papua New Guinea." *Culture, Theory and Critique* 54, no. 3 (2013): 335–354.

LIRNEasia. "Teleuse at the Bottom of the Pyramid 1" 2007. Accessed March 21, 2014. http://lirneasia.net/projects/2004–05/strategies-of-the-poor-telephone-usage/.

Logan, Sarah. "Rausim! Digital Politics in Papua New Guinea." SSGM Discussion Paper 2012/9. Canberra: Australian National University, 2012.

Mariscal, Judith, and Carla Marisa Bonina. "Mobile Communication in Mexico: Policy and Popular Dimensions." In *Handbook of Mobile Communication Studies*, edited by James E. Katz, 65–77. Cambridge, MA: The MIT Press, 2008.

Marvin, Carolyn. *When Old Technologies Were New: Thinking about Electric Communication in the Late Nineteenth Century*. New York: Oxford University Press, 1988.

Miller, Daniel. "The Unpredictable Mobile Phone." *BT Technology Journal* 24, no. 3 (2006): 41–48.

Morauta, Louise. *Beyond the Village: Local Politics in Madang, Papua New Guinea*. Canberra: Australian National University Press, 1974.

Mpogole, Hosea, Hidaya Usanga and Matti Tedre. "Mobile Phones and Poverty Alleviation: A Survey Study in Rural Tanzania." Paper presented at the 1st International Conference on M4D: Mobile Communication Technology for Development, Karlstad University, Sweden, December 11–12, 2008.

Mustafa, Hasrina, Kamaliah Siarap and Koh Boon Suan. "SMS in Romantic Relationship: Effects of User Attachment Styles on SMS Use among University Students in Malaysia." *Media Asia* 36, no. 4 (2009): 223–230.

Nash, Sorariba. "Media Accountability and the New Technology." *Pacific Journalism Review* 3, no. 2 (1996): 27–45.

Pelckmans, Lotte. "Phoning Anthropologists: The Mobile Phone's (Re-)Shaping of Anthropological Research." In *Mobile Phones: The New Talking Drums of Everyday Africa*, edited by Mirjam de Bruijn, Francis Nyamnjoh and Inge Brinkman, 69–91. Leiden: Langaa and African Studies Centre, 2009.

Reilly, Benjamin. "State Functioning and State Failure in the South Pacific." *Australian Journal of International Affairs* 58, no. 4 (2004): 479–493.

Rooney, Dick, Evangelia Papoutsaki and Kevin Pamba. "A Country Failed by its Media: A Case Study from Papua New Guinea." Paper presented at the 13th AMIC Annual Conference, *Impact of New & Old Media on Development in Asia, Bangkok*, Thailand, July 1–3, 2004.

Sanzogni, Louis, and Heather Arthur-Gray. "Technology Leapfrogging in Thailand." In *Encyclopedia of Developing Regional Communities with Information and Communication Technology*, edited by S. Marshall, W. Taylor and X. Yu, 671–676. Hershey, PA: Idea Group Reference, 2006.

Singh, Supriya, and Yaso Nadarajah. "School Fees, Beer and 'Meri': Gender, Cash and the Mobile in the Morobe Province of Papua New Guinea." Working Paper 2011–3. Institute for Money, Technology and Financial Inclusion (IMTFI), 2011.

Souter, David, Nigel Scott, Christopher Garforth, Rekha Jain, Orphelia Mascarenhas and Kevin McKemey. "The Economic Impact of Telecommunications on Rural Livelihoods and Poverty Reduction: A Study of Rural Communities in India (Gujarat), Mozambique and Tanzania: Commonwealth Telecommunications Organisation for UK Department for International Development." London: Department for International Development, 2005. Accessed March 21, 2015. http://www.telafrica.org/R8347/files/pdfs/FinalReport.pdf.

Srivastava, Lara. "The Mobile Makes Its Mark." In *Handbook of Mobile Communication Studies*, edited by James E. Katz, 15–27. Cambridge, MA: The MIT Press, 2008.

Sullivan, Nancy. "Fieldwork Report in Support of an Environmental and Social Management Framework. World Bank supported Rural Communications Fund Project in East Sepik and Simbu Provinces, Papua New Guinea." Madang, PNG: Nancy Sullivan & Associates, 2010.

Suwamaru, Joseph Kim. "Impact of Mobile Phone Usage in Papua New Guinea." SSGM In Brief 2014/41. Canberra: Australian National University, 2014.

Tabinas, Renelle J. A., and Avril A. B. De Guzman. "Effects of Mobile Phone Use on the Livelihood of Market Vendors in Eastern Philippines." Paper presented at the 19th Annual AMIC Conference, Suntec City, Singapore, June 22, 2010.

Tacchi, Jo, Don Slater and Greg Hearn. *Ethnographic Action Research: A User's Handbook Developed To Innovate and Research ICT Applications for Poverty Eradication*. New Delhi: UNESCO, 2003.

Telban, Borut, and Daniela Vavrova. "Ringing the Living and the Dead: Mobile Phones in a Sepik Society." *The Australian Journal of Anthropology* 25 (2014): 223–238.

Temple, Olga. "Tok Ples in Texting and Social Networking: PNG 2010." *Language and Linguistics in Melanesia: Journal of the Linguistic Society of Papua New Guinea* 29 (2011): 54–64.

The Economist. "Less is More." *The Economist* 376 (8434) (2005): 11.

Unwin, Tim. "Development Agendas and the Place of ICTs." In *ICT4D: Information and Communication Technology for Development*, edited by Tim Unwin, 7–38. Cambridge: Cambridge University Press, 2009a.

Unwin, Tim. "Introduction." In *ICT4D: Information and Communication Technology for Development*, edited by Tim Unwin, 1–6. Cambridge: Cambridge University Press, 2009b.

van Dijk, Jan A. G. M. *The Deepening Divide: Inequality in the Information Society*. Thousand Oaks: Sage Publications, 2005.

von Braun, Joachim, and Maximo Torero. Introduction and Overview to *Information and Communication Technologies for Development and Poverty Reduction*, edited by Maximo Torero and Joachim von Braun, 1–20. Baltimore, MD: The Johns Hopkins University Press, 2006.

Walsh, Shari P., Katherine M. White and Ross M. Young. "Young and Connected: Psychological Influences of Mobile Phone Use Amongst Australian Youth." In *Mobile Media 2007*, edited by Gerard Goggin and Larissa Hjorth, 125–134. Sydney: University of Sydney, 2007.

Watson, Amanda H. A. "Early Experience of Mobile Telephony: A Comparison of Two Villages in Papua New Guinea." *Media Asia* 38, no. 3 (2011b): 170–180.

Watson, Amanda H. A. "Mobile Phones and Development in Papua New Guinea: Guiding Principles." SSGM In Brief 2014/42. Australian National University, 2014.

Watson, Amanda H. A. "Mobile Phones and Media Use in Madang Province of Papua New Guinea." *Pacific Journalism Review* 19, no. 2 (2013): 156–175.

Watson, Amanda H. A. "The Mobile Phone: The New Communication Drum of Papua New Guinea." PhD diss., Queensland University of Technology, 2011a.

Watson, Amanda H. A. "Tsunami Alert: The Mobile Phone Difference." *The Australian Journal of Emergency Management* 27, no. 4 (2012): 42–46.

Watson, Amanda H. A. "'We Would Have Saved Her Life': Mobile Telephony in an Island Village in Papua New Guinea." *eJournalist* 10, no. 2 (2010): 106–127.

Weigel, Gerolf. "Part 1: ICT4D Today—Enhancing Knowledge and People-Centred Communication for Development and Poverty Reduction." In *ICT4D— Connecting People for a Better World: Lessons, Innovations and Perspectives of Information and Communication Technologies in Development*, edited by G. Weigel and D. Waldburger, 15–42. Switzerland: Swiss Agency for Development and Cooperation, and the Global Knowledge Partnership, Malaysia, 2004.

Yamo, Henry. "Mobile Phones in Rural Papua New Guinea: A Transformation in Health Communication and Delivery Services in Western Highlands Province." Masters diss., Auckland University of Technology, 2013.

Part II

Self-Determination for Indigenous People through Mobile Technologies

7 Keewaytinook Mobile

An Indigenous Community-Owned Mobile Phone Service in Northern Canada

Brian Beaton, Terence Burnard, Adi Linden and Susan O'Donnell

Introduction

This is the story of Keewaytinook Mobile, a not-for-profit mobile (cellular) phone service built, owned and operated by small, remote, politically autonomous Indigenous (First Nation) communities in northern Ontario, Canada. The people and their ancestors have lived here for thousands of years. The terrain is beautiful and harsh; summers are hot, but in winter, the longest season, temperatures regularly fall below minus 30 degrees Celsius. There are no permanent roads in this region; the Indigenous communities are accessed by small aircraft. The communities generate their own electricity using diesel fuel hauled in on temporary roads built in the winter on the frozen landscape.

Keewaytinook Mobile (KMobile) was created because the Indigenous people in this region wanted it, and they built it themselves because nobody else was going to do it. The KMobile idea began when the leadership of one of the Indigenous communities asked the tribal council Keewaytinook Okimakanak to include mobile services in their network plans. Keewaytinook Okimakanak knew it would be a significant challenge but believed they had the capacity to do it; their telecommunications division KO-KNET had already built and was operating the largest Indigenous-owned telecommunications service in the world.

This story is mostly about the development of the KMobile service infrastructure and why these Indigenous communities own and operate their information and communications technology (ICT) infrastructure to support their capacity development and self-determination goals. KMobile exists in a country in which there are profound and unacceptable social and economic divisions between Indigenous peoples and the non-Indigenous population, a situation condemned by the United Nations (Anaya 2014). Despite Canada's reluctant ratification of the United Nations Declaration on the Rights of Indigenous Peoples in 2010, the federal government has continued its ongoing practice of neglect and underfunding of Indigenous communities, especially communities in the remote northern regions that remain "out of sight out of mind" until the next crisis situation hits the national news. The political and socioeconomic context of KMobile and Indigenous communities in northern Canada is an important part of this story and so that is where it will start.

Settler Colonialism in Canada and OCAP as an ICT Response

In May 2014 James Anaya, United Nations Special Rapporteur on the Rights of Indigenous Peoples, released his damning report on the "distressing socio-economic conditions of Indigenous peoples in a highly developed country" (Anaya 2014, 7). The UN report lists a wide range of human rights issues and crisis situations across Canada. Many Indigenous communities, especially in remote northern regions, are experiencing widespread poverty, high rates of unemployment and chronic diseases, severe housing shortages and overcrowded housing, an underfunded education system influenced by the legacy of the residential school system, unsafe drinking water that poses a serious health risk to residents, and many other crises. For millennia, the Indigenous people in these northern regions survived as hunters and gathers with strong connections to the land and all that it provides; it is only in recent history that they are living on small reserve lands with limited access to the resources needed to develop their communities. In the words of the housing manager of a remote Indigenous community who was interviewed recently by the authors: "We receive just enough to fail".

Despite the challenges, Indigenous peoples are not failing but rather resisting, increasing their resilience and creating resurgence with a strong connection to their lands and resources. Many are building their own community infrastructure—such as KMobile—managing their education and health systems and many other essential services to keep their communities thriving and expanding. The birth rate in Indigenous communities is well above that in non-Indigenous communities across Canada. In the decades leading up to the UN report, confrontations have increased between Indigenous peoples and the Canadian state over land rights, treaty rights, and protection of water and other natural resources. On the intellectual front, analysis of "settler colonialism" is helping Indigenous academics and their non-Indigenous allies explain how it is possible that many Indigenous communities in Canada are experiencing such appalling levels of poverty and underdevelopment. At the community level, the application of the principals of OCAP—Ownership, Control, Access and Possession—across an increasing number of applications is revolutionizing how communities are taking charge of their infrastructure, governance structures, information and knowledge.

Indigenous authors from across Canada are leading the analysis of settler colonialism, including Taiaiake Alfred (2009), Marie Battiste (2013), Jeff Corntassel (2012), Glen Coulthard (2014), Pamela Palmater (2011) and Leanne Simpson (2012). A settler colonialism lens sees that Canadian state policies are designed to remove Indigenous peoples from their traditional lands so that the resources can be extracted for economic gain. Resources taken from Indigenous lands maintain the Canadian economy and the high standard of living experienced by non-Indigenous Canadians, the majority of whom live in large, southern urban centers dependent on these resources.

OCAP—Ownership, Control, Access and Possession—was originally a theory developed by First Nations to apply self-determination to research (Schnarch 2004) and was adopted at the national level by Indigenous leaders in Canada (Assembly of First Nations 2007). As originally conceived, OCAP is an Indigenous response to the role of knowledge production in challenging colonial relations. OCAP principles, or self-determination applied to tele-communications and broadband networks, has at least two implications. First, Indigenous communities must retain access and possession of the capacity and resources to effectively manage the content, traffic and services on their local network. Second, Indigenous communities have a right to own and control the local broadband network in their communities in order to support the flow of information and services (Kakekaspan et al. 2014).

OCAP applied to telecommunications is captured in the "First Mile" and "e-Community" approaches to broadband development in remote and rural Indigenous communities (McMahon et al. 2014). "First Mile" refers to the development of local telecommunications infrastructure that benefits local communities, in contrast to how local infrastructure is often referred to as "last mile" development that benefits centralized, urban-based telecom corporations and governments. Authors Beaton, Seibel and Thomas (2014) have analyzed how community-owned broadband networks and applica-tions are at the core of the social economy in remote Indigenous communi-ties. Ownership and control of the local broadband networks and associated services are vital to self-determination and local community resilience; local ownership of community networks is a critical component of the "First Mile" (McMahon et al. 2011). Implementing First Mile at the Indigenous community level in Canada involves an e-Community approach whereby the communities own and maintain their local broadband infrastructure; the associated benefits such as local employment, revenues and capacity are requirements in local efforts to counter settler colonialism (Beaton and Campbell 2014; Whiteduck 2010).

An important part of the "First Mile" work is to change the policies directing the allocation and management of public funds for broadband infrastructure developments. Canadian governments spend significant pub-lic funds on telecommunications infrastructure in remote and rural regions. However, with rare exceptions, the public funds flow directly to corporate telecommunications companies with little consultation or oversight by the communities the networks are supposed to serve. As has been noted elsewhere: "the direct provision of public subsidies to corporate telecommu-nications companies leaves little recourse for community action" (McMahon et al. 2014, 251).

The problem with giving public infrastructure funds directly to corporations instead of local communities to manage can be illustrated by one recent example. Between 2010 and 2014, the federal and provincial governments spent more than $60 million to build a new fibre transport network in northwestern Ontario, to serve Indigenous communities and mining and other extractive industries in the region. However, rather than funding the Indigenous communities the public funds flowed to the only national telephone company serving this region to build and operate their own new fibre transport network (Philpot, Beaton and Whiteduck 2014). The communities linked by the new fibre network now must purchase their transport services from this provider at a cost far above that charged in urban centers, costs that make it very challenging for the Indigenous communities to deliver affordable services that depend on the network, including the KMobile service, local Internet service, telehealth, school connectivity and other services. In addition, the telephone company left five of the originally proposed remote communities off their fibre network, claiming it required additional public funding to reach them. This is one of many examples of the colonial and capitalist approaches to regional telecom development that enrich corporations and miss an opportunity to build capacity in local communities. This corporate approach is now being challenged by Indigenous groups and their allies through a new non-profit organization, the First Mile Connectivity Consortium, which is making interventions with regulatory bodies so that infrastructure development will truly benefit remote Indigenous communities (McMahon, Hudson and Fabian 2014).

Local Ownership of Telecommunications in the Remote Keewaytinook Okimakanak Communities

The Indigenous owned and controlled KMobile service began in the Sioux Lookout zone of northwestern Ontario, an area about the size of France that supports 26 remote and isolated Indigenous communities. For most of the year, the only way to reach these communities is by small planes. The communities are small, with average populations of about 450 people, and politically autonomous; each is governed by an elected chief and council responsible for not only political governance but also delivering the full range of services and activities necessary for any community to function—from education, public works and health services to justice, policing, recreation activities and more. The terrain is Canadian Shield and tundra—rocks, water, bog and vast forests—rich in wildlife, resources, and thousands of years of Indigenous history. Until a generation ago, the extensive water networks of rivers and lakes were the primary means of transportation, supporting networking, communication and a special way of life deeply connected and dependent on the land and everything it provides. The lands and waterways continue to sustain the lives of the Indigenous people who live here today.

Five of these remote Indigenous communities with year-round residents—Fort Severn First Nation, Keewaywin First Nation, North Spirit Lake First Nation, Poplar Hill First Nation and Deer Lake First Nation—are members of the Keewaytinook Okimakanak tribal council. Keewaytinook Okimakanak ("Northern Chiefs" in the Oji-Cree language) is governed by the chiefs of its member communities. Deer Lake with 1,000 residents is the largest, and the others have resident populations between 400 and 500. Keewaytinook Okimakanak is an intermediary organization; among its functions is to support infrastructure development in its member communities (McMahon et al. 2014). It provides second-level support services, including KO Health, KO Education and notably a series of technology-supported services: the Keewaytinook Internet High School (KiHS) (Potter 2010; Walmark 2010), KO Telemedicine (KOTM) (Williams 2010) and the flagship Kuhkenah Network (KO-KNET), the most extensive Indigenous-owned telecommunications service in the world (Carpenter 2010).

Since its birth in 1994 as a bulletin board service (BBS) to connect students in six remote Indigenous communities with each other, KO-KNET has leveraged strategic funding and partners to create a vast telecommunications network and digital services organization now serving more than 80 Indigenous communities across Ontario. Based in Sioux Lookout, Ontario, KO-KNET services include Internet connectivity, a managed videoconferencing network, the Northern Indigenous Community Satellite Network (NICSN), supporting KiHS (Internet high school) and KOTM (telehealth) and a range of other broadband-enabled services, training and related activities. The remarkable story of KO-KNET has been the focus of five doctoral theses and several university-based research projects resulting in numerous publications (for example: Beaton, Fiddler and Rowlandson 2004; Carpenter 2010; Fiser and Clement 2010; McMahon 2014; O'Donnell et al. 2013). KO-KNET is one of the few examples of an Indigenous organization receiving public funding to build telecommunications infrastructure; as a non-profit, community-run organization, they have ensured that the funding flows directly to and benefits the communities.

KO-KNET has been a pioneer in the First Mile and e-Community approaches, working with the five remote Keewaytinook Okimakanak communities to build community digital infrastructure from the ground up. The e-Community approach envisions broadband as a community-owned infrastructure; decisions are made about broadband connectivity at the community level. Each of the five remote Keewaytinook Okimakanak communities owns a cable network that delivers Internet connectivity to the homes. Four of the communities recently switched to fibre transport that replaced the microwave broadband transport and delivers much faster Internet connectivity. These four communities own their own local fibre network that is now included in their cable network delivering improved connectivity services to the health center (for telehealth) and the school. Fort Severn, the most remote community and satellite served, is currently planning to develop its local fibre network. All five communities run their own local KMobile service in partnership with KO-KNET.

The Development of KMobile as an Indigenous Community-Owned Service

In early 2000, George Kakekaspan was chief of Fort Severn First Nation when the Keewaytinook Okimakanak First Nations won the competition to be the Canadian "Aboriginal Smart Communities" demonstration project (Carpenter 2010). As plans were rolling out to build and operate their broadband community networks, Chief Kakekaspan requested that mobile services be considered as an essential service for his community as part of this project. It would take another seven years before affordable hardware and software became available that supports the OCAP principles and these small Indigenous communities could begin developing their locally owned and managed mobile networks.

In early 2007, Keewaytinook Okimakanak's KO-KNET responded to a request for proposals to develop mobile services in the remote First Nations across northwestern Ontario, Canada. At the time, the federal and provincial governments were searching for cellular providers or groups willing to develop mobile services across this sparsely populated region. This work was to close the mobile gap and create new economic and social opportunities in the existing communities and for new business ventures such as mining and forestry. The government decided to fund KO-KNET to create a pilot project with two remote Indigenous communities to demonstrate the feasibility and operation of the proposed community-owned mobile service. KO-KNET obtained a grant of $1 million from the Ontario government and another grant of $100,000 from the federal government, and each of the partner Indigenous communities contributed their lands, equipment, operators and support for the development of the mobile pilot project.

This pilot project successfully demonstrated that the proposed hardware and software could operate over the existing microwave transport in the case of one of the communities (Keewaywin First Nation) and satellite transport in the case of the second partner community (North Caribou Lake First Nation). The pilot project also helped to establish a strategic development partnership with the regional Dryden Municipal Telephone System (DMTS) to use their mobile/cellular hub to service the Indigenous communities using KMobile. The DMTS partnership with KO-KNET and the use of the KO-KNET mobile hub equipment as a redundant backup system was strategic for everyone working together to build and sustain this new mobile network service.

Initial pilot project meetings with the Indigenous leadership and community members resulted in the decision to double the height of the proposed 100-foot towers. From every person consulted about this new service, the team heard the same first question: "How far will the signal reach?" Community members told story after story about the challenges they faced when working outside their communities on the land and on the water systems and concerns about safety for themselves and their loved ones. The contractor advised the KO-KNET team that by doubling the height of the tower the reach of the signal could potentially be tripled dependent on the signal strength of both the receive and transmit devices. So plans were made to purchase 200-foot towers; tower and telecommunication building sites were selected with the chief and council to address electrical, connectivity, height of land and ease of construction factors.

The two towers and assembled telecommunication buildings were ordered and shipped on the 2008 winter road to Keewaywin and North Caribou First Nations. The winter was unseasonably warm and short; despite concerns that the winter road would not be stable enough to support the transit of the towers and buildings, they made it safely. The two communities worked with the contractors to prepare the sites, erect the towers, position the telecom buildings, mount the antennas and cabling between the antennas and the building and ensure the equipment was operational (Figure 7.1). Next, KO-KNET worked with DMTS, the regional telephone company partner, to provide the telephone lines to the site for local dialing, establish the IP connection from the community headend to the mobile/cellular site and connect the electrical system to the community grid.

Figure 7.1 The new mobile/cellular headend in Keewaywin First Nation, 2008. (Source: KO-KNET).

After that work was completed, KO-KNET worked with the two Indigenous communities to set up the mobile radio equipment and backup electrical storage system. The connections were completed for the local mobile site making it operational by the autumn of 2008. Training sessions were an integral component for this development with workshops delivered by the hardware and software producers held in Sioux Lookout and Dryden for the KO-KNET team and the community partners. KMobile successfully obtained the 850mhz wireless spectrum from the national provider that had been given this space by the federal government but had no plans to use it in this region because the Indigenous communities did not meet their population requirements. Roaming and network agreements were signed to support the operation of their mobile phones in the Indigenous communities.

With the successful demonstration that these technologies could work in these remote communities, the Keewaytinook Okimakanak leadership (community chiefs) directed KO-KNET to construct similar mobile sites in the remaining four Keewaytinook Okimakanak First Nations (Deer Lake, Fort Severn, North Spirit Lake and Poplar Hill). These four sites were turned on by the autumn of 2009 (Figure 7.2).

Figure 7.2 Installing the cell tower in Fort Severn First Nation, 2009.
(Source: KO-KNET).

For much of the four-year period from autumn 2008, when KMobile began operating, the service managed to successfully grow and deliver its mobile cellular service across the region. Throughout this period, KO-KNET created websites to document both the pilot project (KO-KNET 2008) and the KMobile developments (KO-KNET 2014). Many stories about the importance of the service were experienced and shared, and plans were made to put this service into other remote Indigenous communities in the region interested in joining the network. KO-KNET prepared another funding application, and the Ontario government provided nearly $4.8 million for 10 additional Indigenous communities to work with KO-KNET to build their own KMobile services. By the end of this second project in 2012, KO-KNET leveraged this investment to include eight additional remote First Nations for a total of 20 communities operating the KMobile service (Figure 7.3). New challenges were about to emerge for KMobile.

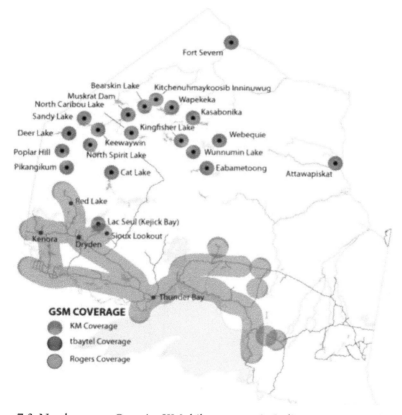

Figure 7.3 Northwestern Ontario: KMobile coverage in Indigenous communities and roaming partner coverage. (Source: KO-KNET).

In autumn 2012, the municipality of Dryden decided to sell its DMTS telephone service to TBayTel. This development meant that KMobile had to decide whether to take on the development and management of the former DMTS hub site that was servicing KMobile or to simply walk away from continuing to provide this service. After consulting with the Indigenous community leaders, everyone agreed to keep the service operating. KMobile completed the transition from DMTS to a fully operated and managed mobile service, including becoming a licensed telecommunications provider capable of supporting roaming agreements with other mobile providers. Operation of the hub site meant additional work for the KO-KNET network team with additional staff being required. KMobile entered new agreements and partnerships with GSM network services and hired a consulting team to assist with the transition and business case for the KMobile operation and maintenance.

The KMobile business model now involves a pay-as-you-go arrangement for customers that includes a $12 per month charge for infrastructure maintenance and development (Figure 7.4). All calls made are charged to the customers with a percentage returned to the Indigenous communities in which the calls originated. Revenues from calls made while roaming outside the communities benefit the network as a whole. Two authors of this chapter, staff members of Keewaytinook Okimakanak, developed the new billing software required to support the KMobile users to use the service both while on the network and roaming on other mobile networks. This software development has saved KMobile hundreds of thousands of dollars by avoiding the need to purchase and manage an existing billing software system. Similarly the KO-KNET team is implementing a new online accounting system for individual users to further support KMobile clients.

Figure 7.4 KMobile SIM card and pay-as-you-go card. (Source: KO-KNET).

Indigenous Community Member use and Experiences of KMobile

The authors have conducted three separate studies to gather Indigenous community member feedback on KMobile and other community-owned telecommunications services. The first included interviews with 42 community members in Fort Severn First Nation in early 2010, shortly after their KMobile service was switched on. Our study described their perspectives and experiences of KMobile in its early days (O'Donnell et al. 2011).

Fort Severn First Nation, the northernmost community in Ontario, is located on the Severn River a few kilometers from where it flows into Hudson Bay (see Figure 7.2). About 400 people live in Fort Severn most of the time, half of whom are adults. Four months after KMobile was operational, about 50 residents had purchased KMobile phones and were buying pre-paid phone cards. More residents may have been using the KMobile roaming agreements with other phone companies, and in addition, many residents (41% of those interviewed) were borrowing mobile phones from others or lending their phones; it was impossible to obtain a definitive number of KMobile "users" in the community. About 20% of those interviewed in those early days used a mobile phone daily, and texting was more popular than voice calls.

The most common reason for using a mobile phone in Fort Severn was safety and security when out on the land—hunting, trapping, fishing or gathering wild food. The range of the KMobile service was of critical interest and importance—for example trappers wondered if the signal would reach their traditional traplines. Other reasons for using KMobile included being easy to contact when moving around in the community or for fun and the novelty of texting people. Reasons given for not wanting a mobile phone were the cost (it was not a priority), not wanting to be easily contacted or not seeing the need for it; for example, one comment was: "For a community like Fort Severn, if I want to talk to someone, I just go over to their house. It's just a small community" (O'Donnell et al. 2011, 670).

In late 2011, the authors conducted an online survey of community members in the Sioux Lookout zone that resulted in 663 responses, including 131 responses from people living in the five remote Keewaytinook Okimakanak communities; the methodology for that survey is described in several publications (Carpenter et al. 2013; Molyneaux et al. 2014; Walmark et al. 2012). In early 2014, the authors conducted a second online survey, this time with responses from 210 residents in the remote Keewaytinook Okimakanak communities; the methodology for the second survey is described in two conference papers (Beaton and Carpenter 2014; Beaton, Seibel and Thomas 2014). The data related to KMobile in the KO communities from these two surveys has not previously been published and will be discussed here.

The 2011 survey was conducted about one year after the KMobile service was switched on in the five remote Keewaytinook Okimakanak communities. At that point, 38% of the respondents were texting and 21% were making a voice call on a mobile phone every day. About 20% were using the Internet on their mobile phone daily—this would have been via a WiFi connection on a smartphone, as KMobile infrastructure supports voice and text only. One respondent commented: "Good job on getting cell service up north, that made my year when I moved back to the community and found cell service, I would have been 'lost' without internet and cell and I probably wouldn't have stayed to live there without it".

Of the 131 remote Indigenous community respondents in the 2011 survey, 50% were current KMobile customers, and 63% indicated they planned to use the service in the next year. In response to the statement "KMobile offers safety and security when out on the land", 61% agreed and 34% did not know. More than half (56%) gave KMobile a good or excellent rating, and 28% rated it fair; 12% responded: "I don't know what this service is". When asked to describe the cell (mobile) phone service in their community; the responses were: Good: sometimes no coverage or dropped calls but good overall (45%); fair (23%); don't know (14%); excellent, always works (11%) and poor, hard to access, many dropped calls (6%). In response to the question: "What do you need to use technology more effectively?" 52% responded: "Better cell (mobile phone) range".

When the 2014 survey was conducted, KMobile had been operating in each of the five Keewaytinook Okimakanak communities for at least four years. In the survey, more than half (54%) of respondents were "very comfortable" using a mobile phone, and 34% were "comfortable". More than a third (39%) were "very comfortable" with a smartphone and the same number were "comfortable". Every day 35% were texting, 24% were making a voice call on their mobile phone and 38% were using the Internet on their mobile phone; again, this would have been through a WiFi connection on a smartphone. In addition, 44% indicated they go online daily using a mobile device.

Similar to the situation in 2011, in 2014 50% indicated they were current KMobile customers, with 58% indicating they planned to use the service in the next year. In response to the statement, "KMobile offers safety and security when out on the land", 68% agreed, an increase from 2011 (61%). Among the many positive comments received about the KMobile service was: "For a person that doesn't use KMobile, I still recommend very strongly to keep it going because a lot of people depend on it in the communities". In the survey, 79% agreed with the statement: "KMobile is an important service in our community". In response to the question: "What do you need to use technology more effectively?" 52% responded: "Better cell (mobile phone) range", and 52% responded "Internet on mobile phone".

Several stories have been documented about how the KMobile service has been vital for medical emergencies. One concerns a woodcutter, Timothy Apetawakeesic, whose calls for help using his KMobile phone were answered after a tree-cutting accident. He had multiple fractures in his leg after a tree fell on him near Weagamow Lake First Nation in late winter 2012. "That's how I survived—the cellphone", said the fortunate man in an article written about the incident (Garrick 2012). The same magazine article mentioned another incident in which a man repairing his vehicle in a remote location caught his hand in a car jack and used his mobile phone to request urgent assistance. Another story documented how during the Ontario provincial election in 2011 the Anik F2 satellite malfunctioned and the telephone services went down in the satellite-served communities, preventing the election officials in Toronto from accessing the polling results using their traditional telephone-based strategies in these remote communities. At the same time, the KMobile phones were working in these communities using a different C-Band satellite, Anik F3, making it possible for election officials to use these local networks to reach the local election officials so their votes could be counted (McMahon 2011).

Conclusion

Two of the authors are writing this chapter in July 2014, during a month-long visit to the five remote Keewaytinook Okimakanak communities. KMobile is obviously a part of the everyday lives of the Indigenous community residents. In each of the communities, essential services continue throughout the summer, staffed by Indigenous community members in the band offices, health centers, community centers, e-Centers and community stores; the maintenance workers are outside keeping the airports functioning, the electricity generators going, the roads graded and the water and sewage systems operating. Almost every worker we speak with uses a mobile phone as part of his or her tasks to keep these services running.

The weather is usually beautiful during the summer months. During this special time of the year, summer community activities include fishing derbies, bible camps, and even the ClearWater Music Festival in Deer Lake First Nation. All of these Indigenous communities are located on rivers or lakes, and fishing is an everyday activity for many community residents; meals of fried fish are often community events. The children swim wherever and whenever they can. Our conversations and interviews with community members are often interrupted by messages and postings arriving on smartphones. Everywhere, mobile phones are in evidence and used by community residents to share news and information about activities, plans and community events with friends, family members and contacts in their own communities and other Indigenous communities and towns and cities far away.

Community residents with smartphones are using them to access the Internet via the many WiFi networks throughout the communities. Facebook is a primary means of communication. Each community has a Facebook page closed to community members and a few outside friends and an active "buy and sell" or auction site that supports the local social economy. Most evenings, several members in each community will use the site to advertise dinner plates for sale. For example, in Poplar Hill First Nation, moose stew and bannock were offered at $15 a plate with 10 plates gone in less than half an hour after the picture of the plates was posted.

The story of KMobile is a successful narrative for these remote Indigenous communities, but many challenges remain. The intent was always to have KMobile support data exchange in addition to voice and text, but to date this goal has proven a significant challenge. The upgrade to 3G—LTE hardware and software in 2012 proved to be a costly experience when the equipment purchased and installed did not work as required despite the due diligence performed beforehand. As a result, alternative solutions are now being researched along with a new business strategy, and the upgrade to LTE is still being planned. The community feedback generated by the research provides additional information for both KO-KNET and the appropriate funding bodies who are required to support these capital investments. Until recently, half of the 20 communities with KMobile sites were served by microwave broadband transport and half by a community-owned C-band satellite link. Many are now linked to the new fibre network that itself was a major fundraising feat for the communities.

Before KMobile existed, the dominant telecommunications companies never considered having a mobile phone service in this region due to their urban-centric business model and network design. Now, mobile technology has become an essential service in these remote Indigenous communities. Additional public investments are now required by the community-owned mobile service to upgrade their network to LTE as well as reach the remaining unserved communities. KMobile is a successful example of public funding flowing directly to an Indigenous organization and Indigenous communities to build and operate telecommunications infrastructure. However, much remains to change government policies created by urban-based bureaucrats to recognize the essential aspect of mobile services in these remote communities and to support community-owned solutions rather than funding solutions from urban-based corporations.

After six years of operating KMobile, the local mobile business case in these remote communities has been demonstrated successfully; the Indigenous business model is all about the public good. Lives have been saved with this tool; people now feel safer traveling on the land and waterways and new economic and social opportunities exist. The operational business models for infrastructure systems such as KMobile involve a complex combination of people, resources, partnerships and support services. The Indigenous communities are investing in the ongoing operation of KMobile with technical staff and resources. These strategic investments in mobile technology are making it possible for everyone to choose where they want to live and raise their families. However, much more work is required to deliver equitable and affordable mobile services in these remote communities.

The five remote Keewaytinook Okimakanak communities we are visiting this summer are resilient. They are experiencing the same challenges as other northern Indigenous communities described in the United Nations report earlier in this chapter—high rates of poverty, unemployment, chronic diseases, overcrowded housing and many other problems typical of remote Indigenous communities in a settler colonial society and economic system based on resource extraction. Surrounding these Indigenous communities are their traditional territories, rich in natural resources. These resources are being extracted for economic gain and to support the comfortable lifestyles of Canadians living in southern cities while the Indigenous communities receive only enough public funding for basic survival. However, in the long history of Indigenous presence in this region, settler colonialism is a recent phenomenon. Indigenous people and communities have always existed across this vast region and will continue to be here for countless generations to come. They are demonstrating their ability, willingness and desire and are quickly developing the capacity and experience to own and operate the infrastructure required to support their lives and livelihoods on their traditional lands.

Acknowledgements

We want to acknowledge and thank our research partner Keewaytinook Okimakanak (KO) and the Nishnawbe Aski who are living in the remote KO First Nations for supporting us in our work. The lead author also acknowledges the traditional, unceded lands and people of the Maliseet Nation where he lives and attends the University of New Brunswick. Thank you to all of the participants and survey respondents who gave their time, ideas and efforts to help move this research forward. In particular we would like to thank Franz Seibel at the Keewaytinook Okimakanak Research Institute and the people who completed the community surveys in 2011 and 2014. The research was conducted as part of the lead author's thesis program in Critical Studies at the Faculty of Education, University of New Brunswick and as part of the First Nations Innovation (FNI) research project—http://fni.firstnation.ca. Project partners include Keewaytinook Okimakanak in Ontario, the First Nations Education Council in Quebec, Atlantic Canada's First Nations Help Desk / Mi'kmaw Kinamatnewey in Nova Scotia, and the University of New Brunswick. The FNI research has been funded since 2006 by the Social Sciences and Humanities Research Council of Canada (SSHRC) with in-kind contributions from the project partners. The lead author's research is also supported by SSHRC.

References

Alfred, Taiaiake. "Colonialism and State Dependency." *Journal of Aboriginal Health* 5, no. 2 (2009): 42–60.

Anaya, James. *Report of the Special Rapporteur on the Rights of Indigenous Peoples, James Anaya, on the Situation of Indigenous peoples in Canada*. United Nations, Human Rights Council, 27th session, May. New York: United Nations, 2014.

Assembly of First Nations. *OCAP: Ownership, Control, Access and Possession— First Nations Inherent Right to Govern First Nations Data*. Ottawa: Assembly of First Nations, 2007.

Battiste, Marie. *Decolonizing Education: Nourishing the Learning Spirit*. Saskatoon: Purich Publishing, 2013.

Beaton, Brian, and Peter Campbell. "Settler Colonialism and First Nations e-Communities in Northwestern Ontario." *The Journal of Community Informatics* 10, no. 2 (2014). Accessed March 18, 1915. http://ci-journal.net/index.php/ciej/article/view/1072.

Beaton, Brian, and Penny Carpenter. "A Critical Understanding of Adult Learning, Education and Training Using Information and Communication Technologies (ICT) in Remote First Nations." Paper presented at the annual meeting of the Canadian Association for Study of Indigenous Education, Brock University, St. Catherines, Ontario, May 26–28, 2014.

Beaton, Brian, Jesse Fiddler and John Rowlandson. "Living Smart in Two Worlds: Maintaining and Protecting First Nation Culture for Future Generations." In *Seeking Convergence in Policy and Practice: Communications in the Public Interest* (Volume 2), edited by Marita Moll and Leslie R. Shade, 283–97. Ottawa: Canadian Centre for Policy Alternatives, 2004.

Beaton, Brian, Franz Seibel and Lyle Thomas. "Valuing the Social Economy and Information and Communication Technologies (ICT) in Small Remote First Nations." Paper presented at the annual meeting of the Association of Social Economy and Non-Profit Research, Brock University, St. Catherines, Ontario, May 28–30, 2014.

Carpenter, Penny. "The Kuhkenah Network (K-Net)." In *Aboriginal Policy Research VI: Learning, Technology and Traditions*, edited by J. P. White, J. Peters, D. Beavon and P. Dinsdale, 119–27. Toronto: Thompson Educational Publishing, 2010.

Carpenter, Penny, Kerri Gibson, Crystal Kakekaspan and Susan O'Donnell. "How Women in Remote and Rural First Nation Communities are Using Information and Communications Technologies (ICT)." *The Journal of Rural and Community Development* 8, no. 2 (2013): 79–97.

Corntassel, Jeff. "Re-envisioning Resurgence: Indigenous Pathways to Decolonization and Sustainable Self-Determination." *Decolonization: Indigeneity, Education & Society* 1, no. 1 (2012): 86–101.

Coulthard, Glen Sean. *Red Skin, White Masks: Rejecting the Colonial Politics of Recognition*. Minneapolis: University of Minnesota, 2014.

Fiser, Adam, and Andrew Clement. "K-Net and Canadian Aboriginal Communities." *IEEE Technology & Society Magazine* 28, no. 2 (2009): 23–33.

Garrick, Rick. "Cellphone Saves Injured Woodcutter." *Sagatay*, Fall (2012): 4–6.

Kakekaspan, Matthew, Susan O'Donnell, Brian Beaton, Brian Walmark and Kerri Gibson. "The First Mile Approach to Community Services in Fort Severn First Nation." *Journal of Community Informatics* 10, no. 2 (2014.). Accessed March 18, 1915. http://ci-journal.net/index.php/ciej/article/view/998.

KO-KNET. "IP Cellular Pilot Project," 2008. Last Modified July 2009. http://meeting.knet.ca/mp19/course/view.php?id=11.

KO-KNET. "Keewaytinook Okimakanak Mobile—KMobile," 2014. Last modified June 2014. http://mobile.knet.ca.

McMahon, Rob. "From Digital Divides to the First Mile: Indigenous Peoples and the Network Society in Canada." *International Journal of Communication* 8, no. (2014). Accessed March 18, 1915. http://ijoc.org/index.php/ijoc/article/view/2456.

McMahon, Rob. "Northern Voices Heard." *Metro News*, October 11, 2011.

McMahon, Rob, Michael Gurstein, Brian Beaton, Susan O'Donnell and Tim Whiteduck. "Making Information Technologies Work at the End of the Road." *Journal of Information Policy* 4 (2014). Accessed March 18, 1915. http://jip.vmhost.psu.edu/ojs/index.php/jip/article/view/146.

McMahon, Rob, Heather E. Hudson and Lyle Fabian. "Indigenous Regulatory Advocacy in Canada's Far North: Mobilizing the First Mile Connectivity Consortium." *Journal of Information Policy* 4 (2014). Accessed March 18, 1915. http://jip.vmhost.psu.edu/ojs/index.php/jip/article/view/175.

McMahon, Rob, Susan O'Donnell, Richard Smith, Brian Walmark, Brian Beaton and Jason Woodman Simmonds. "Digital Divides and the 'First Mile': Framing First Nations Broadband Development in Canada." *The International Indigenous Policy Journal* 2, no. 2 (2011). Accessed March 18, 1915. http://ir.lib.uwo.ca/iipj/vol2/iss2/2.

Molyneaux, Heather, Susan O'Donnell, Crystal Kakekaspan, Brian Walmark, Philipp Budka and Kerri Gibson. "Social Media in Remote First Nation Communities." *Canadian Journal of Communication* 39, no. 2 (2014): 275–88.

O'Donnell, Susan, Lyle Johnson, Tina Kakepetum-Schultz, Kevin Burton, Tim Whiteduck, Raymond Mason, Brian Beaton, Rob McMahon and Kerri Gibson. "Videoconferencing for First Nations Community-Controlled Education, Health and Development." *The Electronic Journal of Communication* 23, no. 1 and 2 (2013). Accessed March 18, 1915. http://www.cios.org/www/ejc/v23n12toc.htm.

O'Donnell, Susan, George Kakekaspan, Brian Beaton, Brian Walmark, Raymond Mason and Michael Mak. "A New Remote Community-Owned Wireless Communication Service: Fort Severn First Nation Builds Their Local Cellular System with Keewaytinook Mobile." *Canadian Journal of Communication* 36, no. 4 (2011): 663–73.

Palmater, Pamela. "Stretched beyond Human Limits: Death by Poverty in First Nations." *Canadian Review of Social Policy* 65/66 (2011): 112–27.

Philpot, Duncan, Brian Beaton and Tim Whiteduck. "First Mile Challenges to Last Mile Rhetoric: Exploring the Discourse between Remote and Rural First Nations and the Telecom Industry." *Journal of Community Informatics* 10, no. 2 (2014). Accessed March 18, 1915. http://ci-journal.net/index.php/ciej/article/view/992.

Potter, Darrin. "Keewaytinook Internet High School Review (2003–2008)." In *Aboriginal Policy Research VI: Learning, Technology and Traditions*, edited by J. P. White, J. Peters, D. Beavon, and P. Dinsdale, 148–57. Toronto: Thompson Educational Publishing, 2010.

Schnarch, Brian. "Ownership, Control, Access, and Possession (OCAP) or Self-Determination Applied to Research: A Critical Analysis of Contemporary First Nations Research and Some Options for First Nations Communities." *Journal of Aboriginal Health* 1, no. 1 (2004): 80–95.

Simpson, Leanne. "Indigenous Perspectives on Occupy." Occupy Talks, Beit Zatoun, Toronto, January 23, 2012. Accessed March 18, 1915. http://leannesimpson.ca/talks.

Walmark, Brian. "Digital Education in Remote Aboriginal Communities." In *Aboriginal Policy Research VI: Learning, Technology and Traditions*, edited by J. P. White, J. Peters, D. Beavon and P. Dinsdale, 141–47 Toronto: Thompson Educational Publishing, 2010.

Whiteduck, Judy. "Building the First Nation e–Community." In *Aboriginal Policy Research VI: Learning, Technology and Traditions*, edited by J. P. White, J. Peters, D. Beavon and P. Dinsdale, 95–103. Toronto: Thompson Educational Publishing, 2010.

Williams, Donna. "Telehealth/Telemedicine Services in Remote First Nations in Northern Ontario." In *Aboriginal Policy Research VI: Learning, Technology and Traditions*, edited by J. P. White, J. Peters, D. Beavon and P. Dinsdale, 159–68. Toronto: Thompson Educational Publishing, 2010.

8 Mojo in Remote Indigenous Communities

Ivo Burum

For many such as I, our lifetime experience has been one of racism at the individual level, reinforced by the powerful institutions of the law [...] and without doubt the media [...] there has been very little challenge to the media about the way in which that institution, practices, promotes and perpetuates racism [...] in our community.

Indigenous Magistrate Pat O'Shane

Introduction

The above statement represents the plight of many marginalized people globally, where media is a carrier of views based on ignorance and prejudice and affected by editorial and financial gatekeeping (Shoemaker 2009). This chapter outlines a process of introducing mobile journalism skills (mojo) and technologies to Indigenous people living in remote communities in Australia. Forming part of a PhD case study conducted in the Northern Territory, in Australia, NT Mojo scoped the degree to which mojo practices—relevant training and tools—might be used to help create a less marginalized Indigenous voice.

At the heart of the NT Mojo project was a skill set and a smartphone kit that included a stabilizing cradle (the mCamLite), a microphone and a small rechargeable light. Developed by Ivo Burum, the project was delivered with the assistance of Batchelor Institute for Indigenous Education. Formal training was held at the Batchelor campus, 100 kilometers south of Darwin, in the Northern Territory. Mojos from five communities flew in for one week of formal training and formed a small community of practice.

Using a case study approach, this research investigates the possibility that mobile journalism (mojo) literacies and technology can create a common digital language that enables a digital bridge across spheres of communication. This chapter explores the degree to which citizens can be taught how to create and publish empowering digital stories using just a smartphone.

The digital era creates possibilities for new cross-cultural online communications where participants with similar social and cultural backgrounds create content and engage in activities concerning local issues and interests of importance to them (Meadows 2005). Potentially, in this communications creation state, the diverse balance of source, content and exposure contributes "to the process of developing well informed citizens and enhancing the democratic process" (Napoli 1999, 9). But meaningful change requires more than access to technology. It needs education to alter the state of consciousness of marginalized or oppressed people to enable them to seek change (Freire 1970). Moreover, the NT Mojo training program discussed here scoped the degree to which Indigenous Australians, trained to use portable mobile technologies, can learn skills to become agents of change.

Methodology

Case-study research expert Robert Yin argues that no one form of evidence "has a complete advantage over all others" and case studies will want to "use as many sources as possible" (Hartley and McKee 2000; Yin 2003, 101). In my investigation, through a series of case studies, I chose to incorporate a variety of evidence available to action research investigators:

- *Documentation*: I included government reviews such as the *Indigenous Broadcast Review* (2010), the *Wilmot Report* (1984) and the documentary film *Satellite Dreaming* (1992) about Indigenous use of media;
- *Interviews*: Because Indigenous people felt more comfortable speaking with me face to face, I conducted short semi-structured before and after interviews;
- *Direct Observation*: The case-study workshops occurred on location providing an opportunity to observe and record how participants dealt with the technical journalistic and social aspects of mojo work in real time.

Background

The NT Mojo project was designed to complement the possibilities created by the deployment of Australia's National Broadband Network (NBN)—a fiber plus Wi-Fi and satellite communications platform. The NBN can potentially provide citizens with opportunities for participation in a diverse mix of information, content and news that's produced locally and sourced across platforms. Both the Indigenous Broadcast Review (Stevens 2011) and the independent review established to examine the policy and regulatory frameworks of converged media and communications in Australia (Boreham 2012) said local content is critical in reflecting and contributing to the development of national and cultural identity.

Maintaining cultural identity was a driver for a series of video workshops run by Dr. Eric Michaels in the 1980s with the Warlpiri, in a remote Indigenous community in Central Australia, called Yuendumu. All the content the Warlpiri transmitted in the 1980s was "locally produced [...] community announcements, old men telling stories, young men acting cheeky" (Michaels 1989, 7). These early transmissions were designed to counteract MSM's gatekeeping approach, which can result from managerial pressures and what media scholar Axel Bruns sees as a journalist's "feel" or desired "news beat" (Bruns 2005, 12). The result is ill-informed news about remote communities that's often obtained from non-Indigenous community workers, or journalists living in larger cities hundreds of kilometers away. This can lack grassroots diversity and shape social image through negative shock portrayal. It commodifies Indigenous people more as news items than as valued contributors to the Australian mosaic (Bell 2008). Whereas "Aboriginal newspapers since the late nineteenth century in Australia, have played a crucial role in the symbolic reclamation of space for an Aboriginal public sphere" (Meadows 2005, 1), the reality suggests these papers have given voice to a select few Indigenous people, heard by a niche group of Australians.

One of the founders of the Central Australian Aboriginal Media Association (CAAMA), Dr. Phillip Batty, says in the film *Satellite Dreaming* (1991) that the camera is one of the most powerful tools used by Indigenous Australians. Marcia Langton, former Chair of Australian Indigenous Studies at the University of Melbourne, who was involved with the early video experiments at Yuendumu and CAAMA, believes that all Indigenous uses of the camera for cultural preservation can be "traced to those early Warlpiri policies of representation" (cited in Michaels 1994, xxxv).

Concerned about the impact of satellite content delivered into remote communities by AUSSAT, the Australian communications satellite launched in 1985, Indigenous people asked for a way to interrupt the signal and insert locally created programs. In 1987 the government recommended the establishment of BRACS (the Broadcast for Remote Aboriginal Communities Scheme). BRACS included a cassette recorder, radio tuner, microphone, speakers, switch panel, VCRs, television set, video camera, two UHF television transmitters, FM transmitter, satellite dish and decoders (Willmot 1984). Based on the Yuendumu participatory model the initiative in 80 remote communities had enormous potential, but because the program failed to provide enough training, some units closed down (Batty 2012). Only those communities, like Yuendumu, that had already developed a response to television, coped with the responsibility of creating local content (Bell 2008, 121). Batty believes success often involved finding funding for an "enthusiastic whitefellah" who would live in community and "work for almost no wages to keep it going" (Batty 2012).

This was the role people like Michaels played. They were, in a sense, cultural gatekeepers. In Michaels' case, in an attempt to further safeguard the local culture against appropriation by visiting ethnographers and MSM, he developed a series of rules—effectively the community's own gatekeeping practices. Those community members who told their story on camera were often criticized by anthropologists and even community members for giving away information. In 1991, I produced *Benny and the Dreamers,* a documentary about the last of the Pintupi to make white contact in the 1950s and '60s. The Australian Broadcasting Commission (ABC) called a few days before the film was broadcast to advise they were not going to air it. They had received complaints from anthropologists about the use of archival footage in the film. However, the Indigenous people, with whom the film was made and about whom it was, had specifically requested that the archival material be used in the film and had vetted every frame. Anthropologists also said the story was incorrect, yet the story about the Pintupi was told by the Pintupi, in their own language and the finished film was screened to and approved by a room full of elders. I advised the ABC their request was another nail in the coffin of self-determination. The ABC broadcast the film in Australia, and it was also shown internationally.

When I arrived in Central Australia in 1990, Warlpiri Media was active and the Central Australian Aboriginal Media Association (CAAMA) was well established with a video unit, a music recording business, radio station and a new RCTS (Remote Commercial Television Services) television station called Imparja. Warlpiri Media and CAAMA employed mainly Indigenous people, but at Imparja scale and technical intricacies meant that staff was predominantly non-Indigenous. This is in line with an observation Michaels made a few years earlier. Once an Indigenous media organization moves away from traditional cultural forms and social organization, "they produce mainstream type programs and need westerners to operate and manage them" (Michaels 1986 in Bell 2008, 82). This was true at Imparja, where an Indigenous owned TV station employed large numbers of whitefellahs. In the main it broadcast aggregated mainstream non-Indigenous content like soap operas, sport, game shows, news and movies. Batty says of Imparja, "it's one of the many tragedies of Aboriginal affairs ... it could potentially offer educational services and health", which he says are "still dysfunctional" (Batty 2012).

Warlpiri Media in Yuendumu and Ernabella Television (EVTV) in Ernabella in the south were creating what Batty had hoped Imparja would—what Meadows refers to as "overlapping public spheres". Communication spaces that "articulate their own discursive styles and formulate their own positions on issues that are then brought to a wider public sphere where they are able to interact across lines of cultural diversity" (Meadows 2005, 38). Batty wanted Imparja to perpetuate an ongoing discussion in an Indigenous media space by acting as the cultural aggregator for community, CAAMA and BRACS content. He also wanted the BRACS communities to take CAAMA/Imparja content (Bell 2008). The aim was to use the Imparja footprint and the BRACS units to create a diverse Indigenous content sphere. Unfortunately politics and responsibilities to advertisers led to programs like *Sale of the Century* being broadcast on Imparja. CAAMA also found producing daily, even weekly, television, much more difficult than radio, and hence it could only deliver a small amount of its own video content. Which vision—Batty's, Michael's or the government's—creates the more diverse cultural sphere can be determined by applying a test for diversity.

For many scholars source diversity sits at the heart of a healthy public sphere (Bruns and Schmidt 2011; Dahlgren 1995; Deuze 2006; Hartley and McKee 2000; Jakubowicz, Goodall and Jeannie 1994; Meadows 2005; Napoli 1999). This is primarily because source is often defined through ownership and "can be measured in terms of diversity in ownership of media outlets, the workforce and content" (Napoli 1999, 12). To this end the originators of BRACS believed that a larger number of outlets—80 units were deployed nationally—would lead to more diverse ownership at source, delivering a diverse content stream that would receive national exposure. According to Napoli (1999), these are the primary components of the diversity equation. However, the lack of training about how to create video programs diminished the potential for BRACS to create diverse video content. "The policeman's wife would take it over and be the program director and bring in all that commercial junk like *Dirty Harry* on a Friday night" (Batty 2012). In those Central Australian communities where training was lacking, many BRACS were reduced to being a radio re-broadcaster or no more than a re-transmitter for the RCTS (Imparja) broadcasts of sport, soaps and commercial DVDs.

Imparja suffered from the same editorial limitations at source when it came to Indigenous content creation and delivery. Exposure can be distinguished by content "sent" or content "received", and seen either from the perspective of the broadcaster or what the audience selects. The assumption being an audience with diverse content options consumes a diversity of content (Napoli 1999, 25), which could result in "exposure to a diversity of views and public issues" (Sustein 1993, 22). But this did not occur at Imparja, where choice of content was restricted primarily to mainstream aggregation, with the odd Indigenous program.

Meadows advises that the utopian goals of an Indigenous public sphere "should not be understood in terms of a non-dominant variant of the broader public sphere" (2005, 38). CAAMA, Imparja, Warlpiri Media and also BRACS need to be viewed as unique offerings in confrontation with the mainstream public sphere, relevant because they potentially enable Indigenous people an opportunity to develop their own counter-discourses, identities and experiences that enable interaction with the wider public sphere (Meadows 2005). Whereas huge gains were achieved with BRACS and other Indigenous media, real opportunities to create a unique Indigenous subfield of communications were missed. "To expand the organization you had to take the Aboriginal people with you and the prime example of that not happening is Imparja TV" (Batty 2012). The type of deliberations Meadows refers to are critical to democratic theory and practice, with new technologies becoming society's "central nervous system" (McChesney 2007, 10). It can be argued that Imparja's early role as a central or sentinel node, disseminating Indigenous information and growing an audience cluster for cultural content, was almost lacking.

In 2005, after more than 25 years of campaigning by organizations such as Warlpiri Media and CAAMA, the Australian government supported the development of National Indigenous Television (NITV), with funding of $48.5 million. Even though NITV was soon delivering more than 350 hours of first-run Indigenous content per annum (Stevens 2011, 92), its first three years were turbulent. Grassroots organizations, including Warlpiri Media and CAAMA, felt the funding could have been better spent at a local community level rather than on urban producers and infrastructure. These are the same criticisms leveled against Imparja 20 years earlier.

In 2010 the Australian government held a review of its investment in the Indigenous broadcasting and media sector (IBR). The key terms of reference were a consideration of cultural benefits for Indigenous Australians and a consideration of the impact of media convergence on the Indigenous broadcasting and media sector. The IBR found that a "one size fits all" centralized aggregation approach would not work given the significant differences that exist between communities (Stevens 2011, 2). At the heart of IBR recommendations is a need to resource and train a more diverse local media especially in remote communities. NT Mojo was designed to do just that: provide training to enable delivery of the type of content that BRACS and Imparja had ostensibly failed to produce.

Parameters for an Evaluation

Case studies such as NT Mojo are often seen as prime examples of qualitative research. Designed to explore communities of practice, NT Mojo investigates a group of trainee mobile journalists (mojos) who engaged in a process of collective learning (Wenger 2015). The NT Mojo Workshop was designed to provide skills and technologies to enable mojos to work as community media practitioners. The action research process situated the researcher as a participant agent of change (McNiff and Whitehead 2006; Wadsworth 2011).

Evaluation occurred in three phases: identification of communities and participants, effectiveness of the training package and immediate impact of the training. However, it's also true that we evaluate "every time we choose, decide, accept or reject" (Wadsworth 2011, 7). In action research we do this by taking a piece of the world and holding it up against a known value: in this case the representation by media mirrored against the use of media by remote Indigenous communities. More specifically the phases of action research as they applied to NT Mojo evaluation are:

- *Noticing:* We bring a set of expectations and values based on previous experience, and we notice the discrepancies between what we are observing and what we expected. For example, in 1990, I found it difficult to find Indigenous people to appear on camera. In 2011, following two decades of being exposed to television, it was easier to find Indigenous people comfortable with appearing on camera.
- *Design:* This involves determining who or what is being researched; who is the research for and what are the outcomes? My background in television and self-shot content meant that professional skills were being taught and outcomes could include job possibilities.
- *Fieldwork:* This occurred around a formal training workshop followed by a further four weeks of supplementary training. Being on the ground I was able to better understand the historical and cultural imperatives (sorry business, family commitments and social structure) that potentially impact this form of story telling (see below).

- *Analysis and conclusions:* These are based on the researcher's immersion and associated observations of mojo reactions during production and the reception to the stories by mojos, community, public and networks.
- *Feedback and planning:* This began on the ground during the workshop. In this case it led to a number of supplementary workshops to determine sustainability, transition and implementation of mojo skill sets into education and MSM.

The NT Mojo Workshop Primer

The program lasted eight weeks and comprised formal training, in-community training and production and follow-up discussions. The various stages of the workshop are outlined below. Whereas many stages overlap, I have tried to list them in chronological order, primarily to provide a guide for researchers wanting to replicate the process and activity:

- *Funding:* The program received government funding primarily because it provided an opportunity to train local people with job-ready media skills, increase digital literacy and create local messaging to augment government's Closing the Gap (CTG) responsibilities. CTG is a government initiative to address the disadvantage faced by Indigenous Australians in life expectancy, child mortality, education and employment.
- *Partnership:* I approached Batchelor Institute for Indigenous Training and Education (BIITE) because I felt they would provide the program with long-term sustainability. At BIITE I was able to train two trainers to help with the in-community training phase. During the training at BIITE we observed that a number of participants didn't like being away from their communities.
- *Community support:* I wrote a project overview and community elders were asked to consider and discuss the project and its requirements—to photograph people and homes and create digital stories for publication online. Elders wanted to use mojo to create positive television and web images to get young people "off the mischief and sniffing and all that is happening here" (Mavis Ganambarr cited in Burum 2012).
- *Permissions:* Beginning the collaborative process by meeting with elders before a camera arrived was important for discussions about cultural restrictions on photography. These can include controlling the mass mediation of photography beyond the face-to-face exchange; class and gender restrictions; the need to account for temporality (what is authorized today is not necessarily authorized tomorrow); and mortuary restrictions (Michaels 1994). Elders spread the word in community, thereby giving their permission for the project to proceed and beginning a selection process based on community osmosis. Mojos provided permission for us to film them and for their work and pictures to be broadcast and used for promotion and research purposes. They required permissions from people they filmed. Requiring mojos to secure permissions meant they had to think about and explain their story.

- *Project Guidelines*: Whereas these may appear restrictive they establish clear parameters and legitimize the project as more than just movie making. The Project Guidelines proved to elders that we had thought about possible issues. The guidelines helped establish an ownership pattern for mojos to help protect their equipment. The guidelines included information on editorial aspects, such as what could be filmed. A selection of the guidelines and their outcomes is listed in Table 8.1.

Table 8.1 Guidelines and outcomes

Guideline	Outcome
G 1.3: Each participant will have to produce one short video during the training phase and a series of 4–8 short videos during the production phase of the NT Mojo project.	All mojos produced a video during training and produced 23 videos during the production phase, an average of 2.5 videos per participant. The lower number was primarily due to a cyclone and local peer pressure.
G 1.4: In keeping with the rules of the NT Mojo project, videos will focus on these themes: health, sport, music, employment, family, my community, art and culture.	We found this broad set of themes covered mojos' desired areas.
G 6.8: During the production phase NTMPT (the NT Mojo Prod Team) representatives will visit participants to further assist with video production.	Mojos found this essential and very helpful. But because of the vast distances between communities and the cyclonic weather, (we could only schedule it in the wet season) this is one of the elements that made this multi-community project difficult. The dry season brings with it the pressures of working outdoors in unbearably hot days.
G 6.9: All participants are encouraged to keep a video journal each day of the production phase.	This could not be policed due to the number of communities and was not a factor in the final analysis. Post workshop interviews were used instead to gauge responses.
G 7.2: Each participant will be responsible for their equipment and keeping it safe and in working order.	We had one breakage, which occurred on the plane en-route from training to community.
G 7.3: This equipment remains the property of the NTMPT until the participant has successfully completed the NTM project after which we will enter a discussion about the possibility of mojos being allowed to keep the equipment.	We decided to give the mojos their iPhone kits because we reasoned they would need them to continue mojo. However, a month later many of the kits were lost, damaged or unaccounted for.
G 8.1: Prior to any recording, participants must make sure that all necessary authorisations and clearances have been obtained.	Participants made sure everyone depicted on camera agreed at least verbally. There were no complaints and hence no transgressions.

- *Participant selection*: The community and elders nominated participants; we spoke with and interviewed them on camera when we arrived back in community. We selected 13 mojos and one support person from each of six communities. One of the support people wanted to make stories as well; this took the number to 14. We based our choices on elders' recommendations and candidates' availability and passion to be involved. "I was praying for this ... and one lady tell me, we got job multimedia job (mojo course), this old lady tell me and I was, *thank God!*" (Participant in Burum 2012). The mojos' ages ranged from 16 to 32, and there were eight male and six female, a spread we felt necessary to assure a diversity of story ideas and content. All mojos spoke some English, a prerequisite we imposed because of time constraints and the level of job-readiness we hoped to achieve.

Community leaders also discussed and chose support persons—teachers or media-related—we decided to train. This provided a friendly face during out-of-community training and an extra hand during in-community training and filming. In Michaels' day, on-camera participation was decided according to skin (moiety) and community standing. In 2011 agreement was based on whether elders thought a candidate could stick it out.

The digital story-telling aspect of mojo interested participants because they saw it as a vehicle for passing on stories, "the mojo workshop it's a new thing for me, it's all about to tell stories ... .it could be a dreamtime story, it could be football ... stories are very important, passed on from generations ... my grandfather always told me stories and I tell my young ones". Others saw mojo as an opportunity to pass on lifestyle messages: "we can teach kids not to do bad things through mojo work" (Participant in Burum 2012). Table 8.2 provides reasons candidates gave for choosing to participate.

Table 8.2 Pre-training interviews

Question	Number of response	Typical response
Why they wanted to be a mojo?	11	They wanted to share positive stories with the rest of Australia and they liked how mojo gave them an opportunity to tell the stories their way.
	7	Wanted to work as journalists
	2	Just wanted to use the skills at school
	4	Wanted to make community messaging
Type of stories they wanted to produce?	4	Wanted to produce stories that supported their cultural practices such as art
	5	Wanted to produce stories on how they live in harmony with their surrounds, such as bush medicine
	2	Wanted to produce sports stories
	3	Were very specific about wanting to promote a healthy life for young people

Question	Number of response	Typical response
Whether they would be able to attend out-of-community training?	14	All said they could.
	6	Were concerned about the time away from community.
We also demonstrated the mojo kit.	14	All mojos took to the technology immediately.
Explanation of the timeframe	14	All were willing and able to commit to the one week of out-of-community training and four weeks in-community filming/training.
Explanation of guidelines and project requirements.	14	Understood the requirements, the need for the guidelines and all were happy to sign and commit to the project.

Selection of Equipment

Change research informs us that the way we define technology at the outset will determine how it is used for generations (Boczkowski 2004). Hence, I developed the neo-journalistic approach and digital story-telling skill set to prove the technology. I also developed a semi-professional production kit (Figure 8.1) consisting of:

Figure 8.1 Mojo Kit 2014.

- The iPhone, which was chosen as the camera because (a) it shoots in high definition (HD) and records digital audio, (b) has great hedonic appeal—no mojo had owned one, (c) it has excellent functionality, (d) it offers the required connectivity, and (f) it worked with the 1st Video Editing App. The fact that it was first a phone, second a camera and then an edit suite made it less threatening and immediately accessible.
- *Edit App:* VeriCorder in Canada developed a commercial-grade editing App enabling multi-track vision and audio editing and mixing. The current iMovie 2.0 is the only other iOS edit App with multiple vision tracks—essential when teaching students how to use a B roll, which is the supplementary footage used to cover edits, or highlight specific interview points.
- *mCAMLite:* stability cradle;
- *Microphone:* small directional mic;
- *Light:* a portable rechargeable light;
- *Tripod:* small and lightweight;
- *Telephony:* Having both 3G/4G and Wi-Fi functions enables the user to choose the fastest upload connection. Wi-Fi is not common in private homes in remote Indigenous communities. We mainly used 3G to upload stories, and each mojo was given an account (Table 8.3).

Table 8.3 3G Upload speeds from remote communities

Upload Test—1 minute file sent using 3G		
Resolution	File size	Upload speed
1080p (High Definition 1)	170 mega bytes	34 min
720p (High Definition 2)	80 mega bytes	16 min 30 sec
360p (Standard Definition)	27 mega bytes	4 min 45 sec
270p (Low quality)	6 mega bytes	1 min 30 sec

Training: Transfer of Knowledge Elements

Training comprised a five-day formal workshop, followed by a four-week block of intensive in-community training and production, with follow-up discussions (Figure 8.2). The aim was to produce what media scholar Lee Duffield (2011, 141) refers to as "start-ready recruits for media jobs backed by a study of contexts". The formal training involved:

- *Basic journalism:* simple research and foundation journalism skills including a community relevant primer on defamation and ethics;
- *Technical recording:* use of the iPhone to record audio and video;
- *Editing:* using the edit App;
- *Publishing:* uploading stories to a website.

I developed a comprehensive mojo training manual that students received with their mojo kit.

Figure 8.2 NT Mojo recording on Bathurst Island: the story tools and skills in Indigenous hands.

Journalism Skills

Participants were introduced to simple story building blocks—actuality (e.g., the police, ambulance or fire brigade attending an accident in the process of doing their emergency work), narration, interviews, overlay (B roll), piece to camera (journalists stand up in front of camera) and music (see iBook). Training involved treating the elements like Legos and learning about their individual shapes before learning how the shapes form story. When learning how to drive a car, we first learn to use the accelerator, then the brake and finally the steering wheel. When these elements are combined in various sequences we get varied driving experiences. Learning to tell stories by learning individual story elements first before learning structure is a similar experience that is less daunting than learning complete story structure from the outset. Story construction or structure, which differs between story styles depending on how elements are constructed, was the most difficult aspect to teach. After training mojos understood the various story elements (PTC, narration, interview, B roll) and had a rudimentary grasp of story structure.

One of the advantages of having all mojos attend formal training was that when learning occurs through engagement in a group participants learn from each other in a community of practice (Wenger 2015). According to Duffield, a community of practice is particularly helpful in the transference phase, where skills are mastered to form new knowledge (2011). For transference to occur, Duffield suggests providing guidelines in the form of readings. In my case, I provided a slide show, a comprehensive training manual, a set of guidelines and an interactive workshop. Group dynamics and working in front of others repetitively enabled transference and helped overcome what Indigenous people call *shame*. Indigenous people from isolated communities often get embarrassed about having to stand out in public or in certain other situations (McChesney 2007, 221), which is due to attention rather than action (Leitner and Malcolm 2007). This can lead to disengagement. There is a fine balance between inviting participation and forcing it. Mojos were encouraged to seek help from each other and to avoid the danger of story choice becoming a reflection of what makes the teacher happy (Worth and Adair 1972).

One aspect of the journalism skill set that was easier to convey was the need for ethical reporting and the care required to avoid defamation. Ethical reporting concepts made sense to mojos who had first-hand experience of unethical treatment by the MSM. Mojos also lived in tight-knit communities where unethical behavior could lead to local cultural issues. An interesting discussion ensued about how mojos should portray edgy stories like alcohol abuse. They reasoned that those types of stories should become an exploration of what is being done about the problem. They decided to use drama reconstruction to overcome the ethical issues associated with portraying specific people as transgressors. This type of story is generally best handled when skill sets are well practiced, yet mojos should be encouraged to make their own story choices from the outset.

iPhone Recording

I left the introduction of the iPhone and camera until participants received basic journalism training. This proved valuable for two reasons: (a) it enabled a focus on the importance of journalism techniques before technology, and (b) it contextualized the use of the iPhone into a device for doing journalism.

Editing

Multimedia story telling can involve a variety of media: actuality, archival, pictures, documents, narration, Tweets and music. Bathurst Island support person Louis Kantilla found the editing technology liberating: "This little gadget, you do everything by snapping the finger, bang you can do it" (Participant in Burum 2012). As indicated in Figure 8.3, the 1st Video Edit App (and iMovie 2.0) has two video tracks. This feature enables more professional checkerboard editing (where narration, interview sync and B roll are structured on separate tracks.

V2		B Roll		B Roll	
V1	Sync		Sync		Sync
A1		Narration		Narration	
A2	Music	FX	Music	Music	Music

Figure 8.3 Checkerboard edit example. *(Note: V1, V2: vision tracks; A1, A2: audio tracks; FX: sound effects).*

Mobile editing happens on location, and this impacts the style of the story and the edit. There should be more actuality, and the feel should be dynamic. This is achieved using a hand-held shooting style and dynamic editing, which creates a news-type story bounce. Editing in general is a way of *thinking* influenced by *ways of seeing* various states of immediate possibility. This is particularly relevant in close-quarter story telling like mojo, where the immediate is an essential story component. It is based on a state that has been described as fluid, where "the relationship between what we see and what we know is never settled" (Berger 1972, 7). Yet we need to impart skills that settle perceptions. In this context mojo praxis and especially editing skills work as a filtration system to make sense of the flowing relationship between two separate unsettled realties—traditional Indigenous and Western worlds. Hence two aspects of the edit process need to be discussed.

First, because the NT mojos came from six different communities and spoke different dialects, mojo potentially became a *common digital language*. Worth and Adair (1972, 44) note that "although the parallel between film and language is not exact, it may be that the manipulation of images ... and structuring in editing is not a random activity". Hence, my premise was that if we taught mojos how to edit (and we did this by introducing an understanding of story elements and structure) and how to sequence stories with a journalistic bounce, they would learn that language style and then "develop patterned ways of filming" (Worth and Adair 1972, 45). I thought these patterned filming structures would be based on their already known culture and languages, which influence not only their "semantic and thematic choice of image", but also the "very way they put images together in a sequence" (Worth and Adair 1972, 45). However, what mostly occurred was that mojos were happy to follow the structural patterns, or story bounce, they had learned during the formal training course. A mojo story that was produced 18 months following the mojo workshop, without my input, was made in exactly the same style and bounce I had taught during workshops.

Second, mojos felt that the story bounce we established was *just like television*. This raised a question as to the relevance of the patterns of journalistic story structure that I introduced, which in one sense seems so different from their traditional story-telling structures. I believe there is, as Worth and Adair (1972) observe, universality involved in working with pictures and music that most people understand and that transcends cultural boundaries. If this is true, then additional training time is required to enable training stories to be deconstructed and restructured in a number of ways to demonstrate different outcomes and editing patterns to take into account other possible structures.

In Community Training and Production Phase

This involved putting into practice what mojos learned during the formal training workshop and trainers moving between communities to assist. Below I have listed my observations about three aspects of this phase.

Story choice

The majority of mojos recorded one of the stories they had identified during the formal training block. Mojos chose their own stories generally based on location, family, local issues and availability of interviewees. The mojos produced 20 stories in four weeks: five health, three sport, four art and culture, three youth, three media, and two local news stories. The news stories that were a response to the flooding caused by cyclonic rains were produced after trainers had left the community. Mojos produced six issue-based stories, which proved difficult to develop in small communities where everyone knew each other. One of the mojos spoke with elders about wanting to produce a story about petrol sniffing. The mojo constructed a drama reconstruction laced with interview grabs from elders so as not to offend locals. It was a smart and effective way of dealing with a sensitive issue and an indication that the workshop was working. It also demonstrated that mojos could think outside the patterned story bounce I introduced.

Production

Because mojos had never before recorded interviews at times they felt shame. This sensitivity was weighed against the importance of their work in the community—mojo as *modern-day storyteller*. All nine mojos who made stories said their biggest difficulty was constructing the story—choosing which bits to shoot and which bits to edit in or out.

Publication

We established a discreet website at *ntmojos.indigenous.gov.au*. Publishing was critical because it showed mojos they had achieved something special. Publishing on the web is seen as being professional and like television. It legitimizes the extra effort required to *get it right*. Second, publishing gets a less marginalized message out to communities and to non-Indigenous people. Finally, it is an opportunity to sell content and for mojos to be compensated for their stories.

The irony was that because of the digital divide many families could only view stories on the mojos' iPhones, at school, at the local media and community centers, or when the stories were broadcast on television. As one of the mojo's mums said, seeing those stories was just the incentive families needed to ask their communities for home computers and computer training.

Outcomes

The success of the NT Mojo's first stage mojo training program can be gauged against a number of criteria.

- *Transference of skills:* and the level of learning and the level of self-sufficiency is a major question arising out of the mojo workshop. Table 8.4 is a summary of what participants achieved.

Table 8.4 Mojo skill audit

Skill	Details	Level achieved
Story choice	Mojos determined their own stories and provided an initial list of interviewees.	Intermediate
Story structure	Mojos decided on the type of material they wanted to cover in discussion with support people. In other instances mojos developed their own story plans and commenced filming independently.	Intermediate
Coverage	Mojos mainly did their own filming. Occasionally trainers would intercede to provide advice or describe technique. Where there were two mojos in a community each would help the other.	Intermediate— more practice required covering and editing sequences
Sound	Mojos recorded their own actuality sound and they recorded their own narration.	High
Writing	Where possible, mojos tried to write their own narration after discussing with trainers what was required. In some instances trainers helped augment and write narration with mojos.	Low—needs more specific and longer training beyond the scope of this workshop.
Editing	All mojos learned to operate the edit App very quickly. Some mojos did their own editing. In other instances trainers offered advice on structure and advanced technique.	Intermediate to high
Story editing	An understanding of story structure and not the mechanical aspect of editing determined the degree to which mojos edited without trainer input.	Low—this was the most difficult of training elements and requires more work.
Render and upload	Mojos learned to render stories and to upload their stories.	High
Ethics and Defamation	Mojos received basic information on ethics, defamation and copyright. There were no issues.	Low—more modules are required, especially given global story reach.

- *Recognition of their achievements*: When two of the mojos won awards, the media and the NT Department of Education wanted to know how to integrate mojo into school. ABC TV news in Darwin and NITV news offered all nine mojos the option of vying for fee-for-service community stringer work. Two of the mojos secured video commissions for the government, and two secured ongoing full time media journalism work in their communities.
- *Improving the model*: Table 8.5 identifies aspects of mojo training and how they might be improved.

Table 8.5 Suggested improvements

Element	Change
Training location	Workshops will need to occur in community or close proximity to enable mojos to feel more at home during training.
Training partners	Could include community media centers, schools, NGOs and MSM (Main Stream Media) DoBCaD (Department of Broadband and Digital Economy) interested in the region. Partnering mojo with existing in-community media and schools provides it with its best chance of sustainability.
Equipment ownership	One of the major issues during the community phase of the workshop was equipment control. We left mojo kits with mojos. Some of those mojo kits are now unaccounted for.
In-community supervision	A more long-term training/production phase is needed which includes a facility to have trainers oversee the program on an ongoing part-time basis.
Ethics and defamation	As discussed above, a more in-depth component is required during formal training. This will be required when mojo is taken to a next, more investigative stage.
Story construction	More time and more exercises are required to enable more stories to be produced during training.
Coverage	More specific coverage exercises are required during formal training. These exercises need to be structured around sequencing so that the focus is on story coverage and not shot-gathering.
Community contact	A local contact person is needed to help organise mojos for stringer and other work.
Payment for stories	Paying mojos for stories will help generate on going interest and sustainability.

Conclusion

In summary, the possibilities of projects like NT Mojo to deliver tools to enable the creation of a dialogue with and between Indigenous people at a grassroots level potentially overcomes the sensitivity that Professor Marcia Langton (1993) believes requires an indirect approach to news interviews and extensive negotiation with communities. When Indigenous communities are able to create and publish their own local stories, bypassing even the more mainstream Indigenous media organizations, an alternative internal cultural process of negotiation is enabled.

Susan Forde (2011, ix) points out that "even though the practices of alternative journalism are older than commercial professional journalism", alternative citizen-generated media remains undervalued. Validating alternative community based journalism across spheres of communication is one challenge at a time that McChesney (2007) describes as a *critical juncture* in communications. This view is supported by Curran (2007), who posits that we need less conformity in journalism and acknowledgment of new communications technologies that sustain audience access to varied viewpoints. I contend that this requires an approach that positions alternative and mobile journalism as a more defined subfield of journalism, so that it is not regarded as just "anything that occurs outside the mainstream news media" (Hirst in Forde 2011, 3). My theory is based on a current development in online media content from raw to more formed citizen-generated web TV media.

According to Meadows et al. (2009) community radio and television is a binding digital glue that creates a sense of belonging and has a "positive impact on individuals in terms of their perceptions of social well-being" (Meadows and Foxwell 2011, 4). Research further indicates that community-based media can assist in managing an individual's state of mind and social isolation by providing what Meadows calls a view of a "positive future" (Meadows et al. 2009, 2011). This is said to be because community journalism is "closer to the complex 'local talk' narratives at community level that play a crucial role in creating public consciousness, contributing to public sphere debate and more broadly the democratic process" (Meadows et al. 2009, 158). Basically community broadcasting has given communities an opportunity to say "we are here" (Meadows and Foxwell 2011, 13). Such journalism has wider democratic implications. It creates what Ewart (2007) refers to as a "series of local public spheres or a community public sphere ... for local level public discussion and debate for those whose voices were typically marginalised or non-existent in mainstream media" (Forde et al. 2002 in Ewart et al. 2007, 181). This research shows that this also impacts positively on content creators working at a local level.

In conclusion, digital literacy training can enable the type of enquiring minds that can, as Paolo Freire (1970) points out, fight against oppression for a more equitable and diverse society. Forde (2011, 2) observes "alternative journalism takes very little account of mainstream journalistic practices and values". Mojo training potentially provides competencies that enable mojos to question the role of journalism and a spoon-fed gatekeeping approach (Lippmann 2008; Singer et al. 2011). This is enabled by practicing a common digital language—a pedagogy that uses skills and technology to form a bridge among school, community and the workplace via digital challenge-based transformative learning programs (Puentedura 2013).

While undertaking the NT Mojo study it became apparent that the role of institutions and educators is instrumental to achieving sustainability. The subsequent Cherbourg Mojo project scoped sustainability with the education system. Whereas the program was successful, sustainability was not an outcome, primarily because there was no budget allocated for it. Therefore sustainability was a critical component of the most current Far West Mojo (FWM) project. Funded by the Red Cross, one of the desired outcomes of FWM is sustainability. Funding is currently being sourced to enable mojos to be paid for their stories for the next couple of years. The Red Cross is partnering with a local Indigenous organization and a youth hub in Ceduna to develop an Indigenous production house to service the 42 welfare Industries working in the region.

A number of workshops were run in schools to test application and pedagogy for delivering better-trained and more aware student citizens (Schofield Clark and Monserrate 2011). The basic premise is that once beyond the subjective hedonic phase, of seeing mojo tools (iPhones) as icons that provide immediate pleasure, the newly acquired skills are used to enter a new phase of behavioral intention. This moral activity and eudemonic philosophy is based on an ethical approach that's designed for communal good. I contend that this state necessitates a more objective approach to personal stories, which can help alternative community-based journalists broaden their audience. As Forde (2011, 118) suggests, appealing to a wider audience may result in being able to make a modest living by creating saleable alternative journalism based on a definition of objectivity, which rejects neutrality in favor of "the quest for truth". This is encouraged through the neo-journalistic approach of mojo that marries foundation journalism skills with new technology and new multimedia skills. As Forde (2011, 119) reminds us, this transformation is not easy, because, "journalists are human beings, everything they produce is subjective", especially community-based alternative journalism. Hence there's a need for structural tools that help content creators find a balance between subjective and objective treatment. The degree to which objectivity impacts a subjective treatment will depend on the mojo, the story, a level of journalistic comprehension and in a commercial situation the brief and/or response from a commissioning body.

A key to this conundrum is the dialectic that occurs between the audience turned journalist and the professional, or between alternative and mainstream media, about the emphasis on telling unbiased stories that are verified. What we need to do is teach citizens and students the skills to understand bias and to tell these types of stories and remind professional journalists that objectivity comes from being accurate, fair and balanced. It is hoped that mojo skills will help professional journalists cross the digital divide and develop in citizen journalists what Forde (2011, 116) calls "a nose for news". One important question that remains is: what is news value and how does this differ across spheres of communication?

References

Batty, Phillip. Interview by Ivo Burrum. In Ivo Burum, PhD diss., Deakin University, 2012.

Bell, Wendy. *A Remote Possibility: The Battle for Imparja Television*. Alice Springs: IAD Press, 2008.

Berger, John. *Ways of Seeing*. Television series. London: BBC, 1972.

Boczkowski, Pablo J. *Digitizing the News: Innovation in Online Newspapers*. 1st ed. Cambridge, MA: MIT Press, 2004.

Boreham, Glen. *Convergence Review: Final Report*. Canberra: Australian Government, 2012.

Bruns, Axel, and Jan-Hinrik Schmidt. "Produsage: A Closer Look at Continuing Developments." *New Review of Hypermedia and Multimedia* 17, no. 1 (2011): 3–7.

Burum, Ivo. *Mojo Working.* Distributed by NITV, Australia, 2012.

Curran, James. "Introduction." In *The Alternative Media Handbook*, edited by Kate Coyer, Tony Dowmunt and Alan Fountain. Routledge, New York, 2007.

Dahlgren, Peter. *Television and the Public Sphere: Citizenship, Democracy, and the Media.* London and Thousand Oaks, CA: Sage Publications, 1995.

Deuze, Mark. "Participation, Remediation, Bricolage: Considering Principal Components of a Digital Culture." *The Information Society* 22, no. 2 (2006): 63–75.

Duffield, Lee. "Media Skills for Daily Life: Designing a Journalism Programme for Graduates of All Disciplines." *Pacific Journalism Review* 17, no. 1 (2011): 141–55.

Ewart, Jacqueline, Susan Forde, Kerrie Foxwell and Michael Meadows. "Community Media and the Public Sphere in Australia." In *New Media Worlds: Challenges for Convergence*, edited by Virginia Nightingale and Tim Dwyer. Melbourne: Oxford University Press, 2007.

Forde, Susan. *Challenging the News: The Journalism of Alternative and Community Media.* UK: Palgrave Macmillan, 2011.

Freire, Paolo. *Pedagogy of the Oppressed.* London: Continuum International, 1970.

Hartley, John, and Alan McKee. *The Indigenous Public Sphere: The Reporting and Reception of Aboriginal Issues in the Australian Media*, 1st ed. New York: Oxford University Press, 2000.

Jakubowicz, Andrew, Heather Goodall and Jeannie Martin. *Racism, Ethnicity, and the Media.* Sydney: Allen & Unwin, 1994.

Langton, Marcia. *Well, I Heard It on the Radio and I Saw it on the Television.* 4th Printing. Sydney: Australian Film Commission, 1993. Accessed March 28, 2015. http://www.screenaustralia.gov.au/cmspages/handler404.aspx?404;https://www.screenaustralia.gov.au:443/getmedia/580b7762-ceb4–4b3f-abad-7ec91b4e6adf/WellIHeard.pdf

Leitner, Gerhard, and Ian G. Malcolm. *The Habitat of Australia's Aboriginal Languages: Past, Present and Future.* Berlin: Mouton de Gruyter, 2007.

Lippmann, Walter. *Liberty and the News.* Paperback ed. Princeton, NJ: Princeton University Press, 2008.

McChesney, Robert. *Communication Revolution: Critical Junctures and New Media.* 1st ed. New York: The New York Press, 2007.

McNiff, Jean, and Jack Whitehead. *All You Need to Know about Action Research.* London: Sage Publications, 2006.

Meadows, Michael. "Journalism and the Indigenous Public Sphere." *Pacific Journalism Review* 11, no. 1 (2005): 36–41.

Meadows, Michael, Susan Forde, Jacqueline Ewart and Kerrie Foxwell. "Making Good Sense: Transformative Processes in Community Journalism." *Journalism* 10, no. 2 (2009): 155–70.

Meadows, Michael, and Kerrie Foxwell. "Community Broadcasting and Mental Health: The Role of Local Radio and Television in Enhancing Emotional and Social Well-Being." *Radio Journal: International Studies in Broadcast & Audio Media* 9, no. 2 (2011): 89–106.

Michaels, Eric. *Bad Aboriginal Art and Other Essays: Tradition, Media, and Technological Horizons.* Minneapolis, MN: University of Minnesota Press, 1994.

Michaels, Eric. *For a Cultural Future: Francis Jupurrurla Makes TV at Yuendumu.* Art and Criticism Monograph 3. Malvern: Art and Text Publications, 1989.

Napoli, Philip. "Deconstructing the Diversity Principle." *Journal of Communication* 49, no. 4 (1999): 7–34.

Puentedura, Ruben. *Technology in Education: A Brief Introduction.* YouTube video. Jan 7, 2013. Accessed March 28, 2015. http://www.youtube.com/watch?v=rMazGEAiZ9c%3E.

Satellite Dreaming. Video Recording. CAAMA, Australia. Distributed by Ronin, Australian Broadcasting Commission (ABC) TV, Australia, 1991.

Schofield Clark, Lynn, and Rachel Monserrate. "High School Journalism and the Making of Young Citizens." *Journalism* 12, no. 4 (2011): 417–32.

Shoemaker, Pamela. *Gatekeeping Theory.* New York: Routledge, 2009.

Singer, Jane B., David Domingo, Ari Heinonen, Alfred Hermida, Steve Paulussen, Thorsten Quandt, Zvi Reich and Marina Vujnovic. *Participatory Journalism in Online Newspapers: Guarding the Internet's Open Gates.* UK: Wiley, 2011.

Stevens, Neville. *Review of Australia Government Investment in the Indigenous Broadcasting and Media Sector 2010.* Canberra: DBCDE, Australian Government, 2011.

Wadsworth, Yoland. *Everyday Evaluation on the Run.* 3rd ed. Crows Nest, NSW: Allen and Unwin, 2011.

Wenger, Etienne. "Communities in Practice: A Brief Introduction." Accessed March 28, 2015. http://wenger-trayner.com/wp-content/uploads/2012/01/06-Brief-introduction-to-communities-of-practice.pdf.

Willmot, Eric. *Out of the Silent Land: Report of the Task Force on Aboriginal and Islander Broadcasting and Communications.* Canberra: AGPS, 1984.

Worth, Sol, and John Adair. *Through Navajo Eyes: An Exploration in Film Communication and Anthropology.* Bloomington: Indiana University Press, 1972.

Yin, Robert K. *Case Study Research: Design and Method.* Thousand Oaks, CA: Sage, 2003.

9 Mobile Technology in Indigenous Landscapes

Coppélie Cocq

Attachment to the land in Indigenous contexts has been emphasized in previous research in relation to identities, traditional knowledge, processes of cultural and linguistic revitalization, place-making and decolonization (Appadurai 1996; Basso 1996; Rydberg 2011). In contemporary Sápmi (Samiland)—the traditional settlement area of the Sami—the close relationship between identity and the land is actualized in struggles over land rights (Cocq 2014). The traditional area of settlement, i.e., the land of the Sami ancestors and where most Sami identities are rooted, is predominantly rural even though most of the Sami in Scandinavia now live in urban areas. In this context, mobile technologies offer not only new opportunities for accessibility in rural and mountainous areas, but also the possibility to include these areas within the "digital landscape" and thus increase the visibility of neglected areas, their inhabitants and their knowledge.

The presumed boundlessness of digital technologies in earlier Internet research (for instance, Meyrowitz 1985) has recently been contested (Lindgren, Dahlberg-Grundberg and Johansson 2013; Moores 2007), and the development of mobile technologies can in fact materialize and concretize our relations and interactions with specific geographical places. Based on the example of an application for mobile devices that reveals a Sami linguistic landscape via augmented reality, this study investigates the potential of mobile technologies for Indigenous empowerment in the context of the Sami community.

The Sami are the only Indigenous people of Europe. Sápmi is a broad area that comprises parts of Sweden, Norway, Finland and Russia. National borders, however, essentially have only political implications because the language areas, modes of land use, etc., span across these borders. Despite the varieties of languages, heterogeneity of livelihoods and diverse conditions and prerequisites for cultural and linguistic vitality, the Sami are one nation with a common flag, a common national hymn and a common national day.

The ten Sami languages are endangered, but efforts at revitalization can be witnessed in many areas in Sápmi. All Sami speak the majority language in each country (i.e., Swedish, Norwegian, Finnish or Russian); however, not all Sami speak or understand a Sami language. Language is a significant symbol of identity in contemporary Sami discourses, whether it is referred to as a mother tongue, as something that has been lost or as something to take back. In the colonial and post-colonial context of Scandinavia, the visibility of Sami culture and languages has been marginalized. Efforts to counteract and question this invisibility and negligence have multiplied (Pietikäinen 2008), and this present study examines one example of such initiatives.

The population of Sápmi has been generally quick to adopt new technologies, from the first generation of cell phones in the 1980s to the modern use of social media in domains as diverse as e-commerce, language acquisition and activism. An early example is SameNet, a social media service that existed long before Facebook. It was used for communication, discussion and e-mailing and as a platform where documents could be shared, for instance, for school and university courses or other educational contexts. SameNet was created in 1997 by the National Association of Swedish Sami (SSR) and was administrated mainly by the Sami education center of Jokkmokk in Sweden. It was shut down in 2011.

The current global technological development that allows for increased accessibility regardless of physical location can, naturally, be witnessed in Sápmi too. Today, the use of digital tools is particularly manifest in education, for instance, with resources that support language learning or with the adaptation and mediation of storytelling in online environments. Mobile technologies are increasing exponentially, from applications for language acquisition and information sharing adapted to smartphones, to the use of Global Positioning System (GPS) technology, to recent projects investigating the possible use of drones in reindeer herding (Lewan 2015). Mobile technology is reshaping not only the media landscape, but also the way we can interact with the environmental landscape. This is the focus of this study, which is illustrated by an application that redefines the linguistic landscape in Sápmi.

Mobility

In this chapter, mobility is approached not only in terms of technological accessibility but also as a practice, i.e., in relation to uses and representations of place, space, and landscape. Thanks to GPS and other navigational techniques, mobile technologies that build on interaction with a physical space have multiplied, including geocaching, mapping tools, check-in buttons in social media platforms, etc. Thus, digital practices related to physical places are common in the parts of the world where Internet accessibility is readily available.

Language is a significant component in our relation to and our representation of places, for instance, through stories tied to the land, through traditional knowledge attached to the flora, fauna and specific phenomena in the landscape or through place names. This aspect has been examined in previous research, e.g., in the field of linguistic landscape studies.

A linguistic landscape is constructed by the combination of "road signs, advertising billboards, street names, place names, commercial shop signs, and public signs on government buildings in a given territory, region, or urban agglomeration" (Landry and Bourhis 1997, 25). In Indigenous contexts, this form of visibility in the landscape is to be approached in relation to a colonial past and to on-going processes of decolonization and linguistic and cultural revitalization. The Sami population and the Sami languages have long been invisible. It is only recently that Sami place names have appeared on road signs and that some cities have set up Sami names on official buildings such as libraries and city halls (Government Offices of Sweden 2009; Språkrådet 2011). The visibility of the Sami languages in the landscape is also subject to hierarchies of languages in multilingual communities, and the presence of Sami in public and private signs in proportion to other languages can differ greatly in different cities and countries (Salo 2012). These differences reflect language policies, questions of power and ideologies that affect attitudes toward minority languages.

These scattered efforts imply that the presence and visibility of the Sami languages are still marginal and marginalized. The history of cartography, when naming places functioned as a tool for colonization (Pettersen 2011, 176), has contributed to shaping the linguistic landscape. Even in cases when local people contributed with geographical information, their presence was erased (Cosgrove and Daniels 1988; Short 2009). Consequently, local places have been renamed or subjected to translation or transcription into the majority language. The linguistic landscape in Indigenous areas is, therefore, very often the one of the colonizer, and the use of official and public signposts in Indigenous languages can be seen as material evidence of the presence of the colonized.

The use of names in mobile and/or digital texts related to specific geographical places also contributes to place making and to the visibility of a linguistic landscape in parallel with the material landscape. Mobile applications also have the benefit of enabling one to interact with the landscape in the Indigenous language and to designate places in their local names. A mobile phone can turn into an object that connects a place with its name. Everyday digital practices often include interacting with geographical places, for example, picking up a smart phone in order to "check in" on a social media platform or to post a photo of one's location are common ways to inform friends about our whereabouts.

The potential of linguistic landscapes is emphasized in the problematization and investigation of the effects of public signs on language behavior. If the "linguistic landscape of a territory can serve [...] an informational function and a symbolic function" (Landry and Bourhis 1997, 25), it might also have effects on ethnolinguistic vitality. Landry and Bourhis suggest "that the presence of private and government signs written in the in-group language may act as a stimulus for promoting the use of one's own language in a broad range of language use domains" (1997, 45). In the contemporary context of Sami linguistic revitalization, it is highly significant to consider the possible effects of digital linguistic landscapes on ethnolinguistic vitality.

Augmented Reality

The application iSikte Sápmi (InSight Sápmi—see List of Websites at end of chapter) illustrates the use of mapping tools and augmented reality in order to (re)create a space where Sami forms the linguistic landscape. Whereas official maps and mapping services provide information about the Sami area only in the national language, digital technologies enable the production of tools that increase the visibility of Sami presence and languages. Although the number of mobile applications for language learning is growing, iSikte is a particularly interesting example in that it focuses on a neglected aspect of language: the language of the landscape.

iSikte Sápmi was produced by Apps Fab AS in 2012 in collaboration with *Gáisi giellaguovddáš* (the Sami language center in Tromsø) and exists for iPhone and, more recently, for Android phones. The application was not designed specifically for Sápmi, and it exists for other countries and regions and in several languages. The application connects a database of place names to a map using GPS technology to identify the location of the user on the map. The user can then, through the phone's camera, get a view of the surrounding landscape with the name of the mountains, lakes and other places. The application has a web 2.0 component meaning that users can add content. Also, it is connected to social media platforms, and pictures of landscapes annotated with the place names can be shared on Facebook.

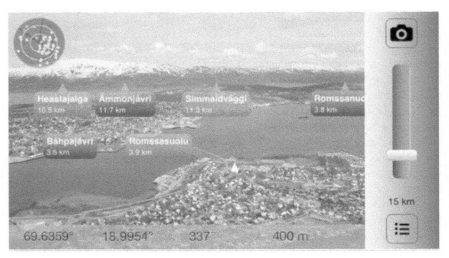

Figure 9.1 Sami place names, view from Tromsø, Norway. (iSikte Sápmi, AppsFab).

The application makes use of augmented reality, i.e., a live view of a physical environment whose elements are supplemented by computer-generated input, in this case text that is added to the landscape as it appears on the screen of a mobile device through the camera lens.

The description of the application by its producers on the Apps Fab website reads:

> *What is the name of the mountain you can see in the distance? What are the lakes you see in front of you? Point there with In Sight—Samiland and you'll see on the screen what they are called. ...*
>
> *In Sight—Samiland shows Sami POIs in Norway. See what's near you when you travel in Samiland: Mountains, Lakes, Places, Buildings, etc. You don't even have to be nearby Samiland: You can manually set your location to a place in Samiland to pretend you are there.*

The app has an offline function that enables the users to select a location and access the data. The accessibility of iSikte Sápmi is enhanced in that it is free of charge, whereas versions of the application for other regions cost 1 to 3 euros. The app has had 2,500–3,000 downloads, a satisfying number in relation to the targeted group, according to the project coordinator (Inger Persson, personal communication, March 4, 2015).

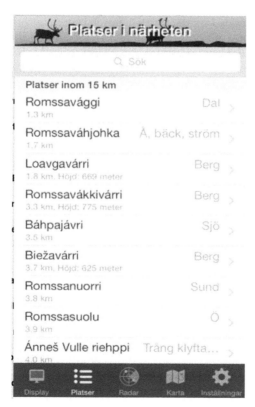

Figure 9.2 List of names of places in the landscape near Tromsø. (iSikte Sápmi, Apps Fab).

The adaptation of the app to Sápmi was initiated by *Gáisi giellaguovddáš*, the Sami language center in Tromsø, which had documented and registered place names in the Tromsø area. Project coordinator Inger Persson (2014) explains:

> *In the spring of 2012, I was asked whether I could get these names published. I even got to choose how I would do it, the only require-ment was that it should be published digitally.*
>
> *After having pondered a while, I decided to go for a mobile solution—an app. I started to look for possible mobile apps, and found iSikte—Norway. I thought: "This is what I want with Sami place names". And I contacted those who produced the iSikte apps, AppsFab AB. AppsFab proved to be positive to my idea of a Sami iSikte. In the fall of 2012, we started working on iSikte Sápmi.*

The data registered by *Gáisi giellaguovddáš* was then connected to the places' coordinates so that they can appear at their proper location in the app. In addition to the data from the language center, place names from other areas in Sápmi were provided by the Norwegian mapping authority (*Kartverket*). The project has the ambition to include Swedish Sápmi if financial support were given (Inger Persson, personal communication, March 4, 2015).

There is also an educational purpose behind the application; it is based on "a project to transmit and increase knowledge about Sami place names and Sami traditional use of the natural environment", and the application implies that "you can learn Sami place names while exploring nature" (iSikte Sápmi). Through the iSikte application, the smartphone becomes a portal of digital signs that creates a Sami linguistic landscape.

A Mobile Linguistic Landscape

Considering the iSikte Sápmi application, what functions does this kind of tool serve? Or, as Landry and Bourhis discuss (1997), what are the informa-tional and symbolic functions of this form of mobile linguistic landscape, and what might the effects on ethnolinguistic vitality be?

The informational function of the application lies in its ability to make accessible comprehensive data about place names. Knowledge about place names in Sami is usually scarce and invisible, with the exception of city names on road signposts in a few municipalities. Sami landscape names are to a great extent descriptive and refer to shapes and/or qualities of inter-est when moving through the landscape, for instance, pasturages, proxim-ity to water, or mythology (Gaup 2005). The functions of landscape terms, and the significance of the knowledge associated to them, are naturally of diverse kinds depending on the relationship to the environment. Reading the landscape was crucial not the least for reindeer herders, and in former times knowledge was based to a great extent on orally transmitted information from generation to generation (for instance, Balto 1997). However, it is doubtful that reindeer herders are the main target group or primary users of the application. Rather, they might be contributors to the database because they retain knowledge about the landscape that might have been lost in other groups of the heterogeneous Sami society.

For Sami speakers in general, the application has an informational function to the extent that it provides terms and names connected to specific places that are hardly accessible otherwise. Reading a database requires a higher degree of engagement than starting an app on a smartphone. Augmented reality in itself implies the direct application of knowledge in the appropriate context. For non-Sami speakers, it would be too ambitious to expect the application to play a significant role in language acquisition. However, the image of the landscape with the Sami names creates a screen where the Sami presence is manifest. Thus, it informs about the history of a geographical area—a history mostly erased by colonialism. Also, the technology positions the visible and manifest presence in a contemporary context; the user is not looking at an old printed map, but at the landscape as it appears from where they are standing looking through the camera lens of a modern technological tool.

The app does not directly increase the visibility of the Sami language in the landscape, and its users are probably people who are particularly interested in the language and in place names. This is indicated by the fact that users play the role of knowledge keepers and suggest corrections when a place name displayed through the app is not correct (Inger Persson, personal communication, March 4, 2015). The Sami presence only becomes visible when a user chooses to activate the application, and the informational character of such a digital linguistic landscape is limited to in-group members. As for the hierarchy of language, Sami gains the same status as Norwegian because both have an app. However, iSikte Sápmi only creates a monolingual landscape; it does not say anything about the power and status relationship between Sami and Norwegian.

The symbolic function of a linguistic landscape complements its informational function (Landry and Bourhis 1997, 27). In the case of a mobile linguistic landscape, the symbolic functions lie partly in the fact that such an application exists. Although applications in Sami languages have multiplied, the development of such products in rare languages cannot be taken for granted. Therefore, such an investment has consequences (not only at a symbolic level) for ethnolinguistic groups whose languages have been previously limited to private domains and spheres. Thus, the presence of the Sami languages through the application can contribute to the development of a positive social identity for the ethnolinguistic groups. This, in turn, is necessary in order to counter and compensate for the former stigmatization of minority languages (Elenius 2006). Ethnolinguistic identity, a significant component in groups "where language is a valued component of identification" (Vincze and Moring 2013, 47–48), implies that language is more than just a tool for communication (see, for instance, da Silva and Heller 2009). It functions as a symbol of identity and a zone of contact for a feeling of community and cohesiveness.

Ethnolinguistic identity is a key for building or reinforcing the language community (Cocq 2015; Giles and Johnson 1987; Tafjel and Turner 1979; Vincze and Moring 2013). From this perspective, a mobile application can contribute to supporting this process by extending the presence and use of the Sami languages to new arenas (a smartphone compared to a map) and taking it from private into public domains. This is enhanced by the social media component of the app that allows users to share their pictures and, consequently, to share evidence of their experience of the landscape through Sami place names.

Such an application might have additional effects on ethnolinguistic vitality in that it can function as "a stimulus for promoting the use of one's own language" (Landry and Bourhis 1997, 25), in this case in relation to the landscape. Its practice on a mobile device implies that the use of the language might emerge in different places, on the move, and in a spontaneous manner—in contrast to a classroom or other specifically limited spaces. As such, the app is closer to a vernacular mode of communication, and, consequently, closer to oral storytelling, which is a traditional mode of transmission of knowledge and identity.

Another effect on language vitality is the positive view of Sami languages that the application might enhance. The status of a language is amplified by an investment in its use in all domains. Revitalization requires changing community attitudes (Grenoble and Whaley 2006), and positive attitudes toward the minority language at the level of society (ideology) and at a group level (how the language is viewed) have also been shown to be necessary for language maintenance (Hyltenstam and Stroud 1991). On an individual level, positive consideration about one's language encourages its use.

Beyond these considerations about the informational and symbolic functions of the iSikte Sápmi application, and in addition to hopeful reflections about its effects on ethnolinguistic vitality, we ought to inquire about what ways and to what extent such tools might contribute to empowerment.

Community Initiatives and Official Signs

The application iSikte Sápmi is the result of a combination of initiatives. The structure of the app was designed by Apps Fab AS and can be applied to any geographical area and language, but it is also the side product of a documentation project produced by *Gáisi giellaguovddáš*. As for areas outside Tromsø, the place names are from the Norwegian mapping authority. The application is, therefore, the result of combined efforts from national authorities and community initiatives.

As mentioned above, the application allows users to suggest and add place names. The possibility to contribute with names implies that the app encourages knowledge exchange. In the case of place names in Indigenous languages, vernacular knowledge is often a rich and unique source because terms are not always well documented in books and rarely in official maps. Furthermore, such initiatives contribute to community empowerment because place names on a digital landscape do not need official approval—unlike road signposts.

The application was chosen with the aim of increasing accessibility to the data (Persson 2014). A digital format was preferred for its ability to disseminate and share knowledge about place names. An easy-to-use system for mobile devices makes the data accessible to more users than a computer database. Whereas a database of place names is of great interest for people searching for specific information or for scholars, the iSikte Sápmi app addresses a broader range of people. Targeted groups are indeed "all Sami, everyone interested in Sami traditional knowledge, youth that are interested in beginning to use Sami place names", but also "others interested in outdoor life, hunters, tourists, etc." (Inger Persson, personal communication, March 4, 2015). An interest in the landscape, in a particular place, or in experiencing an area in a different manner while visiting might be sufficient for motivating the use of the app. A deeper investigation of the actual consumption and use of the product, however, would require further study.

In addition to the advantage of addressing a broader readership than a computer-based format could, the application's greatest benefit is its inherent suitability for mobile devices. Augmented reality means that the database-generated information supplements the real-world environment and appears as an extra layer on the place as it is experienced—and not as a list, a flat map or tables. The information is directly applied to a place, a view or a position a user experiences. The live view of the landscape enhances the significance of the information accessed at the moment—in this case Sami names and terms.

Based on this example, I wish to address the question of the appropriateness and suitability of mobile technologies for interaction with places and landscapes in Sami contexts. Because it can be used on mobile devices and on location, the application illustrates a contribution to a presentation of a digital linguistic landscape. It is not accessible in books or visible to a larger public, but the application enhances the significance of the locations—especially in terms of the traditional use of the natural environment, as a cultural landscape or as places in which stories and histories are embedded.

In Indigenous contexts, the land is a marker of identity connected to a sense of belonging. Colonization processes have implied the erasure of Indigenous presence and history, which is linked to constructions and perceptions of the land as a *terra nullius* (Fitzmaurice 2007; Frost 1981), and this erasure is expressed in the scarcity of Indigenous languages in linguistic landscapes. From this perspective, iSikte Sápmi is an Indigenous response to this ideological encroachment. Naming places becomes part of the decolonization and revitalization processes and is a step toward taking back language and land. A Sami linguistic landscape brings to the fore the cultural landscape and the history attached to it.

What then are the benefits of this form of digital initiative in comparison to mainstream, printed or computer-based media? The application reaches fewer users than traditional media, but it is not the product of a "one-to-many" or "top-down" form of communication, and besides the people of *Gáisi giellaguovddáš* and the registers of *Kartverket*, users can also contribute with locations. The mobile application inherently means that it can be accessed on a portable device. This could of course be compared to a map—which is another form of portable device. However, the application differs from a printed medium because it is mediated through a screen; the landscape appears just as it looks in front of you, which enhances the user's perception of reality. Moreover, it is accessed from a personal item, one's smartphone. It is, therefore, a familiar environment for the users, it is easy to use and it is connected to other familiar functions with "sharing" on social media platforms.

Considering the limited visibility of the place names, which are constrained to a mobile device and require the active initiative of users, what are the effects and effectiveness of such a digital linguistic landscape? In multilingual areas, a linguistic landscape is an indicator of the hierarchy of languages. The material linguistic landscape of, for instance, the Tromsø region, shows a dominant presence of Norwegian and a subordinate presence of North Sami. But in the application, we can see a monolingual landscape in Sami. This is yet another means for strengthening the status of the language and the valuable knowledge embedded in it. In this way, mobile technologies enable a presence of Indigenous languages that is not limited to a private sphere as has been the case previously.

Conclusions

Based on the study of an application that connects Sami place names to geographical places, this chapter discusses place and mobility from an Indigenous perspective. The mobility of mobile technology can indeed reinforce the sense of place because it can function as a form of digital cartography and place-specific knowledge. The emplacement of language in the landscape also implies an emphasis on traditional modes of knowledge transmission through oral storytelling, cultural history and the use of the land.

The benefits of the application for the users are an increased knowledge about language and place names and a strengthened sense of belonging to the place for locals who have a Sami ethno-linguistic identity. The app transforms a colonialized landscape into a familiar landscape. However, even if mobile texts in the linguistic landscape have great potential, their effects are limited in that they are not visible to those who do not already have insight or interest in the language or prior knowledge of the Sami presence in the region.

Indigenous linguistic landscapes can contribute to ethnolinguistic vitality and to the strengthening of identity. From this perspective, a digital linguistic landscape is part of a larger process toward an increased visibility of the Sami languages and cultural and linguistic revitalization. Initiatives such as the iSikte Sápmi application should, therefore, be considered in relation to other community initiatives. Language apps, for instance, also illustrate a mode of bypassing, complementing and providing solutions for lack of efforts at institutional and national levels, but digital tools cannot be studied in isolation from what takes place offline. Contemporary legislation supports the setting up of signposts in Indigenous languages in some municipalities. As for vernacular initiatives, these are expressed in art, cultural events, festivals, etc. Revitalization takes place at the local and group levels and strives for a change of attitudes in society at large.

Looking Forward

The potential and ambitions of mobile technology initiatives should be considered with caution due to the difficulty in measuring their effects. However, the proliferation of initiatives for strengthening the Sami languages and their visibility is promising and might soon provide us with a suitable amount of data to estimate their benefits.

In order to evaluate the role of digital technology for Indigenous languages, there needs to be extensive studies that take into account online and offline efforts and link together the various initiatives in a long-term perspective. Additionally, the accessibility and ease of access to digital tools should take into account issues of digital divides and digital literacies.

List of Websites

(Accessed February 1, 2015)
Apps Fab: https://play.google.com/store/apps/details?id=com.appsfab.insight.sapmi.
iSikte Sápmi (InSight Sápmi): http://www.nrk.no/kanal/nrk_sapmi/1.8862238.

References

Appadurai, Arjun. *Modernity at Large: Cultural Dimensions of Globalization.* Minneapolis: University of Minnesota Press, 1996.
Balto, Asta. *Samisk Barneoppdragelse i Endring.* Ad notam Gyldendal AS, 1997.
Basso, Keith H. *Wisdom Sits in Places: Landscape and Language Among the Western Apache.* Albuquerque: University of New Mexico Press, 1996.
Cocq, Coppélie. "Indigenous Voices on the Web: Folksonomies and Endangered Languages." *Journal of American Folklore* 128, no. 509 (2015): 273–85.
Cocq, Coppélie. "Kampen om Gállok. Platsskapande och Synliggörande." *Kulturella Perspektiv. Svensk Etnologisk Tidskrift* 23, no. 1 (2014).

Cosgrove, Denis, and Stephen Daniels. (Eds.). *The Iconography of Landscape: Essays on the Symbolic Representation, Design and Use of Past Environments.* Cambridge, UK: Cambridge University Press, 1988.

Da Silva, Emanuel, and Monica Heller. "From Protector to Producer: The Role of the State in the Discursive Shift from Minority Rights to Economic Development." *Language Policy* 8, no. 2 (2009): 95–116. doi:10.1007/s10993–009–9127-x.

Dijck, José van. *The Culture of Connectivity: A Critical History of Social Media.* New York: Oxford University Press, 2013, 240.

Elenius, Lars. *Nationalstat och Minoritetspolitik. Samer och FinsksprÅkiga Minoriteter i ett Jämförande Nordiskt Perspektiv.* Lund: Studentlitteratur, 2006.

Fitzmaurice, Andrew. "The Genealogy of Terra Nullius." *Australian Historical Studies* 38, no. 129 (2007): 1–15.

Frost, Alan. "New South Wales as Terra Nullius: The British Denial of Aboriginal Land Rights. *Historical Studies* 19, no. 77 (1981): 513–523.

Gaup, Káren Elle. *Silisávži. Landskap, Opplevelser og Fortellinger i et Samisk-Norsk Område i Finnmark.* Kautokeino: Dieđut, Sámi Instituhtta, 2005.

Giles, Howard, and Pat Johnson. "Ethnolinguistic Identity Theory: A Social Psychological Approach to Language Maintenance." *International Journal of the Sociology of Language* 68 (1987): 69–99.

Government Offices of Sweden (Regeringskansliet). *Lag om Nationella Minoriteter och Minoritetsspråk (2009).* Sweden.

Grenoble, Lenore A., and Lindsay J. Whaley. *Saving Languages : An Introduction to Language Revitalization.* Cambridge: Cambridge University Press, 2006.

Harley, J. B. "Maps, Knowledge and Power." In *The Iconography of Landscape,* edited by Denis Cosgrove and Stephen Daniels, 277–312. Cambridge: Cambridge University Press, 1988.

Hyltenstam, Kenneth, and Christopher Stroud. *Språkbyte och Språkbevarande: Om Samiskan och Andra Minoritetsspråk.* Lund: Studentlitteratur, 1991.

Jenkins, Henry. *Fans, Bloggers, and Gamers: Exploring Participatory Culture.* New York: New York University Press, 2006.

Landry, Rodrigue, and Richard Y. Bourhis. "Linguistic Landscape and Ethnolinguistic Vitality: An Empirical Study." *Journal of Language and Social Psychology* 16, no. 1 (1997): 23–49.

Lewan, Mats. 2014. "Drönare håller koll på renarna." ["Drones Keeping an Eye on Reindeer."] NyTeknik, June 4. Accessed February 1, 2015. http://www.nyteknik. se/nyheter/it_telekom/allmant/article3830955.ece.

Lindgren, Simon, Michael Dahlberg-Grundberg, and Anna Johansson. "Hybrid Media Culture: An Introduction." In *Hybrid Media Culture. Sensing Place in a World of Flows,* edited by Simon Lindgren, 1–15. Abingdon, UK: Routledge, 2013.

Meyrowitz, Joshua. *No Sense of Place: The Impact of Electronic Media on Social Behavior.* New York: Oxford University Press,1985.

Moores, Shaun. "Media and Senses of Place: On Situational and Phenomenological Geographies." *MEDIA@LSE Electronic Working Papers* 12 (2007): 1–22.

Persson, Inger. "Kort presentasjon av arbeidet med iSikte—Sápmi av Inger Persson." May 2, 2014. Accessed February 1, 2015. https://kulturognaturreise.wordpress. com/2014/05/02/kort-presentasjon-av-arbeidet-med-isikte-sapmi-av-inger-persson/.

Pettersen, Bjørg. "Mind the Digital Gap: Questions and Possible Solutions for Design of Databases and Information Systems for Sami Traditional Knowledge."

In *Working with Traditional Knowledge: Communities, Institutions, Information Systems, Law and Ethics*, edited by Jelena Porsanger and Gunvor Guttorm, 163–192. DieDut: Sámi Allaskuvla, 2011.

Pietikäinen, Sari. "Sami in the Media: Questions of Language Vitality and Cultural Hybridisation." *Journal of Multicultural Discourses* 3, no. 1 (2008): 22–35.

Rydberg,Tomas.*Landskap,Territorium och Identitet i Sapmié. Exemplet Handölsdalens Sameby (Landscape,Tterritory and Identity—examples Handöldalens Sami Village)*. Uppsala: Uppsala Universitet, 2011.

Salo, Hanni. "Using Linguistic Landscape to Examine the Visibility of Sámi Languages in the North Calotte." In *Minority Languages in the Linguistic Landscape*, edited by Durk Gorter, Heiko F. Marten and Luk Van Mensel, 243–259. Palgrave Studies in Minority Languages and Communities, 2012.

Short, John Rennie. *Cartographic Encounters: Indigenous Peoples and the Exploration of the New World*. London: Reaction Books, 2009.

Språkrådet. *Språklagen i Praktiken: Riktlinjer för Tillämpning av Språklagen*. Stockholm, 2011.

Tafjel, Henri, and John C. Turner. "An Integrative Theory of Intergroup Conflict." In *The Social Psychology of Intergroup Relations*, edited by William G. Austin and Stephen Worchel, 33–47. Monterey, CA: Brooks/Cole, 1979.

Vincze, László, and Tom Moring. "Towards Ethnolinguistic Identity Gratifications." In *Social Media and Minority Languages : Convergence and the Creative Industries*, edited by Elin Haf Gruffydd Jones and Enrique Uribe-Jongbloed, 47–57. Bristol: Multilingual Matters, 2013.

Part III

Mobiles for Health, Education and Development

10 Using Technology to Promote Health and Wellbeing among American Indian and Alaska Native Teens and Young Adults

Stephanie Craig Rushing, Amanda Gaston, Carol Kaufman, Christine Markham, Cornelia M. Jessen, Gwenda Gorman, Jennifer Torres, Kirsten Black, Ross Shegog, Taija Koogei Revels, Travis L. Lane and Jennifer Williamson

Introduction

An estimated 5.2 million people identified as American Indian and Alaska Native (AI/AN) alone or in combination with other races in 2011, accounting for 2% of the total United States (U.S.) population (Office of Minority Health 2012). One-fifth of AI/AN people live on reservations or other trust lands. Altogether, the AI/AN population is demographically younger than other segments of the U.S. population. Nearly one third of AI/AN people are younger than 18 years old, compared to 24% of the total U.S. population (U.S. Census Bureau, 2010). As a result, adolescent health concerns are of particular interest to AI/AN communities.

High rates of teen birth and chlamydia and gonorrhea infection among AI/AN youth suggest that sexual activity begins earlier than among other U.S. teens (de Ravello, Tulloch and Taylor 2012). Several studies have reported younger than average sexual début and lower rates of consistent condom use among AI/AN youth (Hellerstedt 2004; Kaufman et al. 2004). Despite the compelling need, many Native youth do not receive sufficient reproductive health education. Until recently, few sexual health interventions had been purposefully designed for or rigorously evaluated in AI/AN communities. To fill this void, tribal health advocates across the U.S. have been working to design culturally appropriate sexual health interventions to improve health outcomes among AI/AN youth (de Ravello, Tulloch and Taylor 2012).

In the U.S., the term *mobile health* (m-health) has been used to describe the use of mobile phones and other wireless technologies to educate patients about preventive health services. This chapter will describe the development of six such programs designed to promote AI/AN adolescent sexual health, harnessing the power of laptops, mobile phones, tablets and streaming videos (Table 10.1).

Table 10.1 Technology-based sexual health interventions for AI/AN youth

Intervention	Target Audience	Target Age	Mode of Delivery and Access	Screenshot
Circle of Life	AI youth	10-12	Web-based curriculum searchable on: minorityhealth.hhs.gov Accessed on PCs using broadband and via DVD. Future plans include mobile accessibility.	
Native It's Your Game	AI/AN youth	12-14	Web-based curriculum Accessed via laptops and PCs, on broadband and wireless networks, and was downloaded from USB flash drives.	
Native VOICES	AI youth	15-24	Video Accessed via Facebook and YouTube on desktop and mobile devises, and via DVD.	
Safe in the Village	AN youth	15-24	Video Will be accessible via Facebook, YouTube, and Vimeo on desktop and mobile devises, and via DVD.	
iknowmine	AN youth	15-24	Website, Text Messaging, YouTube, Social Media www.iknowmine.org Accessed via desktop and mobile devises on broadband and wireless networks.	
We R Native	AI/AN youth	13-24	Website, Text Messaging, YouTube, Social Media www.wernative.org Accessed via desktop and mobile devises on broadband and wireless networks.	

Background

Adolescent Sexual Health

In the U.S. age is the single characteristic most closely associated with sexual experience among teens and young adults (Dailard 2006). National Youth Risk Behavior Surveillance System (YRBSS) data indicate that approximately 7% of American youth report sexual début prior to 13; 32% report having had intercourse by 16 years; and nearly 70% of youth experience sexual début by 18 years (Cavazos-Rehg et al. 2009).

Early sexual début is a "significant predictor of both initial and subsequent sexual risk behaviors and related health outcomes, including lack of condom use, multiple and high-risk sexual partners later in life, recurrent STDs, and cervical cancer" (Butler et al. 2006, 378). For these reasons, both abstinence-only and comprehensive sex education programs recommend that youth delay sexual activity. Researchers have identified more than 500 risk and protective factors that influence teens' sexual behavior, representing a range of biologic, environmental and social influences (Kirby 2007). Studies evaluating their relative importance suggest that teens' own sexual beliefs, values, attitudes and intentions are the strongest predictors of sexual behavior (Kirby 2007).

AI/AN Adolescent Sexual Health

AI/AN youth experience numerous inequalities associated with sexual health. In 2012, AI/AN youth 15 to 19 years old had the third highest teen birth rate in the U.S., at 35 per 1,000 (compared with 29 per 1,000 for the nation as a whole) (Martin et al. 2013). AI/AN youth 15 to 19 years old had the highest prevalence of repeat teen birth (at 21.6%, compared to 14.8% among White youth) in 2010 (CDC 2013). AI/AN youth are also disproportionately affected by sexually transmitted diseases (STDs). In 2009, AI/AN people had the second highest rates of chlamydia reported among all races and ethnicities, with the highest rates occurring among young people 15 to 24 years old (CDC and IHS 2012).

The sexual health of AI/AN youth is influenced by a variety of socio-ecological factors, including rural geography, high poverty and substance use, insufficient sex education, poor access to reproductive health services, stigma, sexual violence and historical trauma (de Ravello, Tulloch and Taylor 2012). Native young people's reproductive decisions are also shaped by unique social norms and sexual contexts that include both traditional and contemporary cultural values (Kaufman et al. 2007). The arrival of new life is often viewed favorably by Native communities, regardless of the parent's age. Sexual health messaging is thus highly nuanced in tribal communities and must reflect cultural values, social contexts and health epistemologies to be effective.

Adolescent Media Technology Use

Media technologies are becoming increasingly imbedded in the daily lives of American teens (Rainie 2009). Adolescents now spend an average of seven and a half hours per day using media technologies (Vahlberg 2010). Over three quarters of U.S. teens have their own cell phone; almost half (47%) have smartphones (Madden et al. 2013). About 75% of teens 12 to 17 years old say they access the Internet on their phones or mobile devices at least occasionally, and 25% *mostly go online* using their phone (Madden et al. 2013).

The digital divide that once separated the technology "haves" from the technology "have nots" has almost entirely disappeared in the U.S. (Horrigan 2009). Disparities in use now exist primarily between generations, rather than between racial groups, socioeconomic strata or urban/rural status (Horrigan 2009). A report issued by Native Public Media confirmed this trend in Indian Country, where—despite higher prices for broadband and suboptimal infrastructure—AI/AN adults reported higher rates of media technology use than the national average (Morris and Meinrath 2009). While the speed and quality of Internet access and cell phone coverage is highly variable in tribal communities across the U.S., it is swiftly and steadily improving.

Similarly, urban and rural AI/AN youth (13 to 21 years old) in the Pacific Northwest reported frequent media technology use in 2009, mirroring patterns reported by teens in the general population (Craig Rushing 2010). Seventy-five percent reported using the Internet, and 78% reported using cell phones on a daily or weekly basis. Similar rates of media access and use have been reported by AN youth in rural Alaska (YRBS 2014).

Effective Technology-Based Interventions

The privacy and prevalence of mobile technologies has changed the way teens find health information when they need it (Benight, Ruzek and Waldrep 2008). In 2012, one third of cell phone owners and one half of smartphone owners used their phones to look up health information (Pew Research Center, 2014). Over 75% of urban and rural AI/AN youth in the Pacific Northwest reported going online to get health information in 2009 (Craig Rushing 2010); a rate that has inevitably gone up as smartphone ownership has increased.

In response to these trends, technology-based interventions are being designed to promote adolescent sexual health using a variety of mobile and multimedia platforms, including interactive videos, computer- and web-based curricula, text messaging and social networking (Chavez, Shearer and Rosenthal 2013). Several have reported improvements in cognitive, psychosocial, behavioral and biologic outcomes (Guse et al. 2012; Jones et al. 2014; Vodopivec-Jamsek et al. 2012).

Tailoring Interventions for Cultural-Appropriateness and Community Fit

Integrating cultural values into health interventions has been shown to enhance their effectiveness in diverse populations (Kreuter et al. 2003, 2004). Resnicow et al. (1999) first described cultural sensitivity as containing two dimensions: surface and deep structures. "Surface structure involves matching intervention materials and messages to an observable, 'superficial' characteristic of a target population", including a preferred person, place, language, music, food, location, or clothing (Resnicow et al. 1999, 11). Deep structural adaptations "involve incorporating the cultural, social, histori-cal, environmental and psychological forces that influence the target health behavior in the proposed target population" (Resnicow et al. 1999, 12). This level of sensitivity ensures the intervention is grounded in the popula-tion's core health epistemology, values and beliefs. Messages tailored to the culture of recipients are more likely to be retained, discussed and perceived as personally relevant (Kreuter et al. 2003). Cultural tailoring is particularly important when addressing sensitive topics like sexual health.

Likewise, interventions must align with organizational capacity and community readiness to be sustainably implemented (Peters et al. 2013). Aligning interventions can be particularly challenging in Indian Country, where readiness levels differ from tribe-to-tribe and health services vary from clinic-to-clinic (Indian Health Service 2012).

Community-based Participatory Research

Community-based participatory research (CBPR) offers a useful methodol-ogy to maximize cultural alignment and organizational fit between an inter-vention and its audience. CBPR is a collaborative approach that equitably involves all partners in the research process (Israel et al. 1998). It is deeply committed to ensuring meaningful participation by those impacted by the research during all phases of the study (Minkler and Wallerstein 2003).

There are several distinct benefits to using CBPR in tribal communi-ties, including that it: (a) reflects and acknowledges tribal sovereignty, self-determination and self-governance; (b) allows research to occur in circumstances where it otherwise wouldn't; and (c) better aligns with traditional research approaches. CBPR mirrors the values and strengths of many AI/AN Nations, including respect for community processes and con-sensus, sincere equal partnership, and the ecological view of the individual as intricately linked with family and tribe. In CBPR, equal weight is given to both scientific and Indigenous, community-derived expressions of knowl-edge (Cochran et al. 2008).

All of the interventions discussed in this chapter were developed using extensive CBPR activities with project stakeholders by organizations with long-standing research relationships with the tribes that they serve. This process guided the selection, adaptation and evaluation of each program and was critical to ensuring the fit and usability of resultant interven-tions. Websites, videos and interactive web-based programs were purposely selected by our tribal partners due to their multimedia content that could meet the learning needs of low literacy youth and for their flexible design that could support utilization in diverse urban and rural AI/AN settings (e.g., directly by youth themselves or through caregivers at home or in clin-ics, schools, after-school programs and treatment centers). All six programs are still in the early stages of implementation and evaluation but offer insights on the design and uptake of such programs in tribal communities.

Technology-Based Interventions

Circle of Life

Circle of Life (COL) is a culturally grounded, age-appropriate, comprehensive HIV prevention program for AI/AN youth 10 to 12 years old that was originally developed by ORBIS Associates, an AI/AN-owned nonprofit educational organization. The theoretical foundation for the intervention is the medicine wheel, a heuristic facilitating Indigenous ways of learning and understanding familiar to most tribes in the U.S. (Kaufman et al. 2012). The medicine wheel symbolizes wholeness derived from balance among four quadrants: spiritual, emotional, physical and mental.

COL was adapted into a multimedia format (mCOL) in 2011 and is now available free of charge online and on DVD through the Office of Minority Health Resource Center (see minorityhealth.hhs.gov). mCOL consists of seven online chapters that are supplemented with optional group activities. Three cartoon characters, a teenage boy and girl along with a wise turtle, facilitate the online program. Video clips and games reinforce concepts. The first three chapters, introducing the cultural framework and adolescent development, must be completed sequentially before the sexual health content can be accessed. Lesson plans and a facilitator guide are available on the COL website, and a series of YouTube videos model how to teach each lesson.

Using a group randomized design, the mCOL program is currently being evaluated by a team at the Centers for AI/AN Health at the University of Colorado, Anschutz Medical Campus, in partnership with six Native Boys and Girls Clubs in the Northern Plains (Kaufman et al. 2014). During the evaluation, students accessed the program using PCs with broadband or via DVD. Future plans include reprogramming mCOL for mobile accessibility.

Native It's Your Game

Designed for a slightly older audience, Native It's Your Game (Native IYG) is a 13 lesson, multimedia sexual health curriculum for AI/AN youth (12 to 14 years), adapted from *It's Your Game—Tech* (IYG-Tech) (Shegog et al. 2014). The project was a collaboration of the Alaska Native Tribal Health Consortium, Inter Tribal Council of Arizona, Inc., the Northwest Portland Area Indian Health Board, and the University of Texas Prevention Research Center (Shegog and Markham 2011).

During the adaptation process, AI/AN youth (n=80) and adult stakeholders (n=18) reviewed IYG-Tech, and offered feedback on its use, acceptability and cultural appropriateness. Surface level cultural changes included adding AI/AN youth (role-model videos), traditional music and culturally relevant settings (Kreuter et al. 2000; Resnicow et al. 1999). Deep cultural changes included the addition of a holistic health activity, commentaries from male and female elders to introduce sensitive topics, traditional teaching methods (dance, song, and stories) and more focused content on drug and alcohol use and healthy relationships. After the adaptation process, AI/AN youth (n=45) and adult stakeholders (n=25) re-rated Native IYG on the same usability measures. Most agreed that it was easy to use, appropriately paced, trustworthy, understandable and helpful for making better choices.

A group randomized two-arm nested intervention trial involving over 500 AI/AN middle school students is now underway in Alaska, Arizona and the Pacific Northwest to evaluate the impact of Native IYG on psychosocial and behavioral outcomes. During the evaluation, about one third of the students accessed the program via laptops and PCs, using broadband and wireless networks. Two thirds of the students accessed the program via download from USB flash drives, due to unexpected connectivity issues. Once efficacy has been determined, the research team will disseminate the intervention free-of-charge, with training and technical assistance provided by the three tribal non-profit partners.

Native Voices

Native VOICES was adapted for American Indian teens and young adults (15 to 24 years old), from *Video Opportunities for Innovative Condom Education and Safer Sex* (VOICES), a video-based intervention recognized by the Centers for Disease Control and Prevention (CDC). VOICES is a single-session HIV/STD prevention intervention in which condom use and negotiation skills are modeled by a video and then role-played and practiced by participants (O'Donnell et al. 1998). The project was funded by the Native American Research Centers for Health (NARCH) and carried out by staff at the Northwest Portland Area Indian Health Board, a tribal non-profit organization managed by the 43 federally recognized tribes in Oregon, Washington and Idaho.

The research team used an emergent study design to collect qualitative data and community feedback from tribes in the Pacific Northwest over a three year period (2011–2013). The adapted script was shot on location in Oklahoma City in August 2013. Satisfaction surveys collected at red carpet showings of the video suggest promising results: Over 94% of those surveyed enjoyed the video; 97% found the video to be culturally appropriate for American Indian people; 100% felt the information could be trusted. Nearly 88% felt the video showed real life situations with characters they could relate to, and 94% thought the things the actors did and said about condoms and negotiating safe sex would work for them. After watching the video, 74% felt more likely to get tested for STDs/HIV, and 61% felt more likely to use condoms.

An ongoing randomized controlled trial is evaluating the intervention's cultural appropriateness, acceptability and effectiveness with 800 AI/AN youth in eight urban and rural settings across the U.S. A free Native VOICES

Toolkit has been designed to support the intervention's implementation in diverse tribal settings. The video will also be marketed directly to AI/AN youth on YouTube, Facebook and Twitter, so it can be accessed from cell phones and smartphones.

Safe in the Village

Safe in the Village is a culturally relevant, video-based intervention for Alaska Native youth 15 to 24 years old residing in rural Alaska, which was modeled after *Safe in the City*, a CDC-recognized evidence-based intervention (Myint-U et al. 2010). The project was funded by the Native American Research Centers for Health (NARCH) and carried out by staff at the Alaska Native Tribal Health Consortium (ANTHC).

The project began by carrying out formative research with Alaska Native youth 15 to 24 years old (n=97) residing in rural Alaska Native communities (n=5), using semi-structured interviews and Likert scale surveys. Local site coordinators assisted in recruiting participants and provided key information on their respective communities. After reviewing findings from the formative phase of the study, a local film production company was hired to write the script, cast Alaska Native actors, and film the video in urban and rural locations in Northwest Alaska. A community advisory committee (n=16) provided crucial input to make sure the adapted script was authentic and realistic in its portrayal of village life and culture and used language that was appropriate for the target audience. Committee members also provided guidance on the video's format and how the intervention should be implemented and evaluated. The movie was planned, filmed and optimized in postproduction for mobile device viewing to reach the target audience. The 34-minute movie can be shown alone or with supplemental actor interviews (25 minutes) as a group intervention using a facilitation guide.

An evaluation of the intervention's impact on self-reported knowledge, attitudes, perceptions and behaviors among Alaska Native youth 15 to 19 years old is currently underway. Post evaluation, the video will be released on social media, where it will be uploaded as mini-episodes and in its entirety, so that youth can access it from their cell phones and smartphones.

iknowmine

To address high chlamydia (CT) and gonorrhea (GC) rates among Alaska Native youth, the ANTHC conducted focus groups and in-depth interviews with rural Native youth and community members in 2008. Guided by this insight, the project launched an interactive, multimedia health resource in 2009: www.iknowmine.org. The website featured medically accurate sexual health and drug and alcohol information; Alaska-specific STD screening options; all 200+ testing locations in the Alaska Native Tribal Health system; interactive elements, such as quizzes, polls, games and user-generated content; frequently asked questions; a confidential "ask an expert" section with corresponding 2-way text messaging (operated in collaboration with ANTHC medical staff); linkage to care; a blog produced by ANTHC staff and youth editors; and a free, statewide online condom distribution program at no-cost for Alaska residents. Associated social media channels (Facebook, Twitter, MySpace and YouTube) were publicly launched in December of 2009. In 2010, an online ordering form for free shirts, promotional items and health education materials was added to the site. In 2011, the site began offering free at-home STD tests (Chlamydia, gonorrhea and Trichomonasis) for all Alaskans, in partnership with Johns Hopkins University's *I Want The Kit* (IWTK) program.

In 2012, www.iknowmine.org was re-designed into a comprehensive health resource for Alaska Native youth. The website's content was rewritten by youth and staff and was expanded to include topics such as traditional foods, games, sports and spirituality. To keep up with the increasing use of mobile technologies by AN youth, the site was made compatible with smartphone and tablet use. New social media channels (Instagram and Tumblr) were launched, resource pages for parents and providers were added to the site and non-latex condoms, female condoms and dental dams were added to the online condom distribution program. The re-design also allowed clinics and organizations to order free bulk condoms and health education materials. These services address two important barriers to STD testing and condom use in rural Alaska, where youth are often "too embarrassed" to access them and pharmacies are nonexistent (Leston, Jessen and Simons 2012).

iknowmine Measures and Outcomes

Prior to its re-design, iknowmine had 20,373 site visits and 471 unique condom orders. Engagement was measured through unique user accounts (required to submit a question, order condoms, and upload user-generated content). Since the re-design, www.iknowmine.org has experienced significant growth; total site visits have increased by 679% and unique visitors grew by 400%. The total page-view count since re-design is now over 205,000. Since its launch, iknowmine.org has had 137,197 unique visitors, 157,071 site visits, mailed out over 44,460 male condoms, registered 2,270 iknowmine.org users, enrolled 652 text messaging users and answered more than 250 sexual health questions via its website and text messaging service.

A significant portion of iknowmine's users access it through mobile channels. Approximately 42% of the total website sessions (n = 59,484) are from mobile devices, and 81% of those are from non-tablet devices. The top mobile devices used by iknowmine users were Apple iPhone (all models), Apple iPad (all models) and Apple iPod (all models), cumulatively representing 63% of all mobile device sessions. Social media engagement has seen similar growth in recent years; Facebook likes have increased by 370%, Twitter followers by 234%, and YouTube channel views by 538%.

We R Native

In 2011, *We R Native* was established by the Northwest Portland Area Indian Health Board as a multimedia health resource for Native teens and young adults, promoting positive youth development and healthy decision-making. The service was designed using behavior change theory and extensive formative research with AI/AN youth across the U.S. A media firm was hired to support branding and web development. Since its initial launch through Facebook, additional communication channels have been added, including a website, a text message service (text NATIVE to 24587), a YouTube channel, social media (Twitter, Instagram) and print marketing materials. We R Native includes content on social, emotional, physical, sexual and spiritual health, as well as on AI/AN culture.

The website contains over 330 health and wellness pages, all reviewed by Native youth and experts in public health, mental health, community engagement and activism. Special interactive features include monthly contests, community service grants ($475), an "Ask Auntie" Q&A service and a blog. The text message service and Facebook page alert followers to health tips, new contests, internship opportunities and news stories about Native youth. All channels promote interactivity and reciprocal communication by encouraging feedback and story sharing. A mobile version of the website, launched in 2013, includes the site's health and wellness pages, the "Ask Auntie" Q&A service and monthly contests. The project is funded by the Indian Health Service (IHS).

We R Native Measures and Outcomes

We R Native collects digital metrics recommended by www.digitalgov.gov. During its first two years, the website received over 100,000 page views. The physical health pages on the site accounted for about 8% of the site's total traffic, the mental health pages accounted for 5% of the site's traffic, and the sexual health pages received about 14% of the site's total traffic. Roughly two thirds of the website's users are female; one third are male. The average user visits 3 pages per visit and stays on the site 3:15 minutes.

In 2014, nearly 75% of the site's visitors accessed it from a desktop computer, 19% accessed it from a mobile phone and 8% accessed it from a tablet. Although the mobile phone and tablet are not the primary devices from which users access weRnative.org, mobile and tablet sessions more than doubled from 2013 to 2014 (Mobile: 2,666 to 6,236, Tablet: 850 to 1,726), while desktop sessions increased only 4%. Apple iPhones and iPads are the primary devices used by visitors, accounting for nearly half (45% n=5,666) of the total mobile sessions. In addition, mobile and tablet users have surpassed desktop users as the highest percentage of first time visitors.

Similar user demographics and mobile technology use rates have been observed within our other We R Native communication channels. By October 2014, We R Native's YouTube channel had 250 health and wellness videos with over 14,500 video views. During the last quarter (July 2014 – Sept 2014), over half of the video views came from a desktop computer (57%, n=1,509), one third came from a mobile phone (33%, n=879), and 7% came from a tablet (n=188). The We R Native text messaging service delivers weekly text messages to over 2,000 subscribers located across the U.S. In a recent survey of 350 text message subscribers, 87% reported that they use their cell phone to access the Internet.

At present, the We R Native Facebook page has over 20,900 likes, and over 1,300 Twitter followers. Altogether, We R Native reaches over 32,000 users per week through its various channels, a growing number of whom are accessing the content from mobile devises.

Opportunities and Lessons Learned

Conducting research in AI/AN communities requires a deep understanding of tribal research processes and protocols and adherence to CBPR principles. Approval for this research was obtained from University Institutional Review Boards (IRB), regional and tribal IRBs, tribal councils, tribal health boards/committees and tribal health directors (as appropriate for each study, region and tribe). Study timelines and study protocols must include sufficient time for approval from appropriate review boards and community partners.

In spite of expanded Internet connectivity in tribal communities, many of our study sites had difficulty running online programs. M-health interventions must be prepared to offer technical support throughout the program. Challenges related to IT infrastructure and Internet bandwidth can be minimized by offering programs in multiple formats (streaming online, as a downloadable program, or via DVD or USB flash drive) to meet the unique needs and resources of each setting. While technology-based interventions can *theoretically* be disseminated across large geographic regions, they are critically dependent on having strong implementation plans to achieve their full potential for reach and fidelity.

Likewise, AI/AN tribes vary considerably in their readiness to embrace sexual health programs and in their languages, health epistemologies and coming-of-age teachings. Our resources were designed to supplement and support local teachings, rather than supplant them. Adolescent sexual programs are most effective when they are complemented and reinforced by other health promotion activities at the individual, family and tribal levels and when they are provided in conjunction with appropriate sexual health services.

Once the efficacy of these programs has been established, the research team will use mobile media platforms to market the interventions directly to AI/AN youth, using Facebook, YouTube and websites such as www.iknowmine.org and www.weRnative.org. Steps are also underway to make the programs widely accessible to teachers and facilitators online, as downloadable apps and via DVD or jump drive (depending on the needs of each site). The team is also exploring the possibility of developing compatible versions of the online programs for tablets and smart phones, with the ability to tailor lessons to the age and gender of students accessing them.

Conclusion

Many AI/AN youth receive insufficient culturally relevant sexual health education and as a result experience disproportionate teen pregnancy, STDs and HIV. This situation can be exacerbated by geographic isolation, which makes accessing confidential services and information problematic. Technology-based interventions can help bridge this gap for AI/AN youth in ways that are familiar and inviting.

The systematic community-based approaches described in this chapter provide a roadmap to those interested in making technology-based health programs more meaningful and relevant for AI/AN youth. Much work remains to be done, however, in their adaptation, implementation, evaluation, adoption and dissemination. Internet connectivity and cell phone coverage remain unreliable in many communities. Few randomized controlled trials have rigorously evaluated the effectiveness of m-health programs in rural settings, and evaluating sexual health programs in AI/AN communities is often hampered by measurement challenges associated with small sample sizes.

Fortunately, exciting progress is being made in the field. The mobile and online programs described in this chapter offer culturally appropriate sexual health education across the age span, from pre-teen (10 to 12 years) to tween (12 to 14 years) to young adult (15 to 25 years). By providing this information across multiple media platforms, we hope to reach the greatest number of AI/AN youth, when and where each young person is ready. With continued CBPR partnerships with AI/AN communities across the U.S., effective, accessible technology-based interventions can be designed to reflect traditional and contemporary AI/AN culture, values and experiences.

Acknowledgements

The mCOL project was supported by Grant Number (TP2AH000003) from the HHS Office of Adolescent Health (OAH). The study is registered at www.clinicaltrials.gov (#NCT01698073). The Native IYG study was funded by the Centers for Disease Control and Prevention (CDC) (#5U48DP001949-02) and by the Administration for Children and Families (ACF) (#90AT0013-02-00/CFDA#93.092). The study is registered at www.clinicaltrials.gov (#NCT01303575). Native VOICES and Safe in the Village were funded by Grant Number (U26IHS300289) from the Indian Health Service (IHS), with support from the DHHS Multicultural AIDS Initiative (MAI). The findings and conclusions in this chapter are those of the authors and do not necessarily represent the official position of the Department of Health and Human Services, OAH, CDC, ACF, IHS or MAI.

References

Benight, Charles C., Josef I. Ruzek and Eddie Waldrep. "Internet Interventions for Traumatic Stress: A Review and Theoretically Based Sample." *Journal of Traumatic Stress* 21, no. 6 (2008): 513–20.

Butler, Terry H., Kim S. Miller, Davd R. Holtgrave, Rex Forehand and Nicholas Long. "Stages of Sexual Readiness and Six-Month Stage Progression Among African American Pre-Teens." *Journal of Sex Research* 43, no. 4 (2006): 378–86.

Cavazos-Rehg, Patricia A., Melissa. J. Krauss, Edward L. Spitznagel, Mario Schootman, Kathleen K. Bucholz, Jeffrey F. Peipert, Vetta Sanders-Thompson, Linda B. Cottler and Linda J. Bierut. "Age of Sexual Debut among US Adolescents." *Contraception* 80, no. 2 (2009): 158–62.

CDC (Centers for Disease Control and Prevention). "Vital Signs: Repeat Births Among Teens—United States, 2007–2010." *Morbidity and Mortality Weekly Report*, April 5, 2013. Accessed March 21, 2015. http://www.cdc.gov/mmwr/preview/mmwrhtml/mm6213a4.htm?s_cid=mm6213a4_w.

CDC and HIS (Centers for Disease Control and Prevention and Indian Health Service). *Indian Health Surveillance Report—Sexually Transmitted Diseases 2009.* Atlanta, GA: U.S. Department of Health and Human Services, 2012. Accessed March 21, 2015. http://www.cdc.gov/std/stats/ihs/ihs-surv-report-2009.pdf.

Chavez, Noe R., Lee S. Shearer and Susan L. Rosenthal. "Use of Digital Media Technology for Primary Prevention of STIs/HIV in Youth." *Journal of Pediatric & Adolescent Gynecology* 27, no. 5 (2013): 244–57. doi: 10.1016/j.jpag.2013.07.008.

Cochran, Patricia A. L., Catherine A. Marshall, Carmen Garcia-Downing, Elizabeth Kendall, Doris Cook, Laurie McCubbin and Reva M. A. Gover. "Indigenous Ways of Knowing: Implications for Participatory Research and Community." *American Journal of Public Health* 98, no. 1 (2008): 22–27. doi: 10.2105/AJPH.2006.093641.

Craig Rushing, Stephanie. "Use of Media Technologies by Native American Teens and Young Adults: Evaluating their Utility for Designing Culturally Appropriate Sexual Health Interventions Targeting Native Youth in the Pacific Northwest." PhD diss., Portland State University, Portland, 2010.

Dailard, Cynthia. "Legislating against Arousal: The Growing Divide Between Federal Policy and Teenage Sexual Behavior." *Guttmacher Policy Review* 9, no. 3: Guttmacher Institute, 2006.

de Ravello, Lori, Scott Tulloch and Melanie Taylor. "We Will be Known Forever by the Tracks We Leave: Rising up to Meet the Reproductive Health Needs of

American Indian and Alaska Native Youth." *American Indian and Alaska Native Mental Health Research* 19, no. 1 (2012): i-x. doi: 10.5820/aian.1901.2012.i.

Guse, Kylene, Deb Levine, Summer Martins, Andrea Lira, Jenn Gaarde, Whitney Westmorland and Melissa Gilliam. "Interventions Using New Digital Media to Improve Adolescent Sexual Health: A Systematic Review." *Journal of Adolescent Health* 51, no. 6 (2012): 535–543. doi: 10.1016/j.jadohealth.2012.03.014.

Hellerstedt, Wendy. *Native Teen Voices*, 2004. Accessed March 21, 2015. http://www.ntv.umn.edu/index.shtm.

Horrigan, John. *Wireless Internet Use*, 2009: 48. Washington DC: Pew Internet and American Life Project.

Indian Health Service. *IHS Year 2012 Profile*, 2012. Accessed February 16, 2012. http://www.ihs.gov/publicaffairs/ihsbrochure/profile.asp.

Israel, Barbara A., Amy J. Schulz, Edith A. Parker and Adam B. Becker. "Review of Community-Based Reseach: Assessing Partnership Approaches to Improve Public Health." *Annual Reviews in Public Health* 19 (1998): 173–202.

Jones, Krista, Patricia Eathington, Kathleen Baldwin and Heather Sipsma. "The Impact of Health Education Transmitted via Social Media or Text Messaging on Adolescent and Young Adult Risky Sexual Behavior: A Systematic Review of the Literature." *Sexually Transmitted Diseases* 41, no. 7 (2014): 413–419. doi: 10.1097/olq.0000000000000146.

Kaufman, Carol E., Janette Beals, Christina M. Mitchell, Pamela L. LeMaster and Alexandra Fickenscher. "Stress, Trauma, and Risky Sexual Behavior among American Indians in Young Adulthood." *Culture, Health & Sexuality* 6, no. 4 (2004): 301–18.

Kaufman, Carol E., Jennifer Desserich, Cecelia K. Big Crow, Bonnie Holy Rock, Ellen Keane and Christine M. Mitchell. "Culture, Context, and Sexual Risk among Northern Plains American Indian Youth." *Social Science & Medicine* 64, no. 10 (2007): 2152–64.

Kaufman, Carol E., Anne Litchfield, Edwin Schupman and Christine M. Mitchell. "Circle of Life HIV/AIDS-Prevention Intervention for American Indian and Alaska Native Youth." *American Indian and Alaska Native Mental Health Research* 19, no. 1 (2012): 140–53.

Kaufman, Carol E., Nancy Rumbaugh Whitesell, Ellen M. Keane, Jennifer A. Desserich, Cindy Giago, Angela Sam and Christina M. Mitchell. "Effectiveness of Circle of Life, an HIV-Preventive Intervention for American Indian Middle School Youths: A Group Randomized Trial in a Northern Plains Tribe." *American Journal of Public Health* 104, no. 6 (2014): 106–22.

Kirby, Douglas. Emerging Answers: Research Findings on Programs to Reduce Teen Pregnancy and Sexually Transmitted Diseases. Washington DC: The National Campaign to Prevent Teen Pregnancy, 2007.

Kreuter, Matthew W., Susan N. Lukwago, Dawn C. Bucholtz, Eddie M. Clark and Veta Sanders-Thompson. "Achieving Cultural Appropriateness in Health Promotion Programs: Targeted and Tailored Approaches." *Health Education & Behavior* 30, no. 2 (2003): 133–46.

Kreuter, Matthew W., Debra L. Oswald, Fiona C. Bull and Eddie M. Clark. "Are Tailored Health Education Materials Always More Effective than Non-tailored Materials?" *Health Education Research* 15, no. 3 (2000): 305–15.

Kreuter, Matthew W., Celette S. Skinner, Karen Steger-May, Cheryl L. Holt, Dawn C. Bucholtz, Eddie M. Clark and Debra Haire-Joshu. "Responses to Behaviorally vs. Culturally Tailored Cancer Communication among African American Women." *American Journal of Health Behavior* 28, no. 3 (2004): 195–207.

Leston, Jessica, Cornelia M. Jessen and Brenna C. Simons. "Alaska Native and Rural Youth Views of Sexual Health: A Focus Group Project on Sexually Transmitted Diseases, HIV/AIDS, and Unplanned Pregnancy." *American Indian and Alaska Native Mental Health Research* 19, no. 1 (2012): 1–14.

Madden, Mary, Amanda Lenhart, Maeve Duggan, Sandra Cortesi and Urs Gasser. *Teens and Technology 2013.* Washington, DC: Pew Research Center's Internet and American Life Project, 2013.

Martin, Joyce. A., Brady E. Hamilton, Michelle J. K. Osterman, Sally C. Curtin and T. J. Matthews. "Births: Final Data for 2012." *National Vital Statistics Report* 62, no. 9: 1–68. Atlanta, GA: Centers for Disease Control and Prevention, 2013.

Minkler, Meredith, and Nina Wallerstein. *Community-Based Participatory Research for Health.* San Francisco, CA: Jossey-Bass, 2003.

Morris, Traci L., and Sascha D. Meinrath. *New Media, Technology and Internet Use in Indian Country: Quantitative and Qualitative Analyses.* Flagstaff, AZ: Native Public Media, 2009.

Myint-U, Athi, Sheana Bull, Gregory L. Greenwood, Jocelyn Patterson, Cornelis A. Rietmeijer, Shelley Vrungos, Lee Warner, Jesse Moss and Lydia N. O'Donnell. "Safe in the City: Developing an Effective Video-Based Intervention for STD Clinic Waiting Rooms." *Health Promotion Practice* 11, no. 3 (2010): 408–17. doi: 10.1177/1524839908318830.

O'Donnell, Carl R., Lydia O'Donnell, Alexi San Doval, Richard Duran and Karl Labes. "Reductions in STD Infections Subsequent to an STD Clinic Visit. Using Video-Based Patient Education to Supplement Provider Interactions." *Sexually Transmitted Diseases* 25, no. 3 (1998): 161–68.

Office of Minority Health. *American Indian/Alaska Native Profile,* 2012. Accessed April 15, 2014. http://minorityhealth.hhs.gov/templates/browse.aspx?lvl=3&lvlid=26.

Peters, David H., Taghreed Adam, Olakunle Alonge, Irene Akua Agyepong and Nhan Tran. "Implementation Research: What It Is and How to Do It." *BMJ* 20 (347) (2013): 1–7.

Pew Research Center. *Health Fact Sheet. Internet Project,* 2014. Accessed March 21, 2015. http://www.pewinternet.org/fact-sheets/health-fact-sheet/.

Rainie, Lee. *Teens and the Internet,* 2009. Accessed August 20, 2009. http://www.pewinternet.org/presentations/2009/teens-and-the-internet.aspx.

Resnicow, Ken, Tom Baranowski, Jasjit Ahluwalia and Ronald Braithwaite. "Cultural Sensitivity in Public Health: Defined and Demystified." *Ethnicity and Disease* 9, no. 1 (1999): 10–21.

Shegog, Ron, and Christine Markham. "Partnership To Prevent Teen Pregnancy and HIV/STIs among AI/AN Youth: It's Your Game ... Keep It Real." *The IHS Primary Care Provider* 36, no. 4 (2011): 84–85.

Shegog, Ron, Melissa F. Peskin, Christine Markham, Melanie Thiel, Efrat Gabay, Robert C. Addy, Kimberley A. Johnson and Susan Tortolero. "'It's Your Game-Tech': Toward Sexual Health in the Digital Age." *Creative Education. Advances in Sex Education Special Edition* 5, no. 15 (2014): 1428–47.

U.S. Census Bureau. *The American Indian and Alaska Native Population: 2010 Census Briefs.* U.S. Department of Commerce, 2010.

Vahlberg, Vivian. *Fitting into Their Lives: A Survey of Three Studies about Youth Media Usage.* Arlington, VA: Newspaper Association of America Foundation, 2010.

Vodopivec-Jamsek, Vlasta, Thyra de Jongh, Ipek Gurol-Urganci, Rifat Atun and Josip Car. "Mobile Phone Messaging for Preventive Health Care." Cochrane Database of Systematic Reviews 12, CD007457 (2012). doi: 10.1002/14651858. CD007457.pub2.

YRBS. *2013 Alaska Youth Risk Behavior Survey Results: 2013 Traditional High School YRBS Results.* Alaska Department of Health and Social Services, 2014. Accessed March 21, 2015. http://dhss.alaska.gov/dph/Chronic/Documents/ School/pubs/2013AKTradHS_Graphs.pdf.

11 The Use of Podcasts to Improve the Pronunciation of the Māori Language and Develop Reflective Learning Skills

Lisa J. Switalla-Byers

Introduction

Podcasts were used to improve pronunciation in the Maori language program at a New Zealand primary school as part of an action research project. Action research is a form of investigation designed for use by teachers to attempt to solve problems and/or improve professional practices in their own classrooms. Within an integrated art and storytelling cultural program, children practiced hearing, speaking, reading and writing Māori language. They recorded a speech of greeting (*mihi*) as a podcast, listened, reflected and made changes to it. Another child acted as a critical friend to critique the podcasts and help improve them. They were also encouraged to share their podcasts with their family and friends. The main aim of the podcasts was to listen to their peers since listening is vital for improving pronunciation. All children's learning was supported by a cloak (*korowai*) they had made and in which the gods (*atua*) and their family (*whakapapa*) dwelt, cloaking them in confidence to progress with their language learning. The integration of teaching approaches enabled children to learn cultural (*kaupapa*) aspects related to learning how to greet and introduce themselves in *te reo* Māori (the Māori language). The Indigenous language of New Zealand closely integrates language and culture.

The implementation of this action research was carried out within a State Primary School located in Otago, New Zealand. The learners were from a composite year five and six class. All 25 class members were included within the action research process. Of the learners 10 identified as being of Māori descent, 3 of Pacific Island descent, 12 of Pakeha or European descent. One child was classified by the New Zealand Ministry of Education as "special needs". New Zealand has three official languages including English, Māori and Sign Language. Within the New Zealand Curriculum Document specific references are made to the importance of teaching and learning in relation to all students having the opportunity to acquire knowledge of *te reo Māori me ōna tikanga*, Māori language and culture.

The chapter makes recommendations on how teachers can enhance students' reflective skills and learning strategies within the classroom, encouraging them to self-monitor, be persistent and take personal responsibility for their learning. A glossary of Māori terms is included at the end of the chapter to assist readers.

Background: The Importance of Reflection in Building Learning Skills and the Role of ICT

Establishing a classroom learning culture that encourages and teaches children how to learn will often include a common understanding and shared language of learning. Pohl (2009, 8) states that an essential element in developing a classroom culture of thinking and learning is the explicit teaching of thinking skills to all learners. He states that all teachers need to establish and use an appropriate language of thinking in their classrooms, become familiar with a range of thinking strategies, and use graphic organizers to help learners manage, organize and record their thinking. Pohl is describing not only what teachers do to develop the learning culture but also what students do within that learning culture. Guy Claxton's (2014) Building Learning Power also focuses on what students do within the learning environment. Building Learning Power is about helping children become better learners, developing their portable learning power and preparing students for a lifetime of learning. Claxton states that Building Learning Power is based on three fundamental beliefs. First, the purpose of education is to prepare learners for life beyond the classroom by helping them build mental, emotional, social and strategic resources. Second, education is for all learners and involves helping learners discover personal strengths and passions, and then strengthen their will and skills to pursue them. Third, real world intelligence is something that people can be helped to build up through confidence, capability and passion. Like Pohl it is about creating a culture of learning within classrooms and ideally the school: a culture of learning that systematically encourages and teaches habits and attitudes that empower learners to be resourceful, logical, imaginative, self-disciplined, collaborative and inquisitive.

Pohl (2009, 8) describes a thinking culture as a supportive environment that has the following specific factors working together to reinforce and encourage the explicit teaching of thinking skills:

- Provide learners with the language, tools and strategies to engage in a range of learning tasks.
- Provide opportunities for developing, practicing and refining thinking skills.
- Provide instruction and practice in managing, organizing and recording thinking.
- Assist in the transfer of skills to everyday life as tools for life-long learning.

In addition to what the teacher is doing to develop a culture of thinking and learning, the students or learners also have the responsibilities of learning how to learn and being motivated to become self-regulated life-long learners, an aspiration that involves integrating the teaching of thinking into every teaching and learning activity rather than viewing it as an additional add-on to the planned curriculum of learning.

Reflection as a thinking skill is critical to learning and the transfer of learning concepts and skills to new and different situations. Dewey (1933) describes reflection as being good for developing higher-order thinking skills for learners. Higher-order thinking occurs when learners are reflecting and self-monitoring their actions, attitudes and knowledge, deciding which experiences could be used either collectively or individually in new learning scenarios. Reflection will be discussed in this chapter as a process that involves getting people to talk about their own learning experiences (Kolb 2012). Lord (2008) describes reflection as a never-ending clarifying process, a reflective spiral where you think of new opportunities, refine your language and learning and implement changes. Reflection is a powerful tool for learners to take control of their own learning across a range of contexts such as goal setting, using reflective tools and graphic organizers and peer and self-assessment. Often learners require specific teaching and structured learning activities to reflect fully. It is important that learners understand the purpose of what you are trying to achieve and that reflection skills are explicitly taught. Without explicit teaching learners will often provide surface answers that lack depth and self-awareness.

Reflection is a classroom tool that is often underutilized and probably not completely understood by teachers and learners. Laiken (2002) outlined that often critical reflection is not as valued and therefore it is taught or practiced less in many action-orientated learning environments. Laiken emphasizes that to create a collective learning environment, learners need to be encouraged and taught how to have quality conversations: how to engage in dialogue, state your own views and inquire about others' views with the intention of understanding differences. These are teachable moments where learning occurs. Laiken also outlines the importance of learners practicing self-disclosure through the giving and receiving of feedback through observable behaviors. Each of the above points is an observable behavior that helps to ensure individual and collective responsibility, an expectation that every contribution is valued and everyone can contribute to learning. It also reinforces the creation of a supportive learning environment or class culture that strengthens attitudes and enables learners to take risks and become more confident in their own learning ability.

As teachers the importance of having a structure for feedback and reflection can help maximize the effectiveness of critical reflection. There are many models for reflection available to teachers. Graham and Osborne (2008) outlined a simple and effective model for guiding learners' thinking and learning when participating in providing feedback. It also provides a code of learning, outlining what is valued. It is a method that can be reflected upon either collectively or individually by learners, giving a set of ground rules for establishing a reflective learning culture within a classroom. They note the importance of reflection in action research:

1 Each person has the right and opportunity for reflection.
2 Every idea has value and can contribute to learning.
3 Individual contributions are recognized.
4 Learners are responsible for their own learning.

Technology for Promoting Reflective Learning

Siemens is a researcher who advocates a theory of learning for the 21st Century. His research describes learning theory for a digital age. Within his research he combines learning theories, social structures and technologies. Siemens (2005, 2) quoted Driscoll's definition of learning as "a persisting change in human performance or performance potential (which) must come about as a result of the learner's experience and interaction with the world." Learning needs and learning processes must be reflective of social contexts and social environments. The actual process of learning is as important as what is being learned. Siemens' theory of connectivism is based upon learning and decision-making. New information is constantly being acquired. The learner's ability to make distinctions about what information or knowledge is important and unimportant is critical. The ability to synthesize and recognize connections is a valuable skill. Siemens goes on to describe some significant trends in learning, including how learning occurs:

- Learning in a variety of ways, e.g., communities of practice.
- Learning as a continual life-long process.
- The role of technological tools in shaping our thinking.
- The use of technology to support cognitive information processing.
- The knowledge of where to find information supplementing the acquisition of facts.

He concludes by summarizing that a learner's ability to know what we need for tomorrow is more important than what we know today. Knowledge continues to evolve, and the learner's ability to access knowledge is more important than the knowledge the learner currently possesses.

In a teaching and learning keynote speech Pam Hook (2009) stated that "ICT can make a difference; it's how we use it: learning needs to be multi-structural, relational and extend learning experiences so children can learn how to learn." As classroom teachers try to think of learning from a strengths focus, there is a need to continue to develop an everyday language of reflection, self-monitoring, persistence, listening practice and the development of personal responsibility. Pam Hook encourages a class of learners to be consistently thinking about: What have we put into action? What worked and Why? What didn't work and therefore ... What will we do next? It is evident that Hook places responsibility for learning on the learner and encourages the teacher to be a facilitator of that learning. Hook's questions for reflection are very concise; they are open ended and encourage conversation. These questions could be applied to a range of learning contexts or age groups.

Podcasting is a technology that supports the development of reflective and lifelong learning skills. Dale and Povey (2009) reviewed the use of podcasting for learning, concluding that potential benefits of using podcasting include its ability to both meet a range of learning styles and be used as a tool for developing key skills for critical thinking and reflection. They also found podcasting to be effective for engaging students from diverse social and cultural origins. Podcasting is a tool for teaching and learning that can be used to emphasis and improve pronunciation techniques. Benefits of using podcasting techniques for foreign language learning include accessibility, their ability to be used individually or co-operatively and their ability to be used to target language listening practice. In research by Lord (2008) groups created a podcast channel to upload recordings for group feedback. Participants were required to critically respond specifically to the pronunciation of their peers' podcasts in Spanish. Lord concluded that there was evidence of a more positive attitude toward learning a foreign language after weekly podcasts. Second, Lord concluded that overall there was some degree of improved pronunciation by the end of the semester of teaching.

The Introduction of Podcasting into the Classroom

Podcasting can be used by educators to create authentic listening contexts. In this research learners created podcasts. They were able to listen to and practice speaking, upload a podcast to a podbean site (see www.podbean. com), thus creating a tool for reflection that could be used by learners, peers and their parents. The aim of the podbean site was to create a collaborative community of learners. The primary aim was to help learners to use feedback to improve their pronunciation and fluency of *te reo*. Once the podcast had been uploaded to the podbean site, learners were required to reflect upon their own and two others' podcasts, leaving feedback regarding fluency, pronunciation, accuracy or other elements from a rubric. In addition, following Lord (2008), I also asked the podcaster to verbalize his or her own personal goal and the listener to give feedback in relation to the podcaster's goal. As a teacher I was hoping this would focus the listeners, provide a constructive next learning step and also enable the reflectors to be aware of the podcaster's personal goals as we all learn and progress at different rates and each of the learners within the classroom had different confidence and learning abilities in relation to listening and speaking, even within their first language.

At the start of the school year as the classroom teacher I began to develop a supportive learning environment for the learners in our classroom. I use the word "our" as I wanted the children within the classroom to have ownership over their learning. One of the first tasks was to develop a language nest. Traditionally Language Nests *(Kohanga Reo)* were an immersion-based approach to language revitalization originating in New Zealand in the 1980s as part of the Māori Language Revival. I used the metaphor of a nest as a place to feel safe, be nurtured and grow as a learner. Within the language nest the learners were practicing hearing, speaking, reading and writing *te reo*. The next step was to integrate elements of *tikanga* into the wider classroom programme. Storytelling is a traditional tool used within many Indigenous cultures around the world. Within the Māori Culture oral traditions of *waiata* (song), *whai korero* (speeches), *whakatauki* (proverbs) and storytelling are woven into tradition, safety and everyday learning through art, tool making, carving, weaving and *whakapapa* (genealogy). There is a common expression in *te ao Māori* (the Māori world) *Me ōna tikanaga, me ōna reo*—the language of Māori and the culture of Māori are linked. You learn them together simultaneously; neither should be or can be separated. Together these two parts make a whole: both are important for learning.

An integrated storytelling and art unit was taught around Papa-tua-nuku and Rangi-nui. These two people are the Earth Mother and the Sky Father. Their children are known as *atua* or gods within Māori legends. It is a creation story, a beginning. … There are many pictorial representations of this myth in publication. In class we used imagery to represent each character. We looked at how artists had represented each of them. Predominately Papa-tua-nuku is cloaked in the land, while Rangi-nui is represented with sky imagery. As learners we began to think about how we would represent Papa-tua-nuku and Ranginui on our *korowai* or cloak, which we were making to support our language learning (Figure 11.1). The *korowai* is not only a garment within the Māori world; it is also a metaphor and has many stories linked to it. It, too, is a tool for learning *tikanga* and like a lot of traditional Māori learning this knowledge is handed down and taught to those selected for their interest, talents or place within the family or tribe.

Figure 11.1 Children wearing their *korowai* at a school event. (*Source: Musselburgh School*).

Using the language nest we began to learn our *mihi*, (greetings) *pepeha* (geographical ancestral links to the land where we are from) and *whakapapa* (our genealogy). This is very important within *te ao* Māori as it enables the listener to make geographical, ancestral or tribal links to the speaker: this is called *whanangatangata*. The speaker is introducing and providing information about him- or herself. There is a structure to this speech and sequence that enables the learner to understand what the speaker is saying. It is a great place to begin learning *te reo* as we were beginning with the individual and his or her place in the world. Below is an example of a simple *mihi*:

> *Tēnā koutou, tēnā koutou, tēnā koutou katoa* (Greetings, greetings, greetings to all)
> *Ko Aoraki te maunga* (Mount Cook is my mountain)
> *Ko Waitaki te awa* (Waitaki is my river)
> *Ko Kai Tahu te iwi* (Kai Tahu are my tribe)
> *Ko Hui Rapa te hapu* (Hui Rapa is my subtribe)
> *Ko Araiteuru te waka* (My boat is the Araiteuru)
> *Ko Puke-te-raki te marae* (My marae is called Puke-te Raki)
> *No Otakau ahau* (I am from Otago)
>
> *Ko Joseph tōku te matua* (Joseph is my Father)
> *Ko Rhonda tōku te whaea* (Rhonda is my Mother)
> *Ko Jules rāua Ko Stephani āku teina* (Jules and Stephanie are my sisters)
> *Ko Paora tāku tane rangatira* (Paul is my husband)
> *Ko Liam rāua Ko Quinn āku tamariki* (Liam and Quinn are my sons)
> *Ko Lisa tōku ingoa* (My name is Lisa)
>
> *No reira* (Therefore)
> *Tēnā koutou, tēnā koutou, tēnā koutou katoa* (Greetings, greetings, greetings to you all).

The integration of teaching approaches enabled children to learn cultural aspects related to learning how to greet and introduce themselves in *te reo* Māori. On the inside of the *korowai*, the learners recorded their *whakapapa*, Later that year the children were invited to a regional *hui ako* (regional meeting about learning) where the children spoke *te reo*, shared their *korowai* and were questioned about their learning. Many of them began to make explicit connections to the wider world and were excited that others were taking an interest in their learning. It was very motivational.

The following term involved the explicit teaching of skills around reflection: of themselves as learners and of their learning as a whole. Reflection skills were introduced and modeled across a range of curriculum areas. The reflection process generally included:

- What are you going to do?
- How is it going?
- What do you need to do next?

All children had partners they would reflect with and gain critical feedback from. Sometimes this process was formal, and other times it was learners' choice. After a period of time the learners became better at choosing the criteria for their reflection and with whom they would reflect. The above questions provided a framework for learners to share what they had done, ask for help, plan the next steps and write new learning goals and also to think about the time restraints and whole class timeframes for learning. The purpose for this reflection was to improve our learning and work toward specific learning goals, i.e., an oral presentation of a *mihi* with a specific rubric for assessment. All learners were aware of the assessment requirements, and it was exciting to see some learners become leaders mentoring each other. The rubric also provided a powerful tool for reflection and critical feedback for themselves and their peers. Learners could ask their reflection partner to listen for or provide specific feedback on an element of the *mihi*. They could also ask for increased learning support or practice. Additional tools for learning were available in the form of interactive websites and from audio PowerPoint Slideshows with an emphasis on pronunciation. Other technological tools used included audio stories, and keynotes for pronunciation practice of nouns and simple sentence structures. Podcasting was a valuable tool for reflection in the classroom as there are not a lot of readily available resources to support and model the use of *te reo*. The support of Luana Thomas (the local resource teacher of Māori teaching and learning resources called Taylor Made Resources) was an excellent starting platform.

During the third term the class began to record their *mihi* onto a podcast. By this stage the language nest had been fully embedded within the classroom learning environment. A shared language for reflection was actively being used, refined and practiced throughout the day across several learning contexts. Podcasts were created by individual learners: to do this we began with some whole class teaching and then learners could opt into additional sessions. Learners followed a process of creating their podcasts using the Apple program GarageBand (see www.apple.com/ilife/garageband/). The use of this software enabled audio, visual and weblink material to be embedded in the podcast. The podcasts included a music introduction, images of their *korowai*, their *mihi* and a musical exit. The reflection criteria involved the children reflecting upon their own *mihi* and the *mihi* of two other peers. To do this we established a podbean site and uploaded the *mihi* to it. The introduction of technology was a powerful tool for reflection on their pronunciation as they could listen to what they had recorded. The learners could edit, re-record and make changes before deciding what would be finally uploaded to the podbean site. The reflection process here often involved peer tutoring on how to use the technology as well as feedback upon volume of voice, pronunciation, fluency, phrasing, etc. As a teacher at this stage I found myself giving the children an overview of learning, listening a lot to learning-based conversations, problem solving if the need arose and mentoring the learners to support each other and improve upon what they had already achieved by asking questions, posing challenges, giving some specific feedback and providing encouragement for learning.

Findings and Conclusions

There were three important components to this research. First, creating a classroom culture that promoted thinking and reflection in a safe and meaningful context. Second, developing an understanding of how to teach a second language and transferring this knowledge so English speakers could maximize their learning and improve their pronunciation of a second language. Finally, using podcasting as an m-learning tool to motivate and reflect upon pronunciation and new learning. As a classroom practitioner of year 5 and 6 learners I found the following results:

- Podcasting was a valuable tool for developing learner-generated content.
- Learners were self-motivated on tasks and actively working toward specified timelines.
- Learners enjoyed the use of technology in their learning.
- Learners had ownership over their learning.
- Learners were aware of the purpose of podcasting and reflective feedback to improve pronunciation of *te reo*.
- Learners were engaged in a range of wider, deeper learning skills that were creative, imaginative and practical to help support their learning of a second language.

It was immediately apparent that the learners were engaged in the specified task. This may have been due to the high interest of podcasting being a new and somewhat novel learning tool. It is possible that the higher levels of engagement were due to learners enjoying using technology and having ownership over their own learning. Throughout the process the learners were aware of the purpose of the podcasting and the reflective feedback required of them as learners and participants.

Lord (2008) described reflection as a never-ending clarifying process, a reflective spiral, of thinking of new opportunities, refining language and learning before implementing changes to thinking. This research has determined that good modeling (from the teacher, podbean, *te reo* sites) coupled with opportunities to practice (e.g., talking, listening, reading, writing and games) and to reflect upon our learning does improve pronunciation.

The importance of explicitly teaching reflection skills to learners also became apparent. We established a shared language of learning that was supported and reinforced by specific approaches to learning. The learners in this classroom were actively encouraged to be active life-long learners, aware of their next learning goal and supported to achieve it. Some of the tools used by the classroom teacher included graphic organizers, co-operative learning tasks that encouraged reflection and ways to begin, enter and respond to feedback. Modeling of learning conversations, using assigned roles, and giving participants specific tasks for feedback were also important. I referred to this as "fish-bowling" as the active participants were in the center surrounded by their reflective peers. When learners verbalized their own goal it provided a specific purpose for learning. These goals could have been agreed on by the teacher or via a group conference in advance of sharing them with the class. An advantage of this was that learners were focused on the goal not the individual. If learners were less able, their attempts were still valued by their peers; it removed the fear of failure from the learner. An additional benefit was that listeners could listen with a specific purpose in mind. Finally, it enabled the learner to hear positive feedback, which gave a next learning step that could be enacted or reflected upon again in the near future.

The benefits of creating a shared language for the learners include:

- Being responsible for their own learning: learners were able to recognize their next learning step.
- Creating a shared language and shared knowledge base with their peers.
- Learner-generated content enables an enhanced learning experience to occur as the focus of learning is on process rather than task-based learning.
- Learners were empowered to reflect and improve upon their learning based on reflective feedback from themselves and their peers.
- Time was provided to reflect and enact changes.
- Learners were actively involved with the learning process and became more independent and self-motivated as the year progressed.

The establishment of a routine, the provision of tools for reflection and the integration of learning across the visual arts, oral language, literacy and m-learning through podcasting enabled a variety of children to demonstrate strengths of learning at different times. Throughout the process the learners always knew that they would be aiming to improve their personal pronunciation. The variety of approaches helped to provide time to revisit and to build on existing knowledge. It also helped to encourage personal interest and motivation of learners. Learners had multiple opportunities to present, reflect and revisit their learning, enabling them to make connections and improvements to continually refine their understanding of where they were on their learning journey and what they could do to improve their learning. Outlined below are some of the opportunities learners had for improving and reflecting on their individual pronunciation:

- When the creation myth of Papa-tua-nuku and Rangi-nui was told learners heard a *mihi* woven into the story-telling.
- When they created their *koru* patterns (traditional Māori art patterns) learners were verbalizing the sentences as they wrote, enabling them to practice writing new vocabulary and learn the structure of sentences.
- When sharing their *korowai* learners read and listened to each other's *whakapapa*.
- When presenting their oral *mihi* to the class learners were actively listening to a presentation and reflecting upon a rubric for assessment.

- When at the regional *hui ako* learners heard others presenting and began to understand the importance of their own learning in a wider context.
- When learners began to use the m-learning tool of podcasting they were able to reflect upon their own and others' experiences to improve their pronunciation of *te reo*.

Reflection had a positive impact on the learners in that they began to have direct ownership over their own learning. Learners enjoyed the process of sharing and receiving feedback. Many learners felt that they were sharing good news with their peers. It became a process that motivated learners to improve on what they had done, wanting to please and be a respected peer. Overall the classroom had a different tone or feel, which is hard to explain or measure. It was directly related to the sense of learner well-being. Direct outcomes included increased risk-taking, learners taking direct ownership of their learning by making instant changes to their podcasts if an error in pronunciation was heard. Examples of changes included re-recording to fix mispronunciations or low volume or editing out silences or long pauses, etc. These changes occurred instantly and became part of the podcasting process. Another measurable outcome included learners asking for specific teaching of pronunciation of a phrase or word. Learners were joining practice groups and accessing tools that provided a model for pronunciation. Some learners also became peer-teachers supporting their peers to pronounce words with increased intonation and fluency. Learners were also beginning to seek time throughout the day to practice and improve their pronunciation. I saw this as learners becoming responsible and guiding their own learning. Reflection can occur at any stage of the learning process; it doesn't have to be at the end. Therefore a teacher's role becomes one of facilitating reflection with a class-room climate that encourages a variety of reflection opportunities such as group, individual, teacher-led discussions or student-to-student conversations.

The benefits of creating a shared language for the teacher included:

- The role of the teacher changed to that of a facilitator—shifting the responsibility for learning from the teacher to the learner. The type and quality of feedback given helped the teacher identify progress and areas on which to focus.
- The link between learner and teacher perceptions of thinking and under-standing became more closely aligned. There was an increased focus on learning being cyclic, ongoing and having some repetition so learners could build on existing knowledge.

As a teacher, giving myself permission to step back from a linear action-orientated teaching style was incredibly powerful. A change of mindset, the establishment of the structure outlined by Graham and Osborne (2008) provided a rule of thumb, a way of thinking that initially helped to guide expectations for learning and ways of behaving and responding to each other. Metaphorically, I felt that I was making a two-degree change in the way I structured the learning environment. On the one hand, I was still accountable for ensuring that the learners were provided with quality learning experiences, that this learning could be measured and that a final piece of learning would be completed by all learners. On the other hand, in a way I was a learner too. Together we helped structure the learning activities and learning environment. The cyclic, reflective structure enable me as a teacher to plan and think ahead whilst also giving the learners time to reflect, practice and make changes to improve their pronunciation.

Initially, as a teacher I was very conscious of how I structured the learning experiences, the modeling of reflection processes and the teaching tools used. As the year progressed I found that I was listening more attentively, responding more purposefully and structuring my responses in relation to the specific feedback or needs of the learner. I was facilitating the learning rather than controlling how and when the learning occurred. I could direct learners to various tools, spend some time one-on-one and teach to a specific need for a group of learners. I also found that I was directing learners to specific people, places or tools for learning. Routinely I found myself redirecting myself and the learners toward the collective goal of improving our pronunciation of *te reo*, the Indigenous language of New Zealand. There were times when the process of creating a podcast dominated most of our reflection and new learning. At other times pronunciation was at the forefront of our reflection as we were sharing and giving and receiving direct feedback.

Direct observations that I recorded anecdotally throughout the learning process include the following. In the beginning learners often had a limited skill bank to draw upon when reflecting on their own and others' podcasts. Initially many of the comments fell into the category of passive constructive comments like "I like the … or it sounds good … I can hear you well etc., …" Over time learners were able to make more actively constructive comments when scaffolded by a teacher. Time to practice, modeling by the teacher, the use of sentence stems and providing expectations all helped to improve feedback, making it more robust and relevant for the next learning steps. Another observation included the provision of a process or feedback routine: it really helped the learners articulate and respond to each other based on specific goals. It was at this stage that the most valuable learning occurred. In addition I began to notice that through the reflective process learners were better at determining their own next learning step than their peers, and I wondered why. I concluded the following possibilities: perhaps the learner was taking personal ownership of learning, being self-motivated, goal orientated and focused on improving pronunciation. … Perhaps learners felt empowered to reflect and build on their own next learning steps … or perhaps the focus on the positive feedback and a next learning step led to the peer group perceiving feedback as a possible negative while not wanting to put each other down, thinking that "if I say something critical they may criticize my pronunciation". We had also spent a lot of time thinking and creating a learning environment that encouraged risk taking and supporting a culture of learning; perhaps some learners felt that negative feedback was the same as critical feedback and therefore instead of helping a learner they may have hurt someone's feelings. The ability to focus on the learning rather than the learner may have been hard to discriminate for some learners, especially if that learner was a good friend or had learning difficulties. This is an area that as a class we need to continue to work on so we have more resilience, skills and reflective ability.

Recommendations for Establishing a Reflective Learning Environment Using Podcasting

Establish a class culture that values reflection by:

1 Establishing a common shared language of reflection.
2 Providing regular time to think as individuals and as a group.
3 Structuring lessons to support reflective thinking, inquiry and curiosity.
4 Linking reflection to a social communicative approach, prompting learners to reflect upon what they have done, what they have learned and what they still need to do.
5 Building on the initial reflections by using a reflective process that gives learners opportunities to think about what they know, what they have learned and what they still need to know.

Continue to explicitly teach reflection skills and self-monitoring strategies by discussing:

1 What is the purpose of reflection?
2 What skills do I need to be good at reflecting?
3 What is effective reflection?

Spend some time working on the types of reflective responses children make about their own and others' work. Possible ways to increase children's use of active constructive comments include:

1 Continue to build children's learning power.
2 Highlight strategies children are using so they learn what works and they learn what they can try in the future.
3 Change the focus to one of effort and strategy so it promotes mastery type behaviors.

Establish transference of reflective skills across learning contexts by the following:

1 Provide enough wait-time for learners to reflect on their own and others' learning.
2 Combine with Building Learning Power so reflecting becomes more of an intrinsic process that encourages re-evaluation of learning.

3 Provide time to think.
4 Prompt learners' reflection by asking questions that seek reasons and evidence.
5 Provide some explanations or sentence stems to guide learners' thinking during the reflection processes.
6 Provide social-learning skills to enable learners to see other points of view and to empower learners to focus upon the learning, not the learner, when giving critical feedback.
7 Encourage more robust self-reflection by encouraging learners to give reasons to support what they think, outlining strengths and weaknesses of their own learning.
8 Find ways to reach those children who are above and beyond the cohort.

Continue to use podcasting tools to improve pronunciation because:

1 Instant feedback is a positive tool for developing critical thinking and reflection.
2 Podcasting is a motivational, purposeful tool for teaching pronunciation.
3 Podbean sites can increase wider community feedback from family and community.
4 Podcasting has the ability to meet a range of learning styles and learning experiences.
5 Podcasting has direct benefits for targeting listening practice in order to improve pronunciation.
6 Podcasting can be used to create smaller individual, collective or teacher-directed podcasts within authentic learning contexts.
7 Podcasting is a tool to improve pronunciation through reflective feedback.

To conclude, pronunciation was improved by using the highly motivational m-learning tool of podcasting. Podcasting fostered and enabled learners to become directly responsible for improving their pronunciation and promoted reflection to aid this process. Further longitudinal research into collaborative m-learning experiences using reflection and podcasting to improve language learning is recommended to validate the results of this implementation with larger groups of learners or in different learning contexts. However, the findings are extremely promising for the learning of Indigenous and other languages.

Acknowledgements

I would like to thank the class of year five and six learners who participated in this project. You were a special group of learners who had a collective love of learning that helped make this research possible. I would also like to acknowledge the support of two critical friends whose feedback and advice was valued and appreciated. Lastly, I am grateful to the anonymous reviewers who offered valuable advice for strengthening this article.

Glossary of Māori Terms

Māori	*English*
Atua	Gods
Hui Ako	Regional meeting about learning
Kaupapa	Customary practice
Kohanga Reo	Language nest
Korowai	Cloak
Koru	Traditional art pattern
Mihi	Speech of greeting
Papa-tua-nuku	Earth Mother
Pepeha	Formulaic structure
Rangi-nui	Sky Father
Te Ao Māori	Māori World
Te reo Māori	The Māori language
Tikanga	Traditional lore or custom
Waiata	Songs
Whai korero	Speeches
Whakapapa	Genealogy
Whakatauki	Proverbs
Whanagatanga	Ancestral or tribal links

References

Claxton, Guy. "Building Learning Power." Accessed May 15, 2014. http://www.buildinglearningpower.co.uk.

Dale, Crispin, and Ghislane Povey. "An Evaluation of Learner-generated Content and Podcasting." *Journal of Hospitality, Leisure, Sport and Tourism Education* 8, no. 1 (2009): 117–123. doi:10.3794/johlste.81.214.

Dewey, John. *How We Think: A Restatement of the Relation of Reflective Thinking to the Educative Process.* Boston: D.C. Heath, 1933.

Goodman, Bradley, Amy Soller, Frank Linton and Robert Gaimer. "Encouraging Student Reflection and Articulation Using a Learning Companion." *International Journal of Artificial Intelligence in Education* 9, no. 3–4 (1998): 237–255.

Graham, M., and V. Osborne. "Student Reflection Action Research Project." 2008. Accessed January 2015. http://www.scribd.com.

Hatton, Neville, and David Smith. "Reflection in Teacher Education: Towards Definition and Implementation." *Teaching and Teacher Education* 11, no. 1 (1995): 33–49.

Hook, Pam. "Hooked on Thinking." 2009. Accessed May 15, 2014. http://www.pamhook.com.

Kolb, Alice, and David Kolb. "Kolb's Learning Styles." In *Encyclopedia of the Sciences of Learning,* edited by Norbert M. Seel, 1698–1703. New York: Springer, 2011.

Laiken, Marilyn E. "Managing the Action/Reflection Polarity through Dialogue: A Path to Transformative Learning." NALL Working Paper no. 53–2002 (2002): 1–17. Toronto, Canada: NALL Research Network for New Approaches to Lifelong Learning.

Lee, Mark J. W., Catherine McLoughlin and Anthony Chan. "Talk the Talk: Learner Generated Podcasts as Catalysts for Knowledge Creation." *British Journal of Educational Technology* 39, no. 3 (2008): 501–521.

Lin, Grace Hui Chin, and Paul Shih Chieh Chien. "An Investigation into Effectiveness of Peer Feedback." *Journal of Applied Foreign Languages Fortune Institute of Technology* 3 (2009): 79–87.

Lord, Gillian. "Podcasting Communities and Second Language Pronunciation." *Foreign Language Annals* 41, no. 2 (2008): 364–379.

Pohl, Michael. "Developing a Classroom Culture of Thinking; A Whole School Approach." *TEACH Journal of Christian Education* 5, no. 1 (2009): 8–9.

Siemens, George. "Connectivism: A Learning Theory for the Digital Age." *International Journal of Instructional Technology and Distance Learning* 2, no. 1: (2005). Accessed March 21, 2015. http://www.elearnspace.org/Articles/connectivism.htm.

12 Integrating Multimedia in ODL Materials and Enhanced Access through Mobile Phones

Maria Augusti and Doreen Richard Mushi

Introduction

The use of Information and Communication Technology (ICT) in supporting teaching and learning activities in higher learning institutions is growing tremendously. Advancements in their integration into education systems have made huge contributions in improving quality and reaching institutional objectives in both traditional and open and distance learning modalities of higher education delivery.

Countries around the world are on a quest to improve their educational systems in order to enhance the quality of education and increase access to people in their communities. Extra efforts have been made for Indigenous people, who are normally socially and economically excluded and live in remote areas (UNESCO 2011).

Tanzania has a population of over 43 million with a total of more than 120 ethnic groups (TASAF 2011). Out of these groups, four tribes have been classified as Indigenous. These are the Maasai, Barabaig, Akie and Hadzabe. However, despite this initial recognition, the country is still in the process of screening and determining which other tribes fit into this group. Due to poverty, Tanzania also has a lot of communities that are not grouped as Indigenous but are secluded in their remote locations. In order to ensure education reaches people from all of these communities, the Open University of Tanzania was established as the main university to offer education through an Open and Distance Learning (ODL) mode. UNESCO (2002) introduces ODL to present open access to education by removing constraints of time and space. Remote delivery promotes flexible educational opportunities to learners from various areas and backgrounds. Due to the changing economy and an evolving society, there has been a need to extend ways of education delivery in order to reach a wider user group and reduce inequalities (Isaacs 2012). In Tanzania, current achievements and successes in ODL environments are closely linked to the advancements and developments in the use of ICTs in education (Mnyanyi and Mbwette 2009).

Back to the issue of time and place, the main challenge with ODL in developing countries has been reaching out to large groups of students who are scattered all over the place (UNESCO 2002). When it comes to Indigenous communities, the ODL system has been an effective and efficient mode of delivery. This is mainly due to the fact that people in remote communities can participate in educational activities without being forced to leave their areas. With the movement of education for all, institutions around the world have been looking for solutions for meeting institutional outreach objectives and strategies while maintaining the quality of education delivery and experience (Mbwette 2011). This has meant resorting to the adaption of mobile technologies to support delivery of knowledge to remote and Indigenous communities.

This chapter presents the results of a project that integrated mobile phone technology and multimedia to provide a solution for reaching students from urban to rural and Indigenous communities. Learning materials have been enhanced with multimedia elements in order to improve student engagement and interest. Multimedia technologies have been widely used to supplement the designing and development of learning objects in classrooms and online (McGreal and Elliott 2011).

Tanzania has been using mobile phone technology to alleviate poverty and in other socio-development strategies. This is mainly due to the advantages that mobile phones have brought to our communities. Mobile learning (mLearning) provides a lot of opportunities for users around the world to get access to education despite the limitations of culture, background and status (Traxler and Kukulska-Hulme 2005). In developing countries, the rate of ownership of mobile phones is relatively high compared to the rate of possession of personal computers (Mnyanyi and Mbwette 2009). Furthermore, mobile phone technology is already commonly used by students and cost effective and can be utilized regardless of time or location; therefore, it is quite adaptable (Motlik 2008).

These experiences together with prior studies relate to what has happened at the Open University of Tanzania. The university has been engaged in e-learning activities since 2006/2007 and has been using a learning management system customized from Moodle to support its key teaching and learning activities at the university. The system has been used to offer access to course materials to students scattered all over the country. Despite having regional centers across the country that host computer labs and provide technical assistance to students, the rate of reaching out to a large number of students has still been low. Challenges such as unreliable Internet connectivity, frequent power-cuts and poor infrastructure have been the main contributing factors. Another challenge has been improving learning materials so that they are more interactive and interesting for students to use. One group of students has special needs: the term special needs can be defined as the educational requirements of students with disabilities such as visual, hearing and physical impairments (Collins English Dictionary 2011). Some are from Indigenous and remote communities. The university had to look for ways to allow these groups of students to access learning materials just like all other students. These scenarios are also similar to other universities from developing countries and have made the use of traditional computer technology ineffective for the delivery of education to remote and disadvantaged learners using ODL (Maritim and Mushi 2011). The above challenges have motivated the institutional body to explore the use of mLearning at OUT.

Lack of integrated multimedia in ODL materials due to bandwidth and cost limitations, inaccessibility due to power cuts and unreliable Internet connectivity, together with the many other challenges faced by students with special needs, were seen as issues that could be dealt with through mLearning. Additionally, although multimedia and mobile phone technology have been used in education to enhance teaching and learning in the developed countries for many years, far less has been done in Tanzania, because of many challenges associated with their implementation. By 2011 it was estimated that Africa had 500 million phone subscribers (Rao 2011). Therefore, mLearning and mobile technology has the potential of providing a suitable alternative way for learning at low costs.

OUT Overview

The Open University of Tanzania (OUT) is the largest higher learning institution in Tanzania to offer undergraduate and postgraduate academic programs via the ODL mode. The university operates through 29 regional centers and 69 study centers scattered throughout the United Republic of Tanzania. The student population at OUT is situated in urban, Indigenous and remote areas and includes the visually, hearing and physically impaired. By June 2013, OUT had 44,500 undergraduate students, out of whom approximately 28,000 were active, and 13,500 postgraduate students. These students are mostly from Tanzania, with 47.2% being from the region. This percentage also includes students who are from Indigenous areas, especially Maasai and Sandawe, and from Kenya, Rwanda, Namibia, Uganda and other countries in Africa in addition to some from Europe, USA and Asia who enroll for the Postgraduate Diploma in Curriculum Design and Development (PGDCDD). Currently OUT has Coordination Centres in Kenya, Rwanda , Namibia and soon in Uganda.

The Open University of Tanzania established the Institute of Educational and Management Technologies (IEMT) for managing and administering ICT services to the university. The institute has three main departments: Service Control and Quality Assurance, Information Resource Management and Educational Technology. Educational Technology hosts an E-learning Development and Multimedia Section (EDMS) that oversees the general implementation of e-learning activities at the university. These activities involve management and maintenance of the OUT learning management system (LMS); offering assistance and training for academic staff on the efficient usage of the LMS and online course development; production of learning objects with multimedia elements; and research and further developments in the use of electronic media in education. The department has instructional designers and computer technicians responsible for offering both pedagogical and technical support.

A feasibility study on mobile phone usage undertaken at the Open University of Tanzania revealed that most of the students situated in remote areas possess mobile phones, which confirmed that the use of mobile phone technology can be embedded in the educational process (Mbwette et al. 2011). In light of the mobile phone technology advantages highlighted in the discussion above, Open University of Tanzania recognized the need to implement this technology to assist as many students as possible in a cost effective manner.

Motivation of the Study

Although printed study materials are no longer given to individual students, they are still made available in the mini-libraries located at the regional and coordination centers. In addition, students have been encouraged to access soft copies of study materials prepared by OUT lecturers via Moodle and CDs that are produced for all students enrolled in specific courses. Students have also been encouraged to purchase personal PCs or laptops to ensure wider reading. However the challenge remained that most of the course materials available in the learning management system were in plain text format and hence not accessible by students with special needs. Students who are hearing and visually impaired require study materials integrated with multimedia for accessibility. It has also been a challenge to facilitate the learning process for all categories of students due to the fact that accessing the course materials in the LMS is not always guaranteed due to frequent power cuts, unreliable Internet connectivity and the limited purchasing power of most learners due to poverty. This motivated the need of alleviating problems that hinder students' access to ODL materials.

Target Audience

The target group for the study was the ODL stakeholders scattered nationally and internationally, including students, staff and technologists with special needs and disabilities, females who are marginalized and students from remote and Indigenous communities.

Integrating Multimedia in ODL Materials and Enhanced Access through Mobile Phones

This study was part of a change project sponsored by the Swedish International Development Cooperation Agency (SIDA). The main purpose of the project is to reduce poverty and promote sustainable development through better integration of ICT in education systems. About 12 countries were part of this initiative, Tanzania being one of them. Participants from each country were supposed to come up with innovative project ideas on how they could improve their home countries by using ICT in education.

The Open University of Tanzania took this opportunity to work on a solution that would improve the quality and reachability of learning materials for a wider student group. The two main objectives of the change project from the OUT point of view were first to improve the value of the learning materials by integrating them with multimedia elements and second to enhance access to course materials for all OUT students by using mobile phones.

Integration of Multimedia in ODL Materials

An OUT academic course entitled *Multimedia Technology and Applications*, with the course code *OIT 208*, was used as the pilot course for this project. The course modules were uploaded to Moodle, and their underlying contents such as PowerPoint slides were fully integrated with audio and video elements. The recordings of the instructor's voice were made on a laptop, and the clips were later on integrated into PowerPoint slides using Adobe Master Collection CS3. Furthermore, the recording of a sign language interpreter was embedded in one module to test its usability. A simple Sony digital camera was used for the recording. At the end the course consisted of six modules, each containing PowerPoint slides that had been embedded with audio and video, making a total of 225 slides (Figure 12.1).

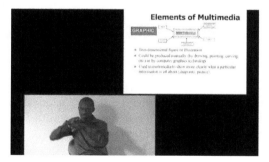

Figure 12.1 A screenshot showing a module within the course in the OUTLeMS with audio and video and sign language interpreter. (Source: C. Chasity 2014).

Enhancing Access of ODL Materials through Mobile Phones

The implementation of OUTmLearning was derived from mTouch-U (2015), an application that enables users to access Moodle from their mobile devices. The application is student centered, making it possible to access and interact with resources and activities that have been constructed by teachers in Moodle. In the case of Open University of Tanzania, OUTmLearning provides a channel for students to access OUT Moodle through their mobile phones. The software was found to be suitable for all modern mobile phones that have Internet features.

The process of deploying the software to OUT considered two key requirements: server and customization requirements. The basic server requirements were installations of Windows Server 2008 R2 or above, IIS7 or above, Microsoft.NET Framework version 4 and ASP.NET MVC3. Customization requirements basically needed files corresponding to the proposed application name, support URL, welcome page content as HTML formatted text, contact e-mail address, Moodle URL, Moodle student accounts enrolled in some courses with content, etc. The customization was done by the Pragma Touch team through teamviewer software installed in the OUT server. The application enabled students to access all the 150 courses through their mobile phones (Figure 12.2).

Figure 12.2 Screenshot of OUTmLearning application. (Source: Field work, 2013).

Evaluation of the Project

The evaluation process was performed to corroborate the usability and feasibility of the developed system. The first phase included expert reviews by technical staff from the E-learning Development and Multimedia Section at OUT. The aim was to involve people who have an established experience with Moodle and its functionalities. The second phase involved conducting a pilot study of the application during the institutional face-to-face sessions. The objective was to obtain feedback from students on the level of accessibility of learning materials and study the challenges associated with using the application. The study was then done in 13 regional centers of OUT with three students randomly selected from each center.

Different types and models of mobile phones were used including Nokia, Samsung, LGE400, Techno N3, Lenovo and HTC. The application was tested accessing course content, downloading a file and engaging in the discussion forum from the OUT Learning Management System (OUTLeMS).

The results were:

- Accessing course content on the LMS was successful for all tested phone types;
- File download from the LMS to the mobile phone was successful for only about 55% of the phone types. This was highly dependent on the memory capacity of the phone;
- Engaging in the discussion forum was successful 95% of the time. Almost all phones were comfortable with the capacity.

Challenges Faced

During the implementation phase of the project the following challenges were encountered:

- Time constraints for the academic staff: academic staff involved in the implementation have their daily work schedules; therefore it was difficult to concentrate on other activities and the project plans revolved around their work and availability;
- Lack of skilled sign language interpreters: sign language experts in ICT subjects are few, making them very expensive to hire.

Furthermore, findings from expert reviews and the pilot project with students revealed the following challenges:

- Readability problems: small screen display of mobile phones failed to engage students. Most students complained about the small text and low resolution due to the size of their phones;
- Poor coverage of mobile networks in some remote areas of Tanzania limited its applicability;
- Awareness of students about the available mobile application for learning was still low.

Implications of the Study to the OUT Community

Having course materials integrated with multimedia and accessed through mobile phones can alleviate problems of student access to ODL materials as nearly all OUT students own mobile phones, although some phones are very unsophisticated. Learning materials with multimedia elements will also be beneficial to students with special needs from various locations inside and outside Tanzania, with the inclusion of Sandawe, Hadza and Maasai people, as they can interact and participate in the educational activities without major limitations.

The use of mobile phones is also anticipated to remove some of the current operational problems of the university ICT infrastructure and services by allowing the potential for agreements with mobile phone companies concerning special rates for academic use. Furthermore, the successful implementation of the project will assist some technophobic teachers to overcome their fear of recording and improve teacher engagement in designing interactive course materials.

Future Work

There are many ways in which this project can be continued. The experiences obtained from this project have also established some new conceptions on how the situation can be improved for learners. For instance the university is exploring the possibility of securing affordable but reliable tablets that will make it easy for students to engage in their courses.

More should be done so that all courses in OUTLeMS can be integrated with multimedia. In this context, there is a need to explore resources that will be required to make the process effective and efficient. There is also a need to provide audio-visual training to technical staff so that they can have proper skills for the production of integrated learning materials. Capacity building is also important and should be made available to OUT staff and students in developing and using multimedia learning materials. Having all ODL materials incorporated with multimedia will enable the entire OUT population to access and use materials regardless of special needs.

There will be a big increase in the number of students from Indigenous and remote communities participating in distance education through mLearning technology. There are growing numbers of students from outside Tanzania due to the accessibility of the ODL materials integrated with multimedia and enhanced teaching and learning anywhere, anytime using the OUT mobile learning application (OUTmLearning). There are now more female students who previously could not go for further studies due to lack of time caused by early marriage/pregnancy and were forced to remain home and take care of the family. The presence of OUTmLearning will enable them to schedule their time and learn through their mobile phones.

A proposition should be made for designing a curriculum in Tanzania to support and provide guidance on the use of mobile learning technology in the teaching and learning processes in the country. The relevance of this fact was discussed by Keegan (2005) in his study focusing on inclusion of mobile learning in the academic mainstream.

Conclusion

The use of mobile phones in supporting delivery of education in academic institutions is clearly on the rise. This is mainly due to the fact that flexible learning opportunities are highly desired by today's learners. Given the fact that most learners in developing countries, especially in East Africa, possess mobile phones (Mbwete 2011), they are very beneficial in imparting knowledge to users from remote and Indigenous communities.

The development and use of multimedia has proved to be successful in enhancing teaching and learning experiences in universities across the world. In the offered academic programs, one has to think of ways of designing resources to best support different learning styles and eventually help students achieve the set learning outcomes. In ODL the implications for the use of instructional technologies in supporting teaching and learning activities are quite extensive (Ally and Tsinakos 2014). The motto for institutions has been to find methods for efficient transfer of knowledge while maintaining the pedagogical requirements.

In developing countries, the main difficulty has been to improve outreach to students who are located in remote areas and belong to Indigenous groups. The use of e-learning technologies to support the transmission of knowledge has not been very successful due to challenges of unreliable Internet connectivity, power cuts and high costs associated with owning personal computers. This has motivated institutions to explore the potential of mobile learning, and there are mLearning success stories from institutions around the world.

The Open University of Tanzania carried out the mobile learning project with the aim of integrating multimedia in ODL materials and enhancing their access through mobile phones. The substantial outcomes of the project were: an academic course developed with modules embedded with video, audio and sign interpretation elements and a mobile learning application called OUTmLearning, which offers students access to the university learning management system.

The testing and evaluation phase of the project presented positive results. The application was well supported by various types of mobile phones. Students from all regions of Tanzania were capable of accessing course content, downloading files and engaging in discussion forums. However, some challenges were identified. Difficulty in readability due to small screens and limited storage specifications were mainly noted. Despite the challenges, overall feedback favored the potential mobile phones have for improving access to remotely located users including Sandawe, Hadza and Maasai people.

This project has provided insights for further developments for the university community in general. Projects designed to overcome the challenges identified in this study are in progress. The use of tablets is an example of anticipated initiatives as are proposals for mainstreaming the use of mobile learning technologies in the curriculum.

An increased number of student enrollments from inside and outside Tanzania is also anticipated. Channeling students through mobile learning will interest a community that has currently been limited by issues of time and space.

References

Ally, Mohamed, and Avgoustos Tsinakos. *Perspectives on Open and Distance Learning: Increasing Access through Mobile Learning*. Burnaby, Canada: Commonwealth of Learning and Athabasca University, 2014.

Collins English Dictionary. 10th ed. Glasgow: HarperCollins, 2011.

Isaacs, Shafika. *Turning on Mobile Learning in Africa and the Middle East: Illustrative Initiatives and Policy Implications.* UNESCO Working Paper Series on Mobile Learning. de Fontenot: United Nations Educational, Scientific and Cultural Organization, 2012. Accessed February 3, 2015. http://unesdoc.unesco.org/images/0021/002163/216359E.pdf.

Keegan, Desmond. "The Incorporation of Mobile Learning into Mainstream Education and Training." In *Proceedings of the 4th World Conference on Mobile Learning,* Cape Town (2005): 1–17.

Maritim, Ezra K., and Honoratha M. K. Mushi. "Mobile Technologies for Enhancing Distance Learning in Tanzania: An Exploratory Study." *Huria: Journal of the Open University of Tanzania* 13 (2012):123. Accessed February 3, 2015. http://www.ajol.info/index.php/huria/article/view/110809.

Mbwete, Tolly S. "E-Learning and Mobile Learning: Some Highlights on Innovative Delivery Methods in Continuing Education." Paper presented at the Conference on Strengthening Universities, Enhancing Capacities—Higher Education Management for Development, Bonn, Germany, November 28–29, 2011.

McGreal, Rory, and Michael Elliott. "Technologies of Online Learning (e-Learning)." In *Theory and Practice of Online Learning,* 2nd ed., edited by Terry Anderson, 115. Edmonton: Athabasca University Press, 2008.

Mnyanyi, Cosmos B. F., and Tolly S. A. Mbwette. "Open and Distance Learning in Professional Development in Third World Countries." In *Proceedings of the 23rd International Council for Open and Distance Education (ICDE) World Conference, Maastricht,* Netherlands, 1–7, 2009.

Mnyanyi, Cosmos B. F., Tolly S. A. Mbwette and Jabiti K. Bakari. "Prospects and Challenges of m-Learning in Open and Distance Learning in Developing Countries." In *Proceedings of the 4th Conference on e-Learning Excellence,* Dubai, 170–178, 2011.

Motlik, Scott. "Mobile Learning in Developing Nations." *The International Review of Research in Open and Distance Learning* 9, no. 2 (2008). Accessed February 3, 2015. http://www.irrodl.org/index.php/irrodl/article/view/564/1039.

mTouch-U. Accessed February 3, 2015. http://mtouch.pragmasql.com.

Rao, Madanmohan. *Mobile Africa Report 2011: Regional Hubs of Excellence and Innovation.* MobileMonday, 2011.

Shuler, Carly. *Pockets of Potential: Using Mobile Technologies to Promote Children's Learning.* New York: Joan Ganz Cooney Center at Sesame Workshop, 2009.

TASAF. *Draft TASAF III Indigenous Peoples Policy Framework.* Dar es Salaam, Tanzania: United Republic of Tanzania, 2011.

Traxler, John, and Agnes Kukulska-Hulme. "Mobile Learning in Developing Countries." Vancouver: Commonwealth of Learning, 2005.

Traxler, John, and Jocelyn Wishart. "Making Mobile Learning Work: Case Studies of Practice." Bristol: ESCalate, University of Bristol, 2011.

UNESCO. *ICTs and Indigenous People.* Policy Brief. UNESCO Institute for Information Technologies in Education, 2011.

UNESCO. *Open and Distance Learning: Trends, Policy and Strategy Considerations.* Paris: UNESCO Division of Higher Education, 2002.

13 Mobile Phones in Rural South Africa

Stories of Empowerment from the Siyakhula Living Lab

Lorenzo Dalvit

Introduction

The uptake of mobile phones on the African continent is the fastest diffusion of technology in the history of humanity. Mobile phones are changing the lives of people from all socio-economic backgrounds across the gender, age, geographical, language and literacy divides. What Castells calls the "Fourth World" is progressively shrinking as more and more people in even the most remote areas become connected (Castells 2000). In this chapter I describe how Information and Communication Technologies (ICT)—and particularly mobile phones—contribute to the empowerment of people in a rural area in South Africa. This work is based on the eight-year experience of the Siyakhula Living Lab, a holistic and multi-disciplinary ICT-for-development project. After a discussion of selected research conducted thus far, I discuss six stories of empowerment through mobile technology.

Mobile in South Africa

The mobile phenomenon can be analyzed through many lenses. In this section I discuss the penetration of mobile phones in South Africa, the issues associated with it and the specific urban/rural divide.

The Mobile Boom

Mobile use in South Africa increased five-fold over the last decade (Global Entrepreneurship Monitor 2012), and three out of four low-income adults in rural and urban areas owned a mobile phone by 2012 (Lana 2014). Uptake of smartphones increases by 20% a year, and by the end of 2015 it is estimated that most phones in South Africa will be smartphones (Jones 2010). As the cheapest model costs approximately ZAR (South African Rand) 550 (US $47), the smartphone experience is coming within reach of a large portion of the population. Donner (2008) characterizes the experience of the Internet by most South Africans as mobile-first and mobile-centric. This has profound implications for how people experience being connected.

The South African government recognizes the importance of universal access for socio-economic development. Through its broadband policy it seeks to achieve universal access by 2015 (My Broadband 2013). The debate around which technology is most suitable to enable South Africa's digital leapfrog is ongoing (Infodev 2012), but mobile phones have a key role to play. The focus on community rather than household and individual access has been blamed for the drop of South Africa to number 80 in the World ranking on Internet connectivity (Goldstuck 2013). At the same time, Goldstuck (2010a, 2010b) notes that up to 60% of adults in peri-urban areas may not be aware of the Internet capability of their phone, and that it may take up to five years for someone to start fully participating online. The National Development Plan (2011) emphasizes the importance of fostering e-skills and sets the goalpost of e-readiness in 2030. For the time being, the function of mobile phones as communication and content creation/consumption devices is as important as network connectivity.

Mobile use among South Africans falls within three categories, associated with the different features of the devices to which they have access. Communication activities such as making calls and sending/receiving SMS are the most common. A survey in a rural area of the Eastern Cape by Dalvit and Strelitz (2013) found that almost everybody uses a phone to communicate. Multimedia activities such as taking pictures, listening to music or watching videos are practiced by 50% of the respondents while approximately 40% of local people perform networked activities such as instant messaging, social networking and browsing the Web. In a study on rural and township[1] learners Gunzo and Dalvit (2012) note that using a phone to listen to music is almost as popular as SMS, while among networked activities only instant messaging is practiced by at least half of the respondents. A study of peri-urban youth in Cape Town (Kreutzer 2009) also notes that mobile phones are often used to take pictures and videos and that instant messaging and the consumption of news are relatively popular.

Coping with Cost

In a country where most of the population lives under the threshold of poverty (StatsSA 2014), the costs associated with mobile phones are a key concern (see Abrahams and Goldstuck 2012). Many low-income phone users have multiple SIM cards to take advantage of promotions by different operators and avoid extra charges when calling somebody on a different network (Dalvit 2014). South African mobile operators allow users to send up to three free call-back messages a day, with ten characters for the sender's name. People often use this space to convey short messages such as "I arrived" or "buy bread". Missed calls, i.e., making the phone ring and hanging up before the receiver picks up, is used to send a pre-set message, e.g., "I have arrived home safely" or to invite somebody who has airtime to call back.

Much has been written about phone sharing as a common practice in Africa in general, and South Africa is no exception (Donner 2008). Explanations range from a gregarious and communitarian orientation of

African culture to a response to material constraints. With reference to Ghana, often considered to be at the forefront of the mobile revolution on the continent, Sey (2009) notes that commercial forms of sharing are quickly disappearing. Research in a South African rural area (Dalvit, Kromberg and Miya 2014) notes an orientation to share one's device out of bare necessity and mainly for communication activities such as calls and SMS. Because of the costs—actual or perceived—of data usage phone sharing for networked activities is kept to a minimum. At the same time, instant messaging is used as a cost-saving mechanism to substitute for more expensive SMS.

Rurality and (Im)Mobility

In Africa, mobile phones play a key role in relation to people's movement, ranging from organizing local trips to keeping in contact with migrant workers. De Bruijn, Nyamnjoh and Brinkman (2009) note the importance of mobile phones for maintaining contact between people in rural areas and migrant workers. Besides maintaining family ties, constant communication sustains a system of remittances, vital for many African rural economies. As noted by Powell (2012), the ongoing process of urbanization on the continent brings more and more Africans within highly connected metropolitan areas. This, however, further marginalizes the majority of the population, which still lives in rural areas. The South African Institute of Race Relations identifies lack of services as the primary driver of the steady outflow of migrants from the rural and peri-urban areas of the Eastern Cape (along with the Limpopo province) to the metropolitan areas of richer provinces. Mobile phones play a crucial role in addressing the tyranny of distance of rural dwellers from opportunities, relatives and services.

Social networking is becoming increasingly popular among South Africans in low-income peri-urban and rural communities. The majority of the nine million Facebook users (Effective Measure 2014) are mobile users. Social networks are used by young people in rural areas to overcome their marginalization, connecting to people in other parts of the country, asking for advice on relationships and discussing political issues. Such conversations take place primarily in African languages, which makes them difficult to capture in research focusing on English (Miya 2014). On the flip side, social networking may entrench social control and pressure to conform in a small and closely knit rural community. Schoon (2012) notes how a hyperlocal social network was used in an impoverished community to gossip and badmouth young people trying to escape their condition.

The Siyakhula Living Lab

The Siyakhula Living Lab is a rural Information and Communication Technology for Development (ICT4D) project spearheaded by two South African universities. In this section I discuss the site, the project and its approach.

The Dwesa Site

The site of the Siyakhula Living lab is a rural area on the Wild Coast of the former homeland[2] of Transkei, in the Eastern Cape Province of South Africa. The area has great but largely untapped potential to attract cultural and eco-tourism. The Transkei is the heartland of the amaXhosa people and the home of former presidents Nelson Mandela and Thabo Mbeki. The area of the project hosts the grave of King Yenza, who fought against the British, and the pool of Nongqawuse, who prophesized that the ancestors would swipe away the settlers if the amaXhosa killed their cattle and destroyed their fields. The area hosts the Dwesa marine reserve, which represents the second successful land restitution claim in post-apartheid South Africa (Palmer, Timmermans and Fay 2002). Project participants from both inside and outside the community commonly refer to the research site as Dwesa because most activities take place in a number of villages adjacent to the Dwesa reserve within the Mbashe municipality.

Figure 13.1 Nqabarha river mouth.

Like many African rural realities, Mbashe is characterized by endemic poverty, lack of services and lack of infrastructure. Most of its 5,000 house-holds rely on subsistence farming, government grants or remittances by family members working in urban areas. Internal migration shaped the demographics of the area. While young men and, to a lesser extent, women migrate to the cities looking for a job, old people, young children and single mothers make up the majority of the local population. Commercial and government services are available in Katiane, the nearest town. The 40 km of gravel road separating it from Mbashe become impracticable after heavy rains, when the rivers overflow and wash away the bridges. Water is available from communal taps, rain-collection tanks and streams or rivers. Until a few years ago only government buildings such as schools and clinics had electricity, but since 2008 households in some villages are connected to the grid. In some areas, however, people still rely on solar panels and generators.

Radio and TV signals are comparatively poor. However, mobile network coverage by either MTN or Vodacom, the two main national operators, is almost universal. A baseline study conducted in 2010 (Pade-Khene, Palmer and Kavhai 2010) suggests that a considerable portion of disposable income is spent on airtime. Recent investigation suggests that the average could be as high as ZAR 160 (US $14) per household against a total income of less than ZAR 1000 (US $86).

Siyakhula—We Are Growing through ICT

Siyakhula means "we are growing" in isiXhosa, the local African language. The name was chosen by the community to signify the development brought about by access to ICT. The project was initiated in 2005 by the Telkom Centres of Excellence in the Departments of Computer Science at Rhodes University (a historically White Institution) and the University of Fort Hare (a historically Black institution). This type of collaboration, still relatively rare at the time, ensured a combination of technical and logistic capability with an understanding of operations in a marginalized context. The original goal was to explore the potential of ICT for socio-economic development. The project soon extended to other academic departments ranging from Communication to Education and African languages, and a wide range of research streams and service development ensued (Dalvit et al. 2007).

Five schools in as many villages were chosen as initial access sites. Schools were the only buildings with electricity and were natural centers for accessing knowledge in the community. Starting with teachers, a train-the-trainer model was adopted to ensure ICT skills trickled down to the community (Dalvit et al. 2006). Over the past eight years, a multidisciplinary group of young researchers from the two universities involved visited the area for one week every month. The project has a car, often used to give lifts to local people, and rents a house in one of the villages (called "the base"). University staff and students are very much part of the community, collecting water from the same communal taps, washing in a basin, using an outside toilet and negotiating the same rough terrain as everybody else. This ensures an understanding of the local context, high presence and frequent interactions. Community members respond with active participation and readiness to provide information on their appropriation of technology, which are key factors to ensure sustainability. This chapter draws on a substantial amount of research on the Siyakhula Living Lab (more than 200 publications).

Cannibalization or Synergy?

Our experience counters the belief that mobile phones replace the need for computers in rural areas. A longitudinal study (Gunzo and Dalvit 2014) explored the relationship between access and use of computers *vis-à-vis* mobile phones among school children in the area. The availability of computers generally correlates with an increase in mobile use. Different categories of activities, i.e., multimedia and networked communication, varied in different ways. For example, availability of computers seemed to increase the consumption of music on a mobile phone, suggesting the download and transfer of content from computers. A significant finding was the increase of networked activities. For example, the number of students using instant messaging on mobile phones doubled over a period of one year. Research by Kavhai, Osunkunle and Dalvit (2010) notes that computers are often used by people in Dwesa to keep in contact with relatives working in the cities via Facebook or Skype. Besides maintaining family ties, this connection is important to ask for remittances. Prior to the arrival of the project, the costs of communication made this prohibitively expensive. Cristoferi and Dalvit (2013) also observed that money is often transferred in the form of airtime, alongside more established practices such as mobile money and SMS notifications, which still require the recipient to travel to an ATM in the nearest town. One could say that networked computers and phones work synergistically to sustain the system of remittances, which is vital for the local economy.

The Red and the Blue

The encounter between tradition in an African rural setting and modern technology proved to be a fascinating area of research. With specific reference to the experience of the Siyakhula Living Lab, Thinyane et al. (2008) argue in favor of unrestricted access to the Internet in rural communities but emphasize the need for contextualized training and services. As an example, a Westerner readily associates a red button with a negation, e.g., "stop", "cancel" or "danger," as opposed to Green, which signifies "go ahead." However, in Dwesa red is the color associated with tradition while blue is associated with modernity.

Slay and Dalvit (2008) explore the red/blue dichotomy further by comparing the effects of a swine flu epidemic in Dwesa and the neighboring Cwebe, focusing on how digital literacy and Internet connectivity empowered the former to respond to the crisis. People from Cwebe, considered less open to innovation and modernity, were not able to put in a claim for government compensation as they did not have access to the relevant information. It would be a mistake to see tradition and technology as contrasting forces. An example of how an Internet-based platform can be used to leverage Indigenous knowledge, preserve it and improve its status can be found in Maema, Terzoli and Thinyane (2013). The authors set up an ontology-based system and populated it with Indigenous knowledge about local herbs, traditional practices, etc.

Figure 13.2 Traditional houses with MTN cellphone tower.

Deconstructing Difference

Dwesa is not a homogeneous community in terms of gender, socio-economic status, level of education or language proficiency. Empowerment of community members through ICT entails overcoming barriers to the outside world as well as within the community. In South Africa, English proficiency is often a requirement for accessing information and services but is not the main language spoken in Dwesa. Since its inception, the Siyakhula Living Lab has made use of software and training material in isiXhosa, which is spoken by almost everybody in the community. Moreover, the experience of receiving computer literacy training in both English and isiXhosa as part of a bilingual intervention proved very beneficial for a group of learners in one of the local schools (Sam 2010). Evidence suggests that this improved understanding of key concepts along with attitudes toward the isiXhosa language and learner's self-confidence in using computers. The ease of creating and sharing content (videos, comments, etc.) in local languages is probably a contributing factor to the uptake of mobile technology in areas such as Dwesa.

Besides epistemological access, through the use of appropriate languages, for instance, it is crucial to ensure physical access. Computer labs are located in schools, and teachers are put in charge of regulating access and training. This places teachers in a position of gatekeepers with respect to ICT, as noted by Mapi, Dalvit and Terzoli (2008). At the same time, these authors note that, countering usual age and gender stereotypes, relatively old women seemed to take the lead in promoting ICT use in the community. A tentative explanation is that mastering an empowering technology is a way for women to acquire status in a patriarchal society. The lesson learned is that disruptive technologies such as computers and mobile phones can either entrench or redefine existing power relationships.

An African Living Lab

The living lab methodology is an established approach to service co-creation within industry in the developed world. The Siyakhula Living Lab (SLL) represents an initial attempt at contextualizing such an approach in an African rural context (Gumbo et al. 2012). The key tenet is the development of contextualized solutions by involving actual users early in the development process. A good working relationship with the community is both the prerequisite and the outcome of shared understandings of goals and challenges. This makes it possible to access data, develop solutions and test services in real time and within real use scenarios.

The SLL seeks to empower the community by promoting the use of computers and mobile phones as tools to support existing activities in domains such as education, tourism and administration but also in traditional affairs and daily life. As noted by Pade-Khene et al. (2011) this implies a holistic approach to community development. Techniques and approaches that aim to promote the effectiveness and sustainability of the ICT project in a rural context relate to social-cultural, institutional, economic and technological dimensions.

The Siyakhula Living Lab initiative attracted positive attention by government and industry, resulting in additional ICT training for teachers and the inclusion of 11 more schools in the project (Dalvit, Siebörger and Thinyane 2012). The project provides free access to wireless Internet, which opens up the potential for the area to be connected to a range of services such as e-Government, e-Commerce, e-Health, e-Education, etc. As an example, a recent study (Cristoferi and Dalvit 2013) found that 8% of respondents in the area claim to do online shopping.

Members of the community are actively involved in shaping decisions concerning the project and in ensuring the security and maintenance of the equipment. Through the project, Dwesa was featured in a breakfast show, evening news and two documentaries broadcast nationally as well as in various national and international publications (*LED*, *De Kat* and *Die Zeit*, etc.). Community members are present alongside young researchers and university students. Seeing their stories featured in the media always provokes great enthusiasm and excitement among Dwesa community members. Consistent with the oral tradition of the area, storytelling seems to be one of the best ways to represent the experiences of Indigenous people. For this reason, six stories have been selected to provide an overview of how ICT, and particularly mobile phones, have empowered the community.

Mobile Stories

In this section I describe six stories relating to the use of ICT—particularly mobile phones—by community members. The stories are arranged to provide a history of the project as well as the different mobile-related activities in the area. The main characters are girls and women of all ages, reflecting the demographics of the area and the generally more positive attitudes noted above.

The Pigs' Story

> *ICT is particularly important for those who are marginalized in so many other ways.*
>
> *(Flor 2001)*

The first story is not specific to mobile but includes a range of technologies, including computers and cell phones, as well as spoken communication. It relates to how access to the Internet is particularly important for those who are marginalized in so many other ways. Shortly before the start of the project, there was an epidemic of swine fever in the Mbashe area. Government officials culled all domestic pigs and promised compensation. After the officials left, members of the local community tried in vain to find out how to claim payment for the lost animals.

When the first computer lab was set up and connected to the Internet and even before ICT training had started, Sisphiwe—one of the local teachers— asked members of the team to help. Within a few hours of being connected, she had found the relevant information on government websites, printed and filled in the appropriate forms and sent them to the correct address. Through word of mouth and phone communication, the news quickly spread through the community. Within a few weeks local farmers received compensation so people could buy new pigs.

This story exemplifies how access to information through the Internet can make all the difference. For members of the local community it would have been too costly and time consuming to travel back and forth to the government offices in the nearest town, negotiate with the clerk at the reception desk the terms of the process, obtain the forms and get the application going. Making this service cheaper and faster ultimately made the difference between a successful use of a service to which people were entitled and yet another unanswered complaint. Direct connection put the matter in the hands of the people concerned, fostering a sense of empowerment and agency. As this happened in early 2006, computers were the only way to connect to the Internet in the area. The current penetration of mobile (and increasingly) smart phones enables many more "pig stories" to emerge.

Figure 13.3 Multi-racial pig family.

... And There Was Light! At a Cost, of Course

Communication tools get socially interesting when they become technologically boring

(Shirky 2008)

In the first few years of the project, only schools and the local clinic were connected to the electricity grid. For a small fee (approximately ZAR 2, which is less than US $0.20), members of the community could charge their phones. Starting in 2008, households in the area were connected to electricity. Most use a prepaid system, i.e., vouchers are bought in town, and a pin code is punched into the prepaid meter box in one's home. Getting the voucher is quite expensive and time-consuming. The bus to town leaves at 5:30 am (an ungodly hour even in the rural areas) and comes back at 2:30, for approximately ZAR 15 (US $1.30) each way. Queues to withdraw money and buy electricity are always long, adding several hours of standing to the cost of bank charges (approximately ZAR 5 (US $0.50). As the average amount per voucher is ZAR 100 (US $9), the additional costs amount to approximately one-third of the cost plus a whole day on the road.

For many, a trip to town is an opportunity to socialize, do shopping and catch up with news, but for those who only need electricity, there is a cheaper option. Nokwaka, an old lady in the Mpume village, sells electricity from her mud-bricks home. For a small fee (ZAR 2, which is less than US $0.20) she collects payment, liaises with the outlet in town and gets a pin code (personal identification number) via SMS on her feature phone. Such a simple service using an established technology provides her with the opportunity to supplement her old-age pension and income as a subsistence farmer while saving the community time and money. Examples of micro-entrepreneurship such as this are particularly significant in an area characterized by economic dependency on the outside world, mainly in the form of grants and remittances.

Finding Mom

> Umntu ngumtu ngabantu – *a person is a person because of other people. (African proverb)*

Family segregation is a sad legacy of the extensive use of migrant work combined with the limitations on the mobility of Africans imposed by apartheid. Nqobile is a young woman born in the Thembisa township of Johannesburg. At the age of four, her parents separated and her father took her back to Dwesa, where she grew up away from her mother. In 2006, Nqobile became a single mother and paid a first visit to her own mother—who had remarried as a second wife into a polygamous family—and her half-sister, whom she had never met. This visit ended with a quarrel with the mother's family, and Nqobile left with only a picture and a neighbor's phone number.

In 2011, once happily married and with a second child, Nqobile tried to re-establish a connection, but the phone number she had was deactivated. After various failed attempts and mistaken identities she tracked down one of her maternal cousins on Facebook and, through a series of negotiations and connections with relatives, obtained her mother's mobile phone number. This enabled Nqobile to get to know about her grandmother's passing and to organize a visit by her half-sister to meet her new family.

Although the relationship with her mother is not yet mended, they keep in regular contact through the half-sister. In this case, the pervasiveness of social networks, accessed by the vast majority of Africans exclusively from their mobile phone, contributed to mending an age-long rift within Nqobile's family. It should be noted that younger members of the family, who are more proficient with technology, re-established and maintain the relationship.

Nqobile also happened to be one of the first champions of the project in the community. Her story as a rural girl from a harsh background but with a passion for computers was featured in a documentary broadcast on national television. Incidentally, she met her husband through the project, and it is probably fair to say theirs was the first relationship in Dwesa to start via instant messaging.

The Chief's Wife

> *The man is the head of the family, but the woman turns the neck.*
> *(South African proverb)*

Many aspects of life in Dwesa are still governed by customary law, and traditional authorities play an important role. The chief of the area is aided by village headmen, called Sibonda after the poles holding up a fence, in such matters as land allocation and administration of justice. In one of the villages, both the Sibonda and his deputy are notoriously difficult to track down. They are both relatively old and uneducated men and do not have a phone. When their services are needed it is customary to just stand outside their home and call them out loud (this is known as the "Dwesa intercom") or send a child to their field or to the local tavern, depending on the time of the day.

When a complex matter requires distance communication, the only option is to call the Sibonda's wife. She is always reachable and always willing to help, from five o'clock in the morning to late in the evening. She acts as a proxy and provides advice on how to approach an issue, when to call and what to say. She actively participates in phone discussions with her husband, providing explanations, corrections and additional information. Although as a woman she is not supposed to interfere in traditional affairs, her control of the mobile phone grants her considerable covert influence and insider's knowledge. In a traditional domain within a patriarchal society, the mobile phone carved a powerful role for this otherwise soft-spoken and humble old lady.

Sharing is Caring, but Keep Your Hands off My Phone!

> *Economics should be about the equitable distribution of wealth, not the equal sharing of misery.*
>
> *(Karl Marx)*

Much has been written about the sharing of technology, particularly mobile phones, in Africa. Whether because of cultural reasons or material constraints, tools and resources are typically shared among family members, neighbors and friends. Yet, not everything is shared and not everything is sharable, even in Dwesa. Vuyolwethu is the last born in a large single-parent household. She received her current phone as a hand-me-down from an older sister and gets small amounts of airtime as a gift from her boyfriend or from older relatives working in Cape Town.

Like most people in the area, she has many different phone numbers, one for each network operator, to take advantage of promotions and avoid charges when calling somebody on a different network. She uses missed calls and call-back messages to invite people to contact her if they are likely to have airtime or just to let them know she is thinking about them if they do not. SMS are considered expensive and she uses them sparingly. She does not have a dream phone, but would like to upgrade to a model supporting the WhatsApp instant messenger in order to save money. She is not aware this entails the use of the Internet so, when asked about features, she is mostly interested in multimedia ones.

For Vuyolwethu, her phone is her radio, CD player and home theater all in one. When she walks around the village, alone or with her friends, the music from her phone is shared with everybody on the street. Her SD card stores more than 500 songs, pictures and short videos. Content sharing is so culturally entrenched that, when asked about copyright infringement, she seems to be blissfully unaware that such a thing exists. When her young niece asks to borrow her phone for a while, Vuyolwethu smiles but denies firmly. Besides the possibility of some adult content being among the videos and pictures, there is clearly something else. She does not mind sleeping in the same bed with two of her nieces and her own small baby, washing in the same room with her older sisters or exchanging clothes with her friends, but the phone is something else. Her phone is probably Vuyolwethu's only truly private possession.

Mama ka Twitter

> The beginning of wisdom is to call things by their proper name. (Confucius)

In his novel "The Heart of Redness", which coincidentally is set in the Dwesa area, South African writer Zakes Mda notes the habit among local people to name their children after artifacts they find interesting. Machintosh is a name as good as any for a boy and Nopettycoat is an acceptable female name. Women with children are respectfully addressed as "mother of" followed by the name of their first born. Mama ka Twitter is a lady in her forties, never married but with a one-year old boy she nicknamed Twitter. When asked why such a name, her first response is a laugh and a reference to the sociable nature of the baby. After some more probing, she adds somewhat cryptically that the baby was brought by Twitter.

Mama ka Twitter (Twitter's mother) does not like to talk about her own business and is very careful about distancing herself from gossip. In a small rural community, just like in the global networked society, reputation is very important. During the interview, she eluded questions on how she got her phone with a laugh and only later admitted she got it from Twitter's paternal grandmother, so she could speak with the child. The security and convenience of having a phone got the best of her, but clearly Mama ka Twitter felt some level of discomfort at being always traceable, especially by the family of the father of her child. Several contradictions indicated she was not willing to share all details of her phone use.

Although the tone remained amicable and relaxed, the interview ended with more questions than answers. This could be interpreted as evidence of mixed feelings toward the interview context but also as indicative of an ambivalent attitude toward mobile technology. On the one hand a fascination that led to picking a technology-inspired name for her child. On the other, a reticence to discuss one's technology use, a kind of reverent detachment. This speaks to the dichotomy between tradition and modernity, resistance and adaptation, trust and suspicion. As Zakes Mda points out in the aforementioned novel, these are deeply intertwined dimensions in the history of South Africa's Indigenous people and inseparable sides of every African soul. This will probably include, in the future, little Twitters.

Conclusions

In this chapter, I discussed the rise of the mobile phenomenon in South Africa, the efforts by government to improve access and challenges such as urbanization and cost. I focused on the experience of the Siyakhula Living Lab in exploring the potential of ICT to empower the community of a small rural area. As part of a holistic approach, mobile phones can empower Indigenous people by responding to the local context.

As many other rural areas in Africa, Dwesa is characterized by lack of services and endemic poverty. People in the area do not have access to relevant information (e.g., on how to enforce their rights) and have to pay more than their urban counterparts for the same service (e.g., buying electricity). The experience of "The Pigs' Story" and "... And There Was Light, at a Cost of Course!" suggest that mobile phones can contribute to accessing funds and avoiding expenditures. While much attention has been dedicated to the provision of new services through e-Government and to income generation, a more efficient use of resources and the ability to cut costs are equally important for low-income rural households.

Technological innovation does not conflict with tradition. The stories "Finding Mom" and "The Chief's Wife" provide examples of how mobile phones can support institutions such as extended families and local authorities. At the same time, age and gender roles within these institutions are challenged and renegotiated through control of the technology. In the former story, Nqobile and her half-sister are maintaining the contact between the two families, a role traditionally performed by the elders. In the latter story, the Sibonda's wife plays an active role in traditionally male affairs. These anecdotes may provide some insight into how mobile phones are changing African rural realities "from within."

People in African rural villages share similar concerns about privacy and ownership of mobile phones as their counterparts in the First World. Dwesa is a closely knit community where life is gregarious and resources are shared with family, neighbors and friends. However, the stories "Sharing is Caring, but Keep Your Hands off My Phone" and "Mama ka Twitter" highlight the complex role of the mobile phone in managing private/public boundaries. On the one hand, connection enables sharing of content and access to help in case of emergency. On the other hand, one must be able to disconnect from others in order to retain control of an important cultural artifact and protect his or her privacy.

The stories presented in this chapter focused explicitly on empowerment. However, one must consider that the pervasiveness of mobile phones has also negative sides, such as putting additional strain on struggling family budgets and potentially increasing the gap between ICT-savvy gatekeepers and those who remain part of the "Fourth World". The discussion ends around the empowering potential of mobile technology for members of marginalized communities and begs a "how" rather than an "if" question, shifting the focus from providing access to understanding use. The account presented in this chapter is by no means exhaustive but provides an overview of the possible role of mobile phones with respect to economic, institutional and socio-cultural empowerment in a small rural area in South Africa.

Notes

1. In Apartheid South Africa, a township was an area of a city or town designated for individuals classified as non-Whites. Townships are still mainly populated by low-income Blacks.
2. Under Apartheid, homelands were partially self-governing areas where individuals belonging to a particular Indigenous group could take residence and nationality.

References

Abrahams, Lucienne, and Arthur Goldstuck. "A Decade of e-Development in South Africa: Sufficient for a Services (R)evolution?" In *National Strategies to Harness Information Technology*, edited by Nagy K. Hanna and Peter T. Knight, 107–152. New York: Springer, 2012.

Castells, Manuel. "The Rise of the Fourth World." In *The Global Transformations Reader* edited by David Held and Anthony McGrew, 348–354. Cambridge: Polity Press, 2000.

Cristoferi, Michele, and Lorenzo Dalvit. "Money-Related Uses of Mobile Phones in a South African Rural Area." Paper presented at the Conference on Mobile Telephony in the Developing World, Jyvaskyla, Finland, May 24–25, 2013.

Dalvit, Lorenzo. "Why Care about Sharing? Shared Phones and Shared Networks in Rural Areas: African Trends." *Rhodes Journalism Review* 34 (2014): 81–84.

Dalvit, Lorenzo, Robert Alfonsi, N. Isabirye, S. Murray, Alfredo Terzoli and Mamello Thinyane. "A Case Study on the Teaching of Computer Training in a Rural Area in South Africa." In *Proceedings of the 22nd Comparative Education Society of Europe Conference,* Granada, Spain, 2006.

Dalvit, Lorenzo, Steve Kromberg and Mfundiso Miya. "The Data Divide in a South African Rural Community: A Survey of Mobile Phone Use in Keiskammahoek." Paper presented at 3rd National South African e-Skills Summit, Cape Town, South Africa, November 17–21, 2014.

Dalvit, Lorenzo, Hyppolite Muyingi, Alfredo Terzoli and Mamello Thinyane. "The Deployment of an e-Commerce Platform and Related Projects in a Rural Area in South Africa." *Strengthening the Role of ICT in Development* (2007): 27.

Dalvit, Lorenzo, Ingrid Siebörger and Hannah Thinyane. "The Expansion of the Siyakhula Living Lab: A Holistic Perspective." In *e-Infrastructure and e-Services for Developing Countries*, Berlin and Heidelberg, 2012, 228–238. New York: Springer, 2012.

Dalvit, Lorenzo, and Larry Strelitz. "Media and Mobile Phones in a South African Rural Area: A Baseline Study." In *Proceedings of the Emerging Issues in Communication Policy and Research conference—Refereed Papers*, Canberra, 70–80. Canberra: News & Media Research Centre, University of Canberra, 2013.

de Bruijn, Mirjam, Francis Nyamnjoh and Inge Brinkman (eds.). *Mobile Phones: The New Talking Drums of Everyday Africa.* Bamenda/Leiden: Langaa Publishers/ African Studies Centre, 2009.

Donner, Jonathan. "Shrinking Fourth World? Mobiles, Development, and Inclusion." *Handbook of Mobile Communication Studies*, edited by James E. Katz, 29–42. Cambridge: MIT Press, 2008.

Effective Measure. "South African Mobile Report: A Survey on Desktop Users' Attitudes and Uses of Mobile Phones." Accessed October 17, 2014. www.sabc.co.za/ wps/ … /South_Africa_Mobile_Report-Mar14.pdf.

Flor, Alexandra G. "ICT and Poverty: The Indisputable Link." Paper presented at the Third Asia Development Forum on Regional Economic Cooperation in Asia and the Pacific, Bangkok, June 11–14, 2001.

Goldstuck, Arthur. "Internet Access in South Africa 2010: A Comprehensive Study of the Internet Access Market in South Africa." Johannesburg, South Africa: World Wide Worx, 2010b.

Goldstuck, Arthur. "South Africa and the Content Revolution: From the Cable Connection to the Mobile Future." 2010a. Accessed May 2014. https://ujdigispace. uj.ac.za/handle/10210/3310.

Goldstuck, Arthur. "The Many Needs for Internet Speed." 2013. Accessed May 2014. http://www.gadget.co.za/pebble.asp?relid=6826.

Gumbo, Sibukele, Hannah Thinyane, Mamello Thinyane, Alfredo Terzoli and Susan Hansen. "Living Lab Methodology as an Approach to Innovation in ICT4D: The Siyakhula Living Lab Experience." In *Proceedings of the IST-Africa 2012 Conference,* Dar es Salaam, Tanzania, 2012.

Gunzo, Fortunate, and Lorenzo Dalvit. "A Survey of Cell Phone and Computer Access and Use in Marginalized Schools in South Africa." In *Proceedings of M4D2012, New Delhi, India* 2012, 232–243. Karlstad: Faculty of Economic Sciences, Communication and IT, Karlstad University, 2012.

Gunzo, Fortunate, and Lorenzo Dalvit. One Year on: A Longitudinal Case Study of Computer and Mobile Phone Use among Rural South African Youth. Paper presented at *International Development Informatics Association 2014 conference*, Port Elizabeth, South Africa, November 3–4, 2014.

InfoDev. "Mobile Usage at the Base of the Pyramid in South Africa." 2012. Accessed May 2014. http://www.infodev.org/infodev-files/final_south_africa_bop_study_ web.pdf.

Jones, Nick. "Gartner Symposium on Innovation. Cape Town International Convention Centre" In Jeremy Daniels "Smart Phones to Rule by 2014, Predicts Gartner." 2010. Accessed June 2014. http://memeburn.com/2010/09/smartphones- to-rule-by-2014-predicts-gartner/.

Kavhai, Mitchell, Oluyinka Osunkunle and Lorenzo Dalvit. "The Impact of Rural ICT Projects in South Africa–A Case Study of Dwesa, Transkei, Eastern Cape South Africa." Masters thesis, University of Fort Hare, 2010.

Lana. "South Africans and their Cell Phones." 2014. Accessed October 17, 2014. http://www.southafricaweb.co.za/article/south-africans-and-their-cell-phones.

Maema, Mathe, Alfredo Terzoli and Mamello Thinyane. "A Look into Classification: Towards Building an Indigenous Knowledge Platform for Educational Use." In *IST-Africa Conference and Exhibition (IST-Africa)*, Nairobi, 1–8. IEEE, 2013.

Mapi, Thandeka Priscilla, Lorenzo Dalvit and Alfredo Terzoli. "Adoption of ICTs in a Marginalised Area of South Africa." *Africa Media Review* 16, no. 2 (2008): 71–86.

Miya, Mfundiso. 2014. "Languages Used on Social Media and Instant Messaging Platforms: The Case of Keiskammahoek." Masters paper, Rhodes University.

My Broadband. "South Africa Connect: The New Broadband Policy." 2013. Accessed October 19. http://mybroadband.co.za/news/government/93243-south-africa-connecHYPERLINK "http://mybroadband.co.za/news/government/93243-south-africa-connect-the-new-broadband-policy.html" t-the-new-broadband-policy. html.

National Planning Commission. "The National Development Plan South Africa (2011)." 2011. Accessed October 17, 2014. www.npconline.co.za/medialib/downloads/home/NPC%20National%20Development%20Plan%20Vision%20HYPERLINK "http://www.npconline.co.za/medialib/downloads/home/NPC National Development Plan Vision 2030 -lo-res.pdf"2030%20-lo-res.pdf.

Pade-Khene, Caroline, Robin Palmer and Mitchell Kavhai. "A Baseline Study of a Dwesa Rural Community for the Siyakhula Information and Communication Technology for Development Project: Understanding the Reality on the Ground." *Information Development* 26, no. 4 (2010): 265–288.

Pade-Khene, Caroline, Ingrid Siebörger, Hannah Thinyane and Lorenzo Dalvit. "The Siyakhula Living Lab: A Holistic Approach to Rural Development through ICT in Rural South Africa." In *ICTs for Global Development and Sustainability: Practice and Applications*, edited by Jacques Steyn, Jean-Paul Van Belle and Eduardo Villanueva Mansilla, 42–77. Hershey, PA: IGI Global, 2011.

Palmer, Robin C. G., Herman Timmermans and Derick Fay. *From Conflict to Negotiation: Nature-Based Development on the South African Wild Coast.* Cape Town: HSRC, 2002.

Powell III, Adam Clayton. "Bigger Cities, Smaller Screens." 2012. Accessed October 17, 2014. http://cima.ned.org/publications/bigger-cities-smaller-screens-urbanization-mobile-phones-and-diHYPERLINK "http://cima.ned.org/publications/bigger-cities-smaller-screens-urbanization-mobile-phones-and-digital-media-trends-afric"gital-media-trends-afric.

Roux, Kobus, Alfredo Terzoli and Marlon Parker. "Rural Innovation Systems in South Africa: Living Labs." In *Enhancing Innovation in South Africa: The COFISA Experience*, edited by Tina James, 74–85. Pretoria: COFISA, 2010.

Sam, Msindisi Scara. "The Development and Implementation of Computer Literacy Terminology in isiXhosa." Master's thesis, Rhodes University, 2010.

Schoon, Alette. "Dragging Young People down the Drain: The Mobile Phone, Gossip Mobile Website Outoilet and the Creation of a Mobile Ghetto." *Critical Arts* 26, no. 5 (2012): 690–706. doi:10.1080/02560046.2012.744723.

Sey, Araba. "Exploring Mobile Phone-Sharing Practices in Ghana." In *info* 11, no. 2 (2009): 66–78.

Shirky, Clay. *Here Comes Everybody: The Power of Organization without Organizations*. New York: The Penguin Press, 2008.

Slay, Hannah, and Lorenzo Dalvit. "Red or Blue? The Importance of Digital Literacy in African Rural Communities." In *Proceedings of the IEEE 2008 International Conference on Computer Science and Software* Engineering, Wuhan, Hubei, 2008, 675–678.

Statistics South Africa. "Poverty Trends in South Africa: An Examination of Absolute Poverty between 2006 and 2011." 2014. Accessed October 17, 2014. http://www.polity.org.za/article/poverty-trends-in-south-africa-an-examination-of-absolute-poverty-between-2006-and-2011-april-2014–2014–04–03.

Thinyane, Mamello, Lorenzo Dalvit, Alfredo Terzoli, and Adam Clayton Powell III. "The Internet in Rural Communities: Unrestricted and Contextualized." *ICT Africa* 13 (2008): 15.

14 Socio-Economic Impacts on the Adoption of Mobile Phones by the Major Indigenous Nationalities of Nepal

Sojen Pradhan and Gyanendra Bajracharya

Introduction

Nepal is a landlocked South Asian country, sandwiched between the two biggest countries of the world, China and India. It has a population of 26.4 million with an area of 147,705 square kilometers. Geographically, Nepal occupies just 0.1% of the earth's surface but is one of the richest countries in the world in terms of bio-diversity because of its unique geographical position and altitudinal variation. The elevation of the country ranges from less than 100 meters above sea level in the Terai, which borders with India, to the highest point on earth, the summit of Mt. Everest at 8,848 meters, all within a distance of about 150 km.

Indigenous people in Nepal predominately inhabit rural areas and are primarily engaged in subsistence agriculture. Traditional occupations continue to be practiced by many Indigenous nationalities in Nepal. Despite the overall predominance of agriculture, there are also huge differences among the 58 Indigenous groups in terms of the varied ways they gain their subsistence in order to survive and the occupations they enter. These range from urban to rural occupations, from hunter-gatherers to merchants and professionals.

This study attempts to understand the relationship between the adoption of mobile phones by the major Indigenous groups and a number of other variables. Population census data is used as the primary source for linking mobile phone ownership with socio-economic variables, literacy, places of residence, the gender of the heads of household (HHs) and occupation. Following the statistical analysis, the results of seven interviews with Indigenous people are presented in order to understand how mobile phones are contributing to Indigenous people's lives in Nepal, in particular the conduct of their occupation. The findings further our understanding of the potential social and economic benefit of mobile phones for developing countries.

Indigenous People in Nepal

The term Indigenous generally indicates people who are the "original" inhabitants of a given territory, i.e., people who were there before the currently dominant ethnic group arrived or before the establishment of state

224 *Sojen Pradhan and Gyanendra Bajracharya*

borders (United Nations 2004). It also refers to people who have been living independently and/or remotely from the influence of the claimed governance of the nation-state, such as those in isolated rainforests. An additional criterion is that such people have maintained, at least in part, their distinct linguistic, cultural and social/characteristics and are differentiated in some degree from the surrounding populations and dominant culture of the nation-state. Another essential factor is that they self-identify as Indigenous, and/or are recognized as such by other groups.

In Nepal they are known as Indigenous nationalities, or in the Nepali language "Adivasi Janajati." Two decades ago in March 1994, the term Adivasi Janajati was defined by the national consultation on Indigenous Peoples of Nepal (Bhattachan and Webster 2005). The term describes those communities that:

- Possess their own distinct and original linguistic and cultural traditions and whose religious faith is based on ancient animism, being worshippers of the ancestors, land, seasons and nature;
- Are existing descendants of the people whose ancestors had established themselves as the first settlers or principal inhabitants within the territory of the modern state of Nepal;
- Have been displaced from their own land for the past four centuries, particularly during the expansion and establishment of the modern Hindu Nation State, and have been deprived of their traditional rights to own the natural resources including *Kipat* or communal land, cultivable land, water, minerals, trading points, etc.;
- Have been subjugated in the State's political decision-making processes and whose ancient culture, language and religion are non-dominant and whose social values have been neglected and humiliated;
- Have a society that is traditionally built on the principle of egalitarianism rather than the hierarchy of the Indo-Aryan caste system and enjoy gender equality with women holding a more advantageous position;
- Formally or informally admit or claim to be "the Indigenous peoples of Nepal" on the basis of the above characteristics.

According to the Population and Housing Census 2011, Indigenous people comprise 35.8% of the total population of Nepal, or around 9.5 million. Nepal has identified and recognizes 58 nationalities through the enactment of the National Foundation for Development of Indigenous Nationalities Act, 2002. According to the Act, "Indigenous refers to those ethnic groups or communities who have their own mother tongue and traditional customs, different cultural identity, different social structure and written or oral history". The four major Indigenous groups in terms of population size are Magar, Tharu, Tamang and Newar, with populations of 1.9, 1.7, 1.5 and 1.3 million respectively. For this study, the population of Nepal has been divided into three broad categories (See Table 1).

Table 14.1 Population of Indigenous and other groups in Nepal. (Adapted from: Population and Housing Census, 2011)

Category	Number	Percent
Major Four Indigenous groups (Magar, Tharu, Tamang, Newar)	6,486,966	24.5
Smaller Groups of Indigenous People (SGIP)	3,000,676	11.3
Non-Indigenous people	17,006,862	64.2
Total	26,494,504	100

The first category consists of the four major Indigenous groups that constitute more than 24.5% of the total population and 68% of the total Indigenous population. The second category consists of Indigenous groups excluding the four major Indigenous groups: this category consists of 11.3% of the total population and 32% of the total Indigenous population, in this chapter referred to as smaller groups of Indigenous people (SGIP). The third category of people consists of those who are not Adivashi Janajati, referred as "Non-Indigenous people," and these constitute 64.2% of the total population (see Figure 14.1).

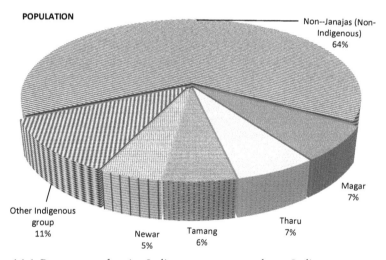

Figure 14.1 Percentage of major Indigenous groups and non-Indigenous nationalities. (Adapted from: Population and Housing Census, 2011).

Methodology

The Central Bureau of Statistics in Nepal (CBS 2014) conducted the Population and Housing Census 2011, which collected information on access to mobile phones at the household level. Publicly available census data, also known as "Public use micro sample" (PUMS) data, was analyzed by the authors to show how mobile phones were adopted by Indigenous households. The information on ethnicity (or caste) was, however, available at the individual level. In order to make the household and individual data consistent for merging, the individual data was confined to the head of household. Place of residence, literacy and occupation are the other major variables used to study the determinants of access to mobile phones. Two-way cross tabulations were prepared to see the relationship between the variables and the access to mobile phones in the household.

In addition, seven interviews were conducted by one of the authors. The interviewees were all Indigenous and were selected to represent a cross-section of occupations, urban and rural localities, and both men and women. Interviews were informal and responses converted to a narrative style.

Current Status and Trends of Using Mobile Phones in Nepal

The International Telecommunication Union (ITU) has predicted that 78% of the world's mobile phone subscribers will soon be coming from the developing countries (ITU 2014). According to the latest data available (2013) through the ITU website, mobile-cellular subscriptions are up to 71.5% in Nepal (ITU 2014). The Management Information Systems (MIS) report from the telecommunication regulator, Nepal Telecom Authorities (dated July 16, 2014), however stated that the penetration rate of mobile phones in the country was 83.2% with mobile phone use rapidly increasing (Figure 14.2 below).

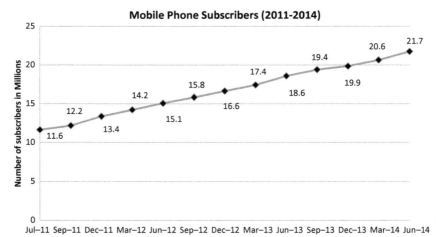

Figure 14.2 Number of mobile phone subscribers in Nepal 2011–2014. (Adapted from: Nepal Telecom Authorities, MIS Report 2014).

From the Population and Housing Census 2011, the total number of households in Nepal having access to mobile phones was 64.6%, being 84% of urban households and 60% of rural households. The number of mobile phones has now almost doubled in the three years from 2011 to 2014 (Figure 14.2). This increasing trend of mobile phone subscriptions in Nepal is due to various socio-economic factors. First, the competitive market place has made mobile phones an affordable commodity rather than a luxury item for only the most affluent. Second, mobile phones are the most reliable, fastest and most convenient means of connecting people. This ease of connectivity has enhanced opportunities for trade, business and commerce. Third, mobile communication technology is more suitable in Nepal because of the availability of wireless infrastructure in complex geographical locations, particularly in the mountainous areas of the country. In many rural areas only wireless media for mobile phone connections (2G or 3G) is available for communication.

Socio-Economic Status of Indigenous People

Socio-economic status (SES) is an overall measurement that characterizes the economic and social position of an individual or family within society. This is often indicated by occupation, level of education, social class, income, wealth and tangible possessions such as houses, cars and jewelry. Table 14.2 presents social indicators such as literacy, gender or sex ratio (SR) and percentage of household heads (HHs) with mobile phones.

Literary Status and Access to Mobile Phones

There is a positive relationship between the literacy rate and households accessing mobile phones in Nepal (Table 14.2). At the national level, both the literacy rate and the percentage of households having access to a mobile phone stand at around 65%. The difference between the literacy status in urban and rural households is approximately 20%. Among the major Indigenous groups, the literacy rate was observed to be the highest among the Newars with 80.1% and lowest among the Tamangs at 62.6%. The proportion of households having access to mobile phones was also correlated with the literacy rate, with the Newars at 82.9% while the Tamangs have the lowest access rate at 60%. The Magars and the Tharus are ranked between the Newars and the Tamangs in terms of literacy rate and the proportion of households having access to mobile phones.

Anecdotal evidence from Nepal suggests that some of the contributing factors for the positive relationship between literacy and mobile phone access are:

• Illiterate people are either unemployed or employed in poorly paying occupations. They are too poor to afford to purchase a mobile phone and then pay for its usage;

Table 14.2 Some social indicators of the four major Indigenous groups in Nepal, 2011. (Analyzed from 2011 Population Census data)

Place/Caste		Literacy	SR	Household Head		% of HHs in Urban	% HHs holding mobile
				Male (%)	Female (%)		
Nepal		65.9	94.2	74.3	25.7	19.3	64.6
	Urban	82.2	104	71.9	28.1		83.9
	Rural	62.5	92.3	74.8	25.2		60.0
Magar		71.1	86.3	65.5	34.5	14.5	64.4
	Urban	83.8	96.5	58.2	41.8		84.6
	Rural	69.2	84.9	66.8	33.2		60.9
Tharu		64.4	96.4	84.5	15.5	11.8	66.3
	Urban	72.2	111.7	82.5	17.5		76.0
	Rural	63.5	94.8	84.8	15.2		65.0
Tamang		62.6	93.7	72.7	27.3	15.1	60.0
	Urban	78.1	101.7	64.3	35.7		83.7
	Rural	60.2	92.5	74.2	25.8		55.8
Newar		80.1	94.5	74.9	25.1	48.5	82.9
	Urban	86.1	97.8	75.0	25.0		89.8
	Rural	74.3	91.5	74.8	25.2		76.4
Other Indigenous Groups		67.7	90.5	69.9	30.1	16.7	61.7
	Urban	81.1	92.5	60.6	39.4		83.2
	Rural	65.2	90.1	71.8	28.2		57.4
Non-Indigenous Groups		64.4	95.5	75.3	24.7	19.0	63.9
	Urban	82.3	107.8	74.3	25.7		83.2
	Rural	60.7	93.3	75.5	24.5		59.3

- Literate people are more eager to adopt the new technology because of their relatively greater exposure to mass media, provided that the adoption is affordable and accessible;
- Literate people are more likely to join paid employment, trade and business sectors of the community that demand the use of mobile phones and the Internet to provide prompt communication, improve access to information and expand business opportunities.

Head of Household and Use of Mobile Phones

The head of household (HHs), according to the census manual, is the key member of the family aged 10 years and above, irrespective of gender, who performs most of the management of the household. The head of household is the decision maker in the family.

The head of household in Nepal is overwhelmingly dominated by males (75%). Tharus have the lowest proportion (15.5%) of females as head of household, while the Magars have the highest proportion (34.5%) (Table 14.2). Both groups have around 65% of household heads with mobile phones. Thus there appears to be no relationship between the gender of the head of household and mobile phone ownership.

Place of Residence and Mobile Phone Adoption

In order to understand the differences in the use of the mobile phones among the Indigenous groups it is useful to look at social, cultural and anthropological aspects. From all recorded history, the Newars are the major inhabitants of the Kathmandu Valley in Nepal (Shrestha 1999). The Kathmandu Valley is the largest economic hub and the most developed part of Nepal. As a result, trade and business provide the major occupations for this Indigenous group. The proportion of the Newars residing in the Kathmandu urban area is approximately half of their population (48.5%), the highest in comparison to any other ethnic group in Nepal. This helps to explain their higher adoption of mobile phones as the latter are of immense use in business (Table 14.2).

Tamangs are the third largest Indigenous group in terms of population size. Most of the Tamangs live in the surrounding districts of the Kathmandu Valley with agriculture being their major occupation. Their literacy rate is also the lowest (62.6%) among the four major Indigenous groups. The proportion of the Tamangs households living in urban areas is 15.1%, and they have the lowest level of mobile phone ownership per household (Table 14.2).

Use of Mobile Phones by Major Occupations of Indigenous Groups

Occupation is one of the most reliable variables to indicate SES status, as it also shows a correlation with other variables (Hauser and Warren 1996), including quality of life (Westover 2009). Since people spend many hours of the day at work, it is one of the most dominant aspects of their lives. Thus, the nature of work impacts on people not only by generating income but also by allowing them to become part of social networks at work.

The International Labor Office (ILO) is responsible for classifying occupations worldwide by adopting the international standard classification of occupations (ISCO). The Population and Housing Census 2011 in Nepal followed these ISCO-08 classifications. Using these criteria the census collected data for 10 major occupations as shown in Table 14.3. Each category of occupation contains more than 200 sub-categories. The population census 2011 also revealed the number of people per ethnic group in each of those 200 occupations.

Table 14.3 Major classification of occupations (ISCO-08). (Adapted from: United Nations, System of National Accounting, 2008)

Type	Occupation Group
0	Armed forces occupations
1	Managers
2	Professionals
3	Technician and associate professionals
4	Clerical support workers
5	Service and sales workers
6	Skilled agriculture, forestry and fishery workers
7	Craft and related trades workers
8	Plant and machine operators and assemblers
9	Elementary occupations

Mobile phone adoption by the five groups of Indigenous people (four major groups and SGIP) was analyzed and compared with the non-Indigenous people (non-Janajatis) of Nepal based on their occupational group (Figure 14.3).

Overall the percentage of mobile phone usage by Indigenous people in Nepal is similar to the usage by non-Indigenous people. Surprisingly, most of the Indigenous groups have *slightly* greater access to mobile phones.

Figure 14.4 presents the mobile phone usage associated with the different Indigenous groups. Recently, businessmen have embraced mobile phones to

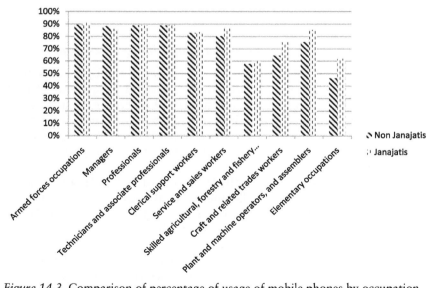

Figure 14.3 Comparison of percentage of usage of mobile phones by occupation groups between Indigenous and non-Indigenous groups in Nepal. (Adapted from: CBS, Population and Housing Census, 2011).

enhance their business opportunities. Tradesmen also are following this trend, allowing their productivity to increase. Many factors have made self-employed or small business owners in Indigenous groups adopt the use of mobile phones.

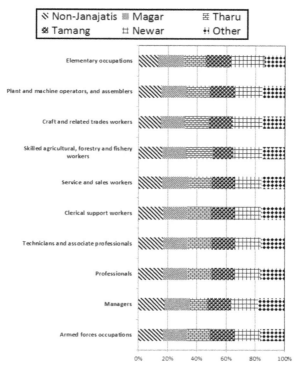

Figure 14.4 Comparison of percentage of usage of mobile phones by occupation groups within Indigenous Groups in Nepal. (Adapted from: CBS, Population and Housing Census, 2011).

Interviews with Indigenous People Regarding their Use of Mobile Phones

A number of interviews were conducted with Indigenous people in Nepal to gain a fuller understanding of the contribution of mobile phones to their lives. An important focus was on how mobile phones provided a benefit to their occupations and businesses.

Two trade workers from the Newar community (occupations of carpenter and metal worker) were interviewed in Lalitpur, the second city in the Kathmandu Valley (Figures 14.5 and 14.6).

Both of them expressed similar benefits of using mobile phones in their occupations. In summary, they were:

- Time savings and easy access—a massive amount of time is saved because there is no need to travel to suppliers and customers as often as before. Mobile phones made it very easy to be in regular contact;

Figure 14.5 Carpenter: Mr Hemdas Silpakar, Patan Wooden Handicraft. (Photo: Siddhi Ratna Shakya).

Figure 14.6 Metal worker: Siddhi Ratna Shakya, Siddhi Metal Handicraft. (Photo: Siddhi Ratna Shakya).

- Fast response to price changes or market condition—when there is a change in market conditions, price movement of raw materials or change in demand or supply, timely information is available through partners and suppliers via mobile phones;
- New customers acquired—mobile phones have added new customers when they made inquiries about products available;
- Mobile phone availability translated to increased sales;
- When workloads increased and additional labor was required, direct phone contact enabled casual workers to be employed quickly.

They were also asked about the negative factors when relying on mobile phones for business. Most mentioned the difficulty of keeping phones charged because of load shedding, also known as rolling blackout. Load shedding is the rotational electrical power shutdown for certain periods of time over different parts of the electricity distribution region and is very common in Nepal due to a lack of sources of electricity.

A local female vegetable grocer from the Tamang Indigenous group, Phool Maya Tamang, aged 37 of Malekhu, in the district of Dhading, 67 kilometers from Kathmandu, was contacted. She has been running the vegetable grocery shop for several years. Three years ago she started using a mobile phone and found it highly beneficial for her business. Now she uses the phone to contact her suppliers and farmers instantly and access up-to-date market information about the shortage and availability of necessities sold in her shop. Sometimes, she is able to contact her son or some friends who collect supplies for her, which allows her more time with her customers. She stated that her business has grown substantially with the mobile phone being the main contributing factor for her success. As an additional benefit, her mobile phone keeps her in regular contact with family and friends.

Ramila Pradhan, a local female high school teacher at Tri-Padma Vidhyashram Higher Secondary School, Lalitpur, was contacted and asked the benefits of owning a mobile phone (Figure 14.7). She indicated that it has changed the quality of her life. Contacting family members, work colleagues or her boss at school for advice and assistance is essential for her. Professionally her phone keeps her in contact with other teachers for either consultation or taking some of her classes when necessary. When political instability occurs, on occasions leading to riots, her mobile phone allows her to be updated about school closures or road closures. These updates are then quickly circulated amongst all work colleagues. Social media is also important, highlighted by her use of Facebook and instant messaging. Being connected to the Internet and making Skype or Viber calls keeps her close to family and friends.

Even in rural areas, mobile phone use is increasing. Indigenous people in rural areas usually work in isolated areas either on their farm or in some cases in the forest, with no proper road access to these areas, which are reached by walking along mountain tracks. Mobile phones have helped these people

Figure 14.7 Teacher using mobile phone outside school. (Photo: Siddhi Ratna
Shakya).

keep in contact with family and friends in their community. For example,
a lot of Nepalese go overseas to seek both skilled and unskilled employment through manpower agencies. An Indigenous farmer interviewed from
a rural area uses his mobile phone to receive phone calls from his brother
who is living overseas. He usually receives a call from his brother to update
him on how he is settling into his new home in Seoul, South Korea. His
brother went to South Korea about one year ago to work at the construction
sites. Other members of the family also get an opportunity to speak to him.
This farmer also uses his mobile phone to keep in touch with other friends
and family members. Mobile connections are the only way to contact family
members in those rural areas.

Older people are using mobile phones to make calls. A farmer couple
from the Tamang group at Sankhu, 16 kilometers north-east of Kathmandu,
were interviewed about the use of their mobile phone. They use their phone
to talk to family members while working on the farm. They have owned
their mobile phone for the last three years but only use it to receive or make
calls. Their children bought the mobile phone for them so that they could be
contacted anytime. Staying connected with family and friends has become a
vital reason for owning a mobile phone, particularly with older generations
in Nepal.

Figure 14.8 Farmer couple talking to their son while working on their farm. (Photo: Siddhi Ratna Shakya).

Conclusion

In a country like Nepal, one of the most mountainous countries in the world with limited road and transportation facilities, mobile phone usage has increased significantly in the last decade, providing family contact, information and the opportunity to expand small businesses.

This study uncovers a number of important relationships between socio-economic variables and access to mobile phones. There is a positive relationship between literacy rate and access to mobile phones. Use of mobile phones is also related to employment and varies largely depending on the type of occupation. Some factors were found to interact with each other, for example, literacy, place of residence and occupation.

Various Indigenous groups have found a wide variety of benefits through the use of mobile phones. This has helped them to fulfill their personal, social and business needs on a daily basis. While interviewing seven people within and outside Kathmandu Valley, Indigenous people from diverse occupations expressed the positive values the use of mobile phones has brought into their lives. For trade workers and business owners the mobile phone has certainly assisted them in increasing their income because of quick access to information or the ability to contact relevant parties. Also, professionals use their mobile phones to communicate, network with their colleagues and get recent traffic updates to send alerts to their friends and family who may be trying to commute to that particular area. Overall, the mobile phone has increased the quality of life of Indigenous people in Nepal.

References

Bhattachan, Krishna B., and Sarah Webster. *Indigenous People, Poverty Reduction and Conflict in Nepal.* Geneva: International Labour Organization, 2005.

CBS (Central Bureau of Statistics), Nepal. "Public Use Micro Sample (PUMS) Data." Katmandu: Planning Commission Secretariat, 2011.

Hauser, Robert M., and John Robert Warren. "Socioeconomic Indexes for Occupations: A Review, Update, and Critique." CDE Working Paper No. 96–01. Madison, USA: Centre for Demography and Ecology, University of Wisconsin-Madison, 1996.

ITU. "The World in 2014, ICT Facts and Figures." Geneva: International Telecommunication Union, April 2014.

Shrestha, Bal Gopal. "The Newars: The Indigenous Population of the Kathmandu Valley in the Modern State of Nepal." *Contribution of Nepalese Studies* 26, no. 1 (January 1999): 83–117.

United Nations. "The Concept of Indigenous Peoples." Workshop on Data Collection and Disaggregation for Indigenous Peoples, January 19–21, 2004. New York: United Nations, Department of Economic and Social Affairs.

Westover, Jonathan H. "Socio-Economic Status and Occupational Differences in the Experience of Mortality." *The Internet Journal of Epidemiology* 8, no. 1 (2009).

Part IV
Cultural and Language Revitalization through Mobile Technologies

15 Cultural Hybridity, Resilience and the Communication of Contemporary Cherokee Culture through Mobile Technologies

Kevin R. Kemper

Introduction

> *All the old things are gone now and the Indians are different.*
> *(Mooney 1995, 397)*

Actually, not all of the old things are gone now, and American Indians are not so different. It is easy to fall into sentimental notions about American Indians, thinking of us as historical artifacts instead of present realities. It is more accurate to see how American Indians have used new technologies to adapt to new cultures and preserve the old culture as much as possible. This produces a contemporary culture that is a mixture of various Indigenous and colonizing influences. There are four types of spaces within this type of hybrid culture—stolen, sacred, self and shared (Kemper 2012, 2013). Stolen space is misappropriated culture; sacred space is protected culture; self space is retained culture; and shared space is disseminated culture, as explained below. Mobile technology facilitates shared space, where anyone inside or outside a culture can access language and other available cultural information.

Despite perpetual changes in technology, the ability to adapt and accommodate language and culture to new technology has helped the GWY, or *Tsa-la-gi* (Cherokee) people to survive and thrive into the 21St century and beyond. For instance, visit <http://www.google.com/webhp?hl=chr> to find the Cherokee language version of Google. In the Cherokee and English languages, you can search for a copy of the first and longest-running American Indian newspaper, *The Cherokee Phoenix,* or you can find applications for mobile technology using Cherokee language. On my iPhone, I have the Cherokee Phoenix app next to CNN, BBC News and ESPN apps as part of my daily news diet. You can get the Cherokee Phoenix on iPad, as well, thanks to the leadership from people like editor Bryan Pollard. There are other digital avenues for reading books in Cherokee to your grandchildren, like I do. Over the past two centuries, the Cherokee people have evolved their language from word-of-mouth to written to wired. Again, language embedded in mobile technology would be within shared space, because anyone can access it. This is most convenient. I remember my father's mother taking my younger sister and me from Ponca City, Oklahoma, to Tahlequah, Oklahoma, to visit *Tsa-La-Gi*, the former name of the present Cherokee Heritage Center (see List of Websites at end of chapter). She had told us that we are part ethnic Cherokee and that we should learn the culture; mobile technology now makes that infinitely faster and more colorful and engaging, though I still prefer to interact personally with traditionalists in *Diligwa*, a current version of a 1710-circa Cherokee village at the Cherokee Heritage Center. When I do, and with their permission, I take photos or make notes by typing into my iPhone.

The Cherokee had developed a syllabary to express language in written form, but modern developments show this syllabary spreading through all kinds of technologies. This chapter provides an ideological critique of expressions of that syllabary in mobile technologies. The thesis is straightforward—there is an ideology of contemporary Cherokee culture that insists that the self spaces of original Cherokee culture should be infused into shared spaces like technology innovations to help Cherokee culture to persevere and thrive. This chapter is somewhat similar to Roy Boney, Jr.'s discussion of "Cherokeespace. com," or how he uses new technologies to combat stereotypes about Cherokee people and culture (Boney 2012, 222). However, this chapter extends Boney's work from an ideological perspective and situates the practical within the theoretical. We need conceptual clarity as to *why* the Cherokee are doing what they are doing and then *how* we view and engage with that.

Research Questions and Methodology

This chapter inquires into the ideological and cultural implications of the use of mobile technologies by the Cherokee people. The methodology is the ideological critique, a form of rhetorical analysis that digs into cultural expressions to unearth cultural meaning (Foss 1996; Cormack 2003). Rather than conducting a detailed qualitative analysis of the language in the cultural artifacts, this chapter reports ways the Cherokee syllabary and culture have invaded mobile technologies and attempts to understand the ideological and cultural implications. Where is the Cherokee in all of it?

This chapter generally engages with cultural expressions from people of the three federally recognized bands of Cherokee—the Cherokee Nation of Oklahoma, the United Keetowah Band of Cherokee in Oklahoma and the Eastern Band of Cherokee in North Carolina, though there are numerous state recognized and non-recognized bands of Cherokee across the country. Those state recognized and non-recognized bands are most controversial among Cherokee people because of disagreements over ethnic and cultural authenticity (Snell 2007). There are countless people like myself with authentic Cherokee ancestry who are not able to be enrolled but who do practice and engage with Cherokee culture.

The cultural artifacts for this chapter are samples of discussions about and examples of the Cherokee use of mobile technology, which had been gathered from searching the Internet, including the English and Cherokee language options at Google and Facebook; the website for the Cherokee Nation of

Oklahoma; the website for *The Cherokee Phoenix*, the official newspaper of the Cherokee Nation of Oklahoma; the website for the United Keetowah Band of Cherokee in Oklahoma; and the website for the Eastern Band of Cherokee in North Carolina. Also, I visited each of the capitals of those bands of Cherokee, where I simply interacted with and internalized some of the culture. There are some personal communications cited in this chapter, but the research is not human subjects research because it uses qualitative analysis of published communications. The analyses are not designed to discover generalizable findings.

Cultural Hybridity and Shared Spaces

An academic term to describe the mixture of cultures is cultural hybridity, which simply means that culture constantly changes and morphs under the influences of competing cultures (see Bhabha 1994; Kemper 2012, 2013; Brennan 2009; Meredith, 1998). That phenomenon creates new culture. In discussions about cultural hybridity and Indigenous cultures, we often are reminded that minority Indigenous culture struggles to survive the onslaught of the cultures of the colonizers (see Cheyfitz 2008, 415; Riding In 2008, 72). When colonizers invade the cultural spaces of other cultures, then culture interacts and merges. In some earlier articles (Kemper 2012, 2013), I have found that at least four spaces are created within cultural hybridity when Indigenous cultures become hybrid with colonizing cultures—*sacred space*, or that which is the most precious and protected; *stolen space*, or that which has been taken without permission of the Indigenous culture; *self space*, or that which the Indigenous culture maintains, but could be available for sharing; and then *shared space*, or that which the Indigenous culture provides in a reciprocal relationship with another culture. This chapter extends that discussion by asking whether Indigenous culture can be maintained in a space where it is in the minority. As we will see, the use of mobile technologies by the Cherokee people functions as a way for reclaiming self space within shared space after their culture and language have been stolen or eradicated.

This could be a function of what some scholars call *cultural resilience*, which has been used to assess the persistence of Aboriginal cultures in Canada, for instance (Tousignant and Nibisha Sioui 2009). In social sciences, the term "resilience" refers to a process of surviving and growing through difficulties (Greene, Galambos and Lee 2003, 78). In the context of Indigenous tribes like the Cherokee, the difficulties encountered during colonization are impossible to quantify fully. "The consequences of many centuries of colonization, repeated trauma, both historical and contemporary, and an explicit national project of ethnocide, cannot be eradicated by a short-term intervention or a well-thought culturally adapted program," say Tousignant and Sioui (2009, 44). This chapter does not operationalize resilience quantitatively, but rather uses the idea of cultural resilience as a focal point for understanding how groups of people like the Cherokee survive and preserve original culture within shared spaces during colonization. Tousignant and Sioui form a definition of cultural resilience that is culturally specific—what works for one group might not work for another—and effective as it works to keep at least some elements of original culture (2009, 46). In fact, they frame of the idea of "a return to the traditional culture of the past as a fundamental path to healing" (2009, 47).

The literature about the purposes and cultural importance of the Cherokee syllabary tends to focus upon the idea of cultural perseverance or the idea of holding onto what is left of original Cherokee culture and language. This appears to be conceptually similar to cultural resilience. For instance, Theda Perdue, a prolific scholar of Cherokee culture, is quick to note how Cherokee culture has persisted through times of technological transformation:

> *The writing of a republican constitution, the establishment of mission schools, the development of commercial agriculture, and the invention and adoption of Sequoyah's syllabary reflect profound changes that forever altered the fabric of Cherokee life. This transformation should not be interpreted simply as the destruction of traditional Cherokee society. Even in the face of intense pressure from whites, Cherokees maintained some control over the evolution of their culture. Although schools, churches, constitutions, and syllabaries were adopted from Anglo-Americans, the Cherokees used them to serve their own purposes. In some respects, the "civilization" of the Cherokees was really cultural revitalization that produced an intense pride in being Cherokee and a sense of Cherokee nationalism previously unknown in the history of the principal people.*
>
> *(Perdue 1989, 73)*

Perdue, one of the leading scholars in Cherokee studies, recognizes the intentionality of the Cherokee people and their governments to take charge of their culture and its use. The syllabary and its subsequent uses have not been historical quirks or accidents, but rather evidence of the adaptability and resilience of the Cherokee people.

Margaret Bender (2002), in a book that explored historical and contemporary implications of the syllabary to the Eastern Cherokee, found at least three major benefits of the syllabary for the Cherokee people: semiotic functioning, self-representation of Cherokee people and social influence upon the use and meaning of the syllabary (Bender 2002, 1). That is, the syllabary functioned as an expression of Cherokee ideology, both historically and contemporaneously (Bender 2002, 3). Though she does not use the language of "cultural hybridity" or "spaces", Bender notes the concept of blended cultures as "a borderland at which a dominant culture meets a resistant one, a borderland characterized by a tension between conscious and unconscious beliefs, between acceptance of the dominant society's values and resistance to them" (Bender 2002, 6–7). This demonstrates basic understandings about cultural hybridity, without using that term, as dominant culture puts enormous pressures upon original culture to conform and adapt. Since adaptability is inevitable, the original culture makes the best of things, as we will see in the example of the Cherokee and mobile technologies.

Also, in the most recent and perhaps most developed historiography of the Cherokee syllabary, Cherokee scholar Ellen Cushman (2011, 187) maintains "that the Cherokee syllabary has played a crucial role in facilitating Cherokees' efforts to maintain a sense of peoplehood through profound social and cultural change". She notes how "the Act Relating to the Tribal Policy for the Promotion and Preservation of Cherokee Language, History and Culture," which had been passed in 1991 by the Cherokee Nation's Legislature, mandates the proliferation of the Cherokee language (Cushman 2011, 188). One provision says the tribe must "[e]ncourage the use of Cherokee language in both written and oral form to the fullest extent possible in public and business settings". This supports the idea that the Cherokee people and governments are assertive with efforts to protect everything about their culture, like with the Cherokee Immersion School started in 2001 (Cushman 2011, 192). Cushman argues that the evolution of "Sequoyan" as a written expression of Cherokee language reflects cultural perseverance: "The more Sequoyan is used, the more Cherokee language, history, reverence for place, and religion can be secured" (Cushman 2011, 216). Cushman, building upon the collective work of Robert Thomas, lists four elements of peoplehood: "language, religion, land, and sacred history …" (Cushman 2011, citing Fink 1998, 121). "The perseverance of Cherokee people has much to do with how this writing system has come to be valued, the way in which it is attached to identity, and its embodiment of spirituality, land, history, and language" (Cushman 2011, 217). In discussing Clifford (2001, 479), she also says, "Cultural perseverance, then, would be viewed as a place where Native cultures, for example, enact part of their sovereignty—a process that allows them to name who they are, what practices count, what structures govern, and what technologies allow for adaptation".

However, it seems Cushman sets up a zero-sum game at times where either white culture or Cherokee culture wins or loses; rather, it is cultural hybridity, with Cherokee persistence in the middle of it all. Therefore, it stands to reason that younger generations require social media and mobile technology that help them to continue the Cherokee and also interact with English. It is common for scholars—or even those casually glancing at the issue—to assume the Cherokee want to survive as a people. The deeper truth is that the Cherokee survive by using up-to-date technology. As Cushman (2011, 186) said, "This ability to adapt to many media while retaining crucial links to linguistic information has proven especially important for the proliferation of Sequoyan in modern times". I have friends who use Cherokee fonts in their Facebook pages. On one hand, it invites you to learn what they mean, but on the other hand, the need for meaning is subsumed by the dominant language and culture. This hybrid use of language is, in and of itself, a unique culture.

This historiography of the Cherokee syllabary at times tends to create a false illusion that Cherokee people somehow have been passive victims who manage to hold onto culture and language by a proverbial thread, though there are some exceptions. This chapter, on the other hand, asserts that the Cherokee have been, are now and always will be assertive in expanding culture and language. That does not mean there are not pressing threats that could drive the language and syllabary into practical extinction, but it does mean that the Cherokee are neither passive nor victims.

The Ideology of Shared Spaces

Historical Background of Cherokee Language and Technology

Again, there is an ideology of contemporary Cherokee culture that insists that the self spaces of original Cherokee culture must be infused into shared spaces like technology innovations to help Cherokee culture to persevere and thrive. There are ample historical illustrations of this throughout pre-contact and post-contact Cherokee history. You could take a couple of days to read the excellent, full-length books about the evolution of the Cherokee syllabary until present-day (Bender 2002; Cushman 2011); you quickly could check out Roy Boney, Jr.'s, (2011) succinct, illustrated comic book at the *Indian Country Today Media Network*. It is essential to note that the first words in the first text bubble and the first word of the title are in Cherokee syllabary, which tells us that the Cherokee language comes first, and then the English. This reasserts the primacy of Cherokee culture for Cherokee people. Using some humor to make a serious point, the second piece of art in the comic book shows Sequoyah changing into a live, present-day image. Sequoyah says, "But first, I must do a little **adjusting** ... because we can't go much further if we don't adjust". This reflects a commitment to cultural progress, a common theme in the historiography of the Cherokee syllabary (Figure 15.1).

Figure 15.1 Sequoyah's syllabary in the mobile era. (Source: Roy Boney, Jr. 2011).

People in the United States often have heard of Sequoyah, the silversmith who created the Cherokee written syllabary, but few know that there are hints that the Cherokee had written language long before Sequoyah but destroyed the documents and technology because of concerns it might bring harm to the people. I do not take a scholarly stance for or against the stories that I have heard and read in various places, but I do share a few of them as they represent the long-standing efforts by the Cherokee people to use and preserve language.

For instance, an apocryphal story raises questions of whether the Cherokee had writing technology before contact with Europeans. As one rendition goes, "God gave the red man a book and a paper and told him to write, but he merely made marks on the paper, and as he could not read or write, the Lord gave him a bow and arrows, and gave the book to the white man" (Bender 2002, 26,). Bender also quoted the first Cherokee editor, Elias Boudinot, to finish the story:

> *They have it handed down from their ancestors, that the book which the white people have was once theirs; that while they had it they prospered exceedingly; but that the white people bought it of them and learned many things from it, while the Indians lost credit, offended the Great Spirit, and suffered exceedingly from the neighboring nations; that the Great Spirit took pity on them and directed them to this country.*

Sequoyah, or George Guess, retains the "official" honor of having invented a syllabary to express Cherokee language in writing and print, which is commonly thought to be the original written Indigenous language in North America, but a deep look at the historiography of the syllabary provides more questions than answers to all of that (see Foster 1885; Foreman 1973). In fact, Traveller Bird (1971), who claimed to have been a descendant of Sequoyah, argued that the stories about Sequoyah basically were lies perpetrated by whites and white-friendly Cherokee. Traveller Bird said that Sequoyah believed in the syllabary as "their most effective and defensive tool" against the whites (Traveller Bird 1971, 93–94), but also that Sequoyah and his wife Eli were punished by disfigurement because of accusations of witchcraft for using the syllabary (Traveller Bird 1971, 106–107). Scholars do note the accusations of witchcraft (see Bender 2002, 26), but questions remain as to the historical accuracy of Traveller Bird's work (see Wadley 2014).

Margaret Bender (2002, xii) argues that the written language empowered by the syllabary "itself is ideological, enacting categories that structure the Cherokee social world." That is, she does not think that external ideologies were superimposed upon the syllbary. Bender (2002, 43) primarily sees culture code embedded within the syllabary; this theory would explain the persistence until today. She says, "The concept of a syllabary as code helps to reveal the coherence of a set of beliefs, behaviors, and material products, including literacy practices and performances—reading, writing, possessing, and wearing materials in syllbary; literacy learning and teaching practices in syllabary; and the perceived nature of syllabic texts of various forms". That is, the syllabary represents deep cultural and ideological meanings of the Cherokee who use it, instead of functioning as a static system for symbols (Bender 2002, 129).

Again, there is serious concern that the Cherokee language would be subsumed by the dominant culture. This chapter uses academic terms like "cultural hybridity" to describe how the Cherokee language functions ideologically. Even those involved with the Cherokee over the years have seen this problem. As one 19th century missionary wrote, "As matters now are, the Cherokee language itself must, in the nature of things, soon give place to the English and Se-quo-yah's alphabet and Se-quo-yah's people will no longer be separated from the great mass of the American people, but blend into one and thus fade away" (Timothy Hill, quoted in Foster 1885, xvii).

In the historiography of written communications by Cherokee in Cherokee language, the prominent people are Cherokee: Sequoyah, who created the syllabary, and Elias Boudinot, who first edited and published the newspaper, *The Cherokee Phoenix*. However, white missionaries—particularly the Reverends Evan Jones and Samuel Worcester—were key people in applying the syllabary to typesetting, which was necessary for the printing of Bibles, religious tracts, newspapers and other documents. We must not forget how influential whites have been and still are with Cherokee culture.

Even the prospectus of *The Cherokee Phoenix*, first published in the early 1800s, reflects cultural hybridity, as Boudinot earnestly desired to help the Cherokee people by preserving original culture and adapting to new culture:

> As the great object of the Phoenix will be the benefit of the Cherokees, the following subjects will occupy its columns.
>
> 1 The laws and public documents of the Nation.
> 2 Account of the manners and customs of the Cherokees, and their progress in Education, Religion and the arts of civilized life; with such notices of other Indian tribes as our limited means of information will allow.
> 3 The principle interesting news of the day.
> 4 Miscellaneous articles, calculated to promote Literature, Civilization, and Religion among the Cherokees (Boudinot, in Perdue 1996, 90).

From the invention and implementation of the Cherokee syllabary to today, we cannot escape the dominant culture, but we cannot escape the Cherokee culture.

Shared Spaces and Mobile Technology

A pertinent and interesting introduction to Cherokee language embedded within mobile technology, as well as an extended explanation of methodology, would be a description of how I access Cherokee language as part of my own life and for writing this chapter. I speak very little Cherokee, and instead I study Choctaw because my tribal ethnicity and identity is more Choctaw than Cherokee. However, this project has expanded my connections with and understandings of my Cherokee ancestry and culture. I will not say that this discussion includes all of the technical applications or environments or that these still will be valid by the time you read this chapter. You will get a sense of the exciting possibilities of contemporary Cherokee culture on social media and other mobile technologies as it evolves every second. I challenge you to see if you can integrate Cherokee—even momentarily— into your mobile technologies.

Because of the first-person narrative, some of this discussion could be considered autoethnography, but it would be simpler to say that I dove into use of Cherokee language with mobile technologies so I could understand the ideologies and the concepts in a personal way. For instance, I have a Google app on my iPhone, but I have not been clear as how to change the language function to Cherokee. However, I can go to Google on my MacBookPro and go to "Settings" in the lower right hand corner of the screen and hover to get "Search Settings." Then, I can select "Language" on the left side of the screen and then scroll down and select GWY (Cherokee).

To translate on Facebook, you can go to <https://www.facebook.com/?sk=translations> and select "Cherokee" from the dropbox. A helpful guide is available at <https://www.facebook.com/translations/app_guide>. Under "Settings" and then "General"; you can edit language and select "Cherokee". I did this on my regular account and on my Facebook app. You also can get Cherokee language apps at iTunes. This creates a hybridity of Cherokee language and syllabary with English language and alphabet.

From iTunes, I had downloaded the free app for *The Cherokee Phoenix,* though the stories are not presented in Cherokee language. I purchased a CD of Cherokee hymns at the Cherokee Heritage Center in Tahlequah, Oklahoma, the location of the capitals of the Cherokee Nation of Oklahoma and the United Keetowah Band. These were loaded into my iTunes on my MacBook Pro, which then can be loaded onto my iPhone. Also, by searching "Cherokee language" at iTunes you can find and purchase numerous apps, songs, podcasts and books.

If you do not have Apple technology, never fear. You can find Cherokee games for Android at Web sites like Cherokee Language Lessons (Games), or Cherokee language apps at the Google Store (Cherokee Language Lessons). There is a Cherokee language pack for the newest Microsoft Windows systems at Windows Language Packs Web page.

You could go to the Museum Store of the Museum of the Cherokee Indian to get a copy of Bo Taylor's "Rebuilding the Fire: Traditional Songs of the Eastern Band of the Cherokee." I bought this CD at the Museum of the Cherokee Indian in Cherokee, North Carolina, so I could put the traditional dance songs on my iPhone.

Everywhere I now go, I can learn, read and listen to Cherokee language and culture, if I wish. Yet, it is mashed together with the English language and Euro-American cultures. Since I also have been studying Choctaw diligently at places like *Chahta Anumpa Aiikhvna* (School of Choctaw Language) and with Choctaw hymns and Bible passages on my iPhone and in print, my brain is both stimulated and overwhelmed when I attempt to think in three cultures at once. Yet, all of those languages and cultures reflect some of my own background. It is *my* culture, all wrapped into one. Are all of those technological innovations truly Cherokee language and culture? Or something different? And, why does that matter? All of that is cultural hybridity.

"Speaking a language means we have a culture", a Cherokee elder was quoted as saying. "There is a big difference between people who have a culture and people with a history" (Cherokee Preservation Foundation). This culture can be found in all kinds of popular media. In a copy of the United Keetowah Band's official newspaper, *The Gaduwa Cherokee News* (November 11, 2011), you are invited to like the band's Facebook page, which can be found at <https://www.facebook.com/pages/United-Keetoowah-Band-of-Cherokee-Indians-in-Oklahoma/269228928301>. However, except for some Cherokee script in the band's official seal, or a greeting from Miss Keetoowah Cherokee, you will see little Cherokee language, which is ironic since the Keetoowah band formed in part because of a concern to preserve traditions.

The use of mobile technologies functions as an intentional way for the Cherokee people to decolonize their experience. It almost functions like "reverse colonization", which Steve J. Stern (1998, 53) defines as:

> *the entire array of responses whereby Indians and other subalterns colonized apparatuses of colonial control and profit-taking—the numerous ways that peoples condemned to colonization ended up invading and putting to their own use the colonial state's legal rules and political alliance games, or appropriating colonial markets and production niches in ways that furthered their own life strategies while undercutting colonial monopolies and revenues, or engaging and redeploying the meanings of Catholic religion and sacred patronage, or incorporating written decree, memory, and genre into a sense of group self, right, and destiny.*

Simply put, the intentional insertion of Cherokee syllabary and language into mobile dialogue flips colonization on its head; if Cherokee culture and other Indigenous cultures have to be infected with Euro-American culture, then Euro-American culture can expect cross-contamination with the cultures it colonizes.

"We are using the tools of mass communication & making them our own by using them our way on our own terms", Roy Boney, Jr., wrote in a Facebook comment when I informally asked my Facebook friends about what they thought of the premise of this chapter. If you go to his personal website, you will find examples of his graphic art, writing and use of the Cherokee syllabary within the context of English language and American culture. Boney, who serves the Cherokee Nation in its Language Technology Program of the tribe's Education Services Group, invades the Euro-American cultural space with Cherokee syllabary: the header for this Web site is his name and primary occupation, but note that the Cherokee syllabary comes first: "DB ᎣhᎦ AWoᎥ: ᎫCGᏸᎠWᎤ-ᏆᎤᏸ Roy Boney, Jr.: Artist." In his blog Boney inserts Cherokee syllabary in surprising places. In fact, on September 7, 2012, he posted a modified copy of his Cherokee Nation citizenship card, which included his art in place of a photograph, and used syllabary for his name. On your mobile device, you can download Boney's (2007) animation video, "On a Spring Day", at YouTube. The spoken words of the Cherokee are in Cherokee, while the subtitles and some of the spoken words of whites are in English, as the story explains about the Trail of Tears, in which the Cherokee people and other southeastern tribes were forcibly removed from the southeastern United States to present-day Oklahoma. Boney serves as the primary narrator.

In an interview by Boney (2012), Cushman noted the significance of the syllabary in the digital environment, which activates various senses: "The digital makes possible the close unification of sound place, action, image and Sequoyan". This is itself a form of hybridity, bringing together different media and expressions into a cohesive message. This is reflected in numerous technological advances. For instance, the Cherokee language was the first Indigenous language from North America to be used in Gmail (Cherokee Nation News Office 2012). These advances are not just creative serendipities—the Cherokee intentionally participate in what will further the language. The Cherokee Nation is a liaison member of the Unicode Consortium, which helps integrate the language on all kinds of computers and mobile technologies.

The ideological critique simply looked to see how much the Cherokee language is used in mobile technologies and communications. The findings show that the Cherokee language is found more on Facebook or Twitter than in official publications of the various bands, though you do find Cherokee language there. Within these shared spaces, you find surprising instances of the language, making you wonder what the script and words mean. Those who are Cherokee—and even some who are not—might be motivated to take a Cherokee language course or at least ask someone what the language means. This does not appear to result in expanded fluency, but it does draw those interested in fluency into mobile technology. The language expands and evolves through instantaneous use.

The location of cultural survival is in all spaces—stolen, self, shared, and sacred. Cherokee language and culture embedded in mobile technologies are in shared space, offered to anyone willing to learn and use the language. Even in this chapter, you have been confronted with a bit of Cherokee syllabary, an intentional use for more than just academic accuracy. It is usable, it survives and it propagates. Imagine the explosion of Cherokee syllabary across mobile technologies, where Instagrams get shared instantaneously to Cherokee *and* non-Cherokee.

Conclusion

After writing this chapter and experiencing Cherokee language and culture, I find myself drawn to absorbing more and more—what is that word, how does that relate to who I am, how can I use that in real life, etc.? I have decided to use the Cherokee language some on my Facebook page, at least for now, so I can continue this experience. That is the point of those who have merged Cherokee language and culture with mobile technologies. The Cherokee language and culture exhibits resilience in the face of colonizing influences.

Future research needs to examine whether use of the Cherokee syllabary in newer technologies results in a quantitative increase in the numbers of Cherokee speakers and whether that exposure translates into fluency. The Cherokee Preservation Foundation found in one survey less than 500 fluent Cherokee speakers. In August of 2014, I saw a large billboard in Cherokee, N.C., which claimed 216 fluent speakers on that reservation. There are just about 50 monolingual Cherokee speakers left in the Cherokee Nation of Oklahoma (Eaton 2014).

After finishing this chapter, I intend to visit family and friends and otherwise enjoy the Cherokee National Holiday over Labor Day Weekend in Tahlequah, Oklahoma (http://www.cherokee.org/AboutTheNation/NationalHoliday.aspx). I can use mobile technology to find my way there, look up the meaning of Cherokee words in Cherokee syllabary, listen to Cherokee dance music, learn more about my ancestors and the Cherokee culture that passed along to me, take photographs of activities and help keep Cherokee culture alive for another seven generations and beyond. All that with just an iPhone. The American Indians might be a little different, thanks to cultural hybridity and general progress, but the old things are not gone.

List of Websites

(All accessed March 23, 2015)
Cherokee Heritage Center. http://www.cherokeeheritage.org.
Cherokee Language Lessons (Games): http://www.cherokeelessons.com/Games/.
Cherokee Nation of Oklahoma. http://www.cherokee.org.
Cherokee Phoenix. http://www.cherokeephoenix.org.
Eastern Band of Cherokee (North Carolina). http://nc-cherokee.com.
Facebook. http://www.facebook.com.
Google. http://www.google.com.
Google Store (Cherokee Language Lessons). https://play.google.com/store/apps/developer?id=Cherokee+Language+Lessons.
Museum Store, the Museum of the Cherokee Indian: http://cherokeemuseum.mivamerchant.net/mm5/merchant.mvc.
Roy Boney, Jr., blog. http://boneyart.blogspot.com/.
Roy Boney, Jr., Web page. http://royboney.com/.

United Keetowah Band of Cherokee in Oklahoma. http://ukb-nsn.gov.
Windows Language Packs. http://windows.microsoft.com/en-us/windows/
language-packs#lptabs=win8.

References

Bhabha, Homi. *The Location of Culture*. New York: Routledge Classics, 1994.
Boney, Jr., Roy. "'Character' Study: Author Ellen Cushman is Fascinated with Cherokee
Writing." *Indian Country Today Media Network*. March 25, 2012. Accessed
March 23, 2015. http://indiancountrytodaymedianetwork.com/2012/03/25/
character-study-author-ellen-cushman-fascinated-cherokee-writing-104649.
Boney, Jr., Roy. "Cherokeespace.com: Native Social Networking." In *American
Indians and the Mass* Media, edited by Meta G. Carstarphen and John P. Sanchez,
221–26. Norman: University of Oklahoma Press, 2012.
Boney, Jr., Roy. "Exclusive: Artist Roy Boney's Special Graphic Feature on the Cherokee
Language." *Indian Country Today Media Network,* September 20, 2011. Accessed
March 23, 2015. http://indiancountrytodaymedianetwork.com/2011/09/20/
exclusive-artist-roy-boneys-special-graphic-feature-cherokee-language-54344.
Boney, Jr., Roy. On a Spring Day. Video. Jan. 30, 2007. https://www.youtube.com/
watch?v=OoJwtn2-lRg&noredirect=1.
Brennan, Niall P. "Representing National Culture, Values, and Identity in the Brazilian
Television Mini-Series." *Networking Knowledge: Journal of the MeCCSSA
Postgraduate Network* 2, no. 1 (2009): 1–21. Accessed March 23, 2015. http://
ojs.meccsa.org.uk/index.php/netknow/article/view/43.
Cherokee Nation News Office. "Cherokee Nation Joins International Language
Consortium." December 14, 2011. Accessed March 23, 2015. http://www.
cherokee.org/News/Stories /32858.aspx.
Cherokee Nation News Office. "Gmail Now Available in Cherokee." November 19, 2012.
Accessed March 23, 2015. http://www.cherokee.org/News/Stories/33269.aspx.
Cherokee Preservation Foundation. "Overview." Accessed March 23, 2015. http://
cherokeepreservation.org/what-we-do/cultural-preservation/cherokee-language/.
Clifford, James. "Indigenous Articulations." *Contemporary Pacific* 13, no. 2 (2001):
468–90.
Cormack, Michael. *Ideology*. Ann Arbor: University of Michigan Press, 2003.
Cushman, Ellen. *The Cherokee Syllabary, Writing the People's Perseverance*, Vol 56,
American Indian Literature and Critical Studies Series. Norman: University of
Oklahoma Press, 2011.
Eaton, Kristi. "Tribes Draw Knowledge from Monolingual Speakers." *Associated
Press*, in *The Cherokee Phoenix*, April 8, 2014. http://www.cherokeephoenix.org/
Article/Index/8120.
Fink, Kenneth. "Riding Behind with a Pillow Strapped On." In *A Good Cherokee,
A Good Anthropologist*, edited by Steve Pavlik, 119–27. Berkley: University of
California Press, 1998.
Foreman, Grant. *Sequoyah*. Norman: University of Oklahoma Press, 1973.
Foss, Sonya. *Rhetorical Criticism: Exploration & Practice*. Long Grove, Ill.:
Waveland Press, 1996.
Foster, George E. *Se-quo-yah: The American Cadmus and Modern Moses*.
Philadelphia: Office of the Indian Rights Association, 1885. https://archive.org/
details /sequoyahamerican00fost.

Greene, Roberta R. *Resilience: Theory and Research for Social Work Practice.* Washington, DC: NASW Press, 2002.

Greene, Roberta R., Colleen Galambos and Youjung Lee. "Resilience Theory: Theoretical and Professional Conceptualization." *Journal of Human Behavior in the Social Environment* 8, no. 4 (2003): 75–92.

Kemper, Kevin R. "Sacred Spaces: Cultural Hybridity and Boundaries for Visual Communication about the Hopi Tribe in Arizona." *Visual Communication Quarterly* 19 (2012): 216–31.

Kemper, Kevin R. "You Have to EARN Access: A Case Study of Arizona Tribes and Reporting About Indigenous Religion Around the Pacific Rim." *Asia Pacific Media Educator* 23, no. 1 (2013). doi: 10.1177/1326365X13510095.

Meredith, Paul. "Hybridity in the Third Space: Rethinking Bi-Cultural Politics in Aotearoa/New Zealand." Paper presented at Te Oru Rangahau Maori Research and Development Conference, Massey University, 1998. Accessed March 23, 2015. http://lianz.waikato.ac.nz/PAPERS/paul/hybridity.pdf.

Mooney, James. *Myths of the Cherokee.* New York: Dover.

Perdue, Theda. *The Cherokee.* Indians of North America, edited by Frank W. Porter III. New York: Chelsea House Publishers, 1989.

Riding In, James. "Presidential Address American Indian Studies: Our challenges." Wicazo Sa Review, 23, no. 2 (2008, Fall): 72. Accessed March 23, 2015. http://www.jstor.org/stable/30131262.

Snell, Travis. "Non-Recognized 'Cherokee Tribes' Flourish." *The Cherokee Phoenix,* January 19, 2007. Accessed March 23, 2015. http://www.cherokeephoenix.org/19212/Article.aspx.

Stern, Steve J. "The Decentered Center and the Expansionist Periphery: The Paradoxes of Foreign-Local Encounter." In *Close Encounters of Empire,* edited by Gilbert M. Joseph, Catherine LeGrand, and Ricardo D. Salvatore, 47–68. Durham, NC: Duke University Press Books, 1998.

Tousignant, Michel, and Nibisha Sioui. "Resilience and Aboriginal Communities in Crisis: Theory and Interventions." *Journal of Aboriginal Health* (2009): 43–61. Accessed March 23, 2015. http://www.naho.ca/jah/english/jah05_01/V5_I1_Resilience_03.pdf.

Traveller Bird. *Tell Them They Lie: The Sequoyah Myth.* Los Angeles: Westernlore Publishers, 1971.

Wadley, Ted. "Sequoyah (ca. 1770-ca. 1840)." *New Georgia Encyclopedia.* Last modified 2014. http://www.georgiaencyclopedia.org/articles/history-archaeology/sequoyah-ca-1770-ca-1840.

16 eToro

Appropriating ICTs for the Management of Penans' Indigenous Botanical Knowledge

Tariq Zaman, Narayanan Kulathuramaiyer and Alvin W. Yeo

Introduction

The preservation, management and sharing of Indigenous knowledge (IK) is crucial for social and economic development in rural and Indigenous communities. Access to information and communication technologies (ICTs) to digitize Indigenous knowledge and cultural heritage is becoming increasingly important. However, Indigenous people face several challenges as they attempt to use technologies for preservation and management of their IK and cultural heritage (Duncker 2002). One of the issues is the gap of understanding among information system designers, knowledge engineers and Indigenous communities (users of information systems and domain experts of IK). Although IK is oral and performative in nature, ICT tools for IK management are designed in the same way as the tools for text based knowledge documentation and dissemination, such as e-readers with annotation facilities, meta-data and semantic search facilities, text-based knowledge sharing platforms and distributed databases.

Despite many attempts at modeling Indigenous cultures and knowledge systems (Hunter 2005; Ara Irititja Team 2011), neither theoretical understandings nor practical outcomes have influenced mainstream technology design. This becomes even more problematic if the purpose of the technology is to support the preservation of IK itself. IK is a highly contextualized body of knowledge linked to locations, situations and cultural, social and historical contexts and community activities (Van Der Velden 2002).

Therefore researchers have underscored the need for a holistic approach to design Indigenous knowledge management systems (IKMS) (Kargbo 2006). Winschiers-Theophilus et al. (2012a), working in the African context, accept "interconnectedness", "holistic view", "local community as co-designer" and participatory processes as key factors in designing ICT tools for IKMS. These principles can equally be applied to the design of IKMS in other parts of the world.

The Context

In this chapter we present a case study of designing and developing ICT tools, including mobile devices, for the Penan's Botanical IKMS. Penans are one of the Indigenous communities living in Sarawak, Brunei and Kalimantan (Secrombe 2008). The Sarawak Penan population in 2010 was estimated to be 16,281, of whom about 77% had settled permanently. The remaining 20% are semi-nomadic while 3% still live as nomads (Lyndon et al. 2013). The Penan community in Sarawak are divided into three groups: the Penans living in the upper reaches, middle reaches and lower reaches of the Baram river area. This reflects the nomadic Penans' practice of relating to the rivers and rainforests where they roam.

In early 2012, researchers from the University of Malaysia Sarawak (UNIMAS) with the collaboration of the Penan community at Long Lamai initiated the eToro project for documentation and preservation of Indigenous Botanical Knowledge of the Penans. Based on a participatory approach, the researchers worked with the community to thoroughly study and document the local structures of Penan Botanical knowledge management, day-to-day plant-use activities, localized context and the potential role of ICTs to support this system. The project activities included exploring the specific needs of the community and establishing cultural protocols for supporting creative explorations to develop ICT tools. As part of this process the roles of the stakeholders were clarified using process flow diagrams that were developed with the active participation of community.

The Community and Ngerabit eLamai

Long Lamai is one of the oldest settlements of the Indigenous Penan community in the upper reaches of Baram, Miri Sarawak. The village is accessible by flying from the nearest town Miri to Long Banga and taking a one and a half hour boat ride to Long Lamai, or one can drive eight hours along logging roads and hike an hour through dense forest (Figure16.1). There are approximately 500 Penans living in Long Lamai (Siew et al. 2013).

The community is egalitarian in nature, and strong community cohesion is reflected in their daily activities and interactions. The village is a true picture of remoteness: it has no road access, limited electricity, no proper water supply and no telephone connectivity. The community has very limited communication with the outside world and lacks basic health facilities. The available infrastructures at Long Lamai consist of a Penan school, a church and a telecenter called Ngerabit eLamai.

UNIMAS is one of the active implementers of telecenter projects in Malaysia. In 1999, UNIMAS initiated the multi-award winning eBario project with the aim to bridge the digital divide in order to stimulate socioeconomic development in the Bario community in the Kelabit highlands

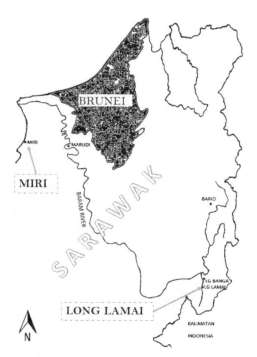

Figure 16.1 Location of Longa Lamai and Miri in the State of Sarawak, Malaysia.

(for details see Yeo et al. 2011). Given the success of eBario, the Ministry of Science Technology and Innovation (MOSTI), Malaysia, provided funding to research and replicate another four telecenters at Ba'Kelalan and Long Lamai in Sarawak state, and Kpg Buayan and Pulau Larapan in Sabah state, both states of Malaysia.

Ngerabit eLamai is the telecenter of the Long Lamai community and also the first replication site of the eBario project. After a one and a half year negotiation and community engagement process the community built the infrastructure for the telecenter by *gotong-ryong* (collective action of community). The telecenter is the only source of telecommunication at the village. It is equipped with three networked PCs, three laptops, a printer and a scanner and is powered up by solar energy. The telecenter also provides other facilities such as a telephone connection, the Internet, printing and photocopying services. When the telecenter was installed, the researchers conducted training courses (Figure 16.2) of basic computer and information technology skills for the local community members, including "train the trainer" sessions (Juan et al. 2010). Meanwhile, several community members also have their own devices such as laptops, tablets and smartphones connected via Wi-Fi installed in the telecenter or using the low mobile coverage that can be accessed in a specific location in the village.

Figure 16.2 Participants in the train the trainer session at the telecenter in Long Lamai.

After providing access to ICTs, the next step was to understand the community needs holistically and to identify the potentials for using ICTs in the local context.

The Problem

Under the agenda for development and modernization established by the Malaysian/Sarawak government in the 1950s, the Penans were encouraged to shift from hunting and gathering to cultivation. In this new environment it is increasingly difficult to fall back on the forest for daily needs or trading products. The semi-nomadic and settled lifestyle of today has changed their customary routine. They are now more dependent on cash income in their new economic life compared to their nomadic life. They now have less means of income from jungle produce, such as *rattan* (a species of palm) or agarwood (*gaharu*, a fragrant resinous heartwood). The community is also facing challenges to their social and cultural lifestyle. It is becoming more difficult to keep the Penan Indigenous languages alive and to retain the Indigenous knowledge about plants and jungle life as modern life overtakes their traditional lifestyles. Unfortunately, the deprivation of their forest territory and the loss of their elders over time are making the situation worse. The traditional social and cultural life of the Penan is gradually dying and the community is facing a rapid loss in their cultural assets.

The Penan of Long Lamai view the entire rainforest as their home and living a lifestyle that does not exploit or harm their home is important. They rely on these forests for food, medicine and the materials they use to build their homes. In the early 1990s, the Long Lamai community lost their primary forest and the wild plants surrounding their village in a forest fire that resulted in the loss of knowledge about plants, birds and animals. Many of the older generation forgot parts of their Indigenous knowledge. During a visit to Long Lamai by the university researchers, there was a meeting with the telecenter manager, who is an elder from the local community. He showed his interest in pursuing the use of ICTs to nurture the transfer of cultural knowledge from the older generation to the younger generation. The main reason was his observation that the youth of Long Lamai were taking more interest in ICTs while the elders wanted to educate them about the traditional lifestyle.

He led the authors on a visit to a nearby forest and within ten minutes' walk inside the jungle identified more than 10 different plants with local names and uses (Figure 16.3). However, we noticed that sometimes he referred to his notebook as he himself was also forgetting this rich knowledge.

Figure 16.3 A forest walk with Garen Jengan who is the Ngerabit eLamai Telecenter Manager and a member of the local community.

eToro: The Solution

In response to the problem of rapid knowledge loss, the researchers from UNIMAS and the local community of Long Lamai collaboratively developed the concept of the eToro project. Our research goal was to develop tools that would enable the Penans of Long Lamai to digitize their own knowledge in a manner that they consider to be meaningful and acceptable. The community goal for the project is to document their botanical knowledge and to bridge the intergenerational knowledge gap within community.

We followed a community-based co-design approach, built on principles of participatory action research aimed at mutual learning (Winschiers-Theophilus et al. 2012b). Ideally the tools become part of the existing knowledge system of the community, thereby ensuring a continuous knowledge collection and local curation process. This is important as we acknowledge that IKMS are dynamic and holistic by nature and can hardly be captured and represented by limited outsider visits.

Understanding the Local Knowledge Structure

The Penans' Indigenous botanical knowledge (IBK) management has a complex structure governed by a social and cultural belief system, assimilated with community activities and both the tacit, or implicit, and explicit knowledge bases. The Penan youth no longer accompany the elders in *Toro*, an instructive walk in the forest. They do not listen to the elders' stories or participate in activities that have been the common mode of knowledge transfer. *Toro* is a joint activity of a Penan family and also works as an activity-based knowledge sharing and mentoring journey in the forest that links community elders to members of the younger generations in grooming future guardians of the rainforest. Mentoring includes lessons on livelihood combined with a notion of stewardship, incorporating concepts of conservation ethics and ownership. The Penan elders embody local practices and knowledge with a strong desire to preserve it for the next generations.

The proper representation of this complex system in ICT-based IKMS faces the challenges of knowledge gaps between researchers and local community, language barriers and the lack of equivalent concepts and terminologies on both sides. Some of these challenges can be handled by developing the common understanding of the local knowledge system. To understand the activities in the *Toro* journey and to capture the essentials from a local perspective, we engaged the local community members in developing an agreed-upon sketch of the *Toro* journey (Figure 16.4).

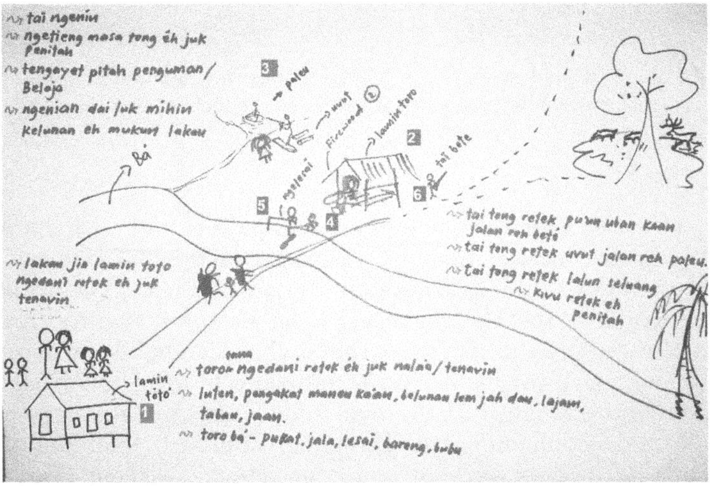

Figure 16.4 Drawing depicting activities in the *Toro* journey.

The development of the drawing was a method used for exploring the representations of local concepts and to inform the researchers. The figure shows that the activities start by leaving the *lamin toto* (house in the village) (labeled 1) and finding a place in the jungle that has enough food, such as fruit trees, fish in the nearby river, sago plants, and animals for hunting. When a family finds the place, they establish their *lamin toro* (traditional temporary hut) (labeled 2). The subsequent activities include extracting sago (3), cooking food (4), fishing (5) and hunting (6). An entire family—parents and children—complete the journey together. There is a gender-based division of labor as well as knowledge in performing the *Toro* activities. These drawings were complemented by a set of individual drawings depicting what is significant to the Penans within a given context. Although there is a good rapport between UNIMAS and the community that was granted the telecenter project, we observed that community elders needed another round of defining cultural protocols and putting them into an explicit form, given the sensitive nature of the project, which deals with the botanical information. For Penans forest is "everything". They have diverse procedures, rules and "regulations" in relation to their land and plants, which regulate their interactions within the community, with other communities and with the environment upon which they depend. The importance of these rules becomes more significant when they engage with external actors. However, outsiders often do not understand these cultural protocols as they are mainly unwritten and exist as customary practices. To avoid any issue, the community and researchers documented these cultural protocols in a written form as a guidebook for researchers and visitors of the Long Lamai community (for details refer to Zaman et al. 2014). The cultural protocols emphasize community ownership rights and governance with respect to knowledge resources. For example, the practice of *molong* deals with conservation ethics and stewardship of resources. In the Penan social system, one can *molong* a resource (such as a tree or land) by making a declaration. The declaration process includes marking the tree and clearing away around the tree. The declaration works as a public statement that there is a steward of this specific tree who will take care of it and in return gain rights to use the tree. The rights of the steward are protected by the system of *molong*. The process of documenting the cultural protocols fostered the engagement process, development of mutual understanding and respect for customary laws, values and the decision-making process.

The Tools

The IKMS tools have been co-designed with the local community members to support the collection and management of Indigenous plant knowledge (Siew et al. 2013). First, questions were explored such as: what type of information does the community want to collect, where do they want to store it, in which format, and who will have access rights to it? Researchers, students from Long Lamai living in Kuching (the capital city of Sarawak) and Long Lamai community representatives participated in this process. Several sessions were organized to verify the data requirement specifications such as language accuracy, the types of data fields and the attributes for data items. These activities resulted in a document with 26 datasets in local and English languages that were approved by the community representative team.

The most common problems associated with the effective implementation of ICT in rural areas are usability, maintenance of software and hardware systems and inconsistent electric power supply (Kwacha 2007). We designed our software system for mobile tablet devices due to their function-alities and robustness that allow them to be easily carried everywhere. They have a screen size that is not too small, consume less power than a desk-top computer, require less maintenance and have the multiple functions of picture, video, audio and GPS-coordinate capturing capabilities in a single device. In addition, it was noted that the structure and user interface of the tablet PCs are friendly and easy to use for Indigenous communities (Rodil, Eskildsen and Rehm 2011).

To capture the text, images, videos and GPS coordinates of the plants, we used an Android-based tablet and Open Data Kit (ODK 2015). The data collection form was manually designed by the community and mapped into a digital ODK survey form for mobile devices. During discussions with a community elder, we realized the complexity of interlinked plant knowledge and the weaknesses of form-based entries and organization.

The captured data can be accessed within an Indigenous Content Man-agement System (iCMS). The iCMS architecture is based on a front-end and back-end distribution. The application runs on the main system at the front-end while the data is stored on the externally attached hard drive at the back-end. The designed iCMS provides a secured data protection mechanism for the collected content. To address community requirements, the iCMS provides profile-based access rights to view or update information to users of the system. The collected information is protected and not available publicly. Once the rules are created and implemented the system accommodates an accountabil-ity mechanism for the user's activities. The IK Manager can generate reports of user activities performed during login sessions. On the base of the activities the manager can make a decision to moderate the user's role and activities.

In addition to data protection, data storage in a web based or local repository is a significant factor of digital IKMS. To ensure maximum con-trol, iCMS stores the data on external hard drives following an *e-insitu* approach. The *e-insitu* approach is the facility for the community to have physical, *insitu* control of the data and storage device, in addition to logical (electronic) data protection mechanisms. Under the *e-insitu* approach the hard drive is kept in the custodianship of a community-appointed member.

Following the *Toro* journey in Figure 16.4, the documentation activities consist of three phases. In the first phase, the young community members travel to the forest with the elders and use tablets and the ODK to col-lect knowledge about plants in the form of text, images, video and GPS coordinates. In the second phase, the collected information is manually verified in community meetings, and in the third phase iCMS is used to manage the data.

We explored the usability and usage possibilities of the tablet-based soft-ware system and iCMS prototypes with three elder men, three women and two young Penans. All the participants easily recognized the ODK form entries and were able to enter text data and pictures of plants and to record video. Participants found the categorization of form entries peculiar but the selection and swipe gesture intuitive. However the gestures requiring two fingers were only properly managed by the young and computer literate participants. The elders also had to be open to change. Not all community members were privy to all the data collected. Thus, with the tablet-based system and iCMS, the elders now had to control the access rights to documented information.

In addition to the elders' expectation that the youth be able to list all important plants in the surroundings is the expectation that the youth be able to collect the necessary information about the plants. This expectation, if realized, will foster knowledge transfer.

Creating Value for the Community

Based on snapshots of knowledge gathered using the eToro system, the community gains access to traditional knowledge that can be visualized and re-purposed for a number of innovative applications. The collected information is fully under the control of the community, which decides on the level of sharing and the resulting gains that can be translated into economic benefits. The community then plans the workflow for knowledge sharing and plots each trajectory, whether it is used for learning within community, brainstorming with researchers or packaging for tourists. With the IKMS developed, the next step is to determine if there are other innovative uses for the information gathered. By using resources that are indexed by geographical coordinates, topology maps can then be designed. These maps can be used for the learning trajectory for the youth within a guided environment. The same maps generated are now being used to map the trajectories for tourists and to produce unique travel maps. An illustration of the tourist trail for a unique offering of eco-tourism exploration can be seen in Figure 16.5.

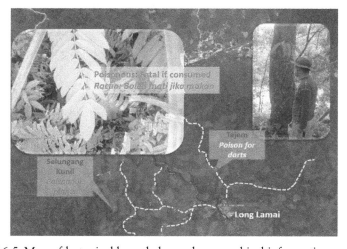

Figure 16.5 Map of botanical knowledge and geographical information.

These offerings preserve the broad, holistic, Indigenous view of life while fully harnessing and recontextualizing traditional knowledge. The old ways of life are now brought alive through the rich contextual modeling within an encompassing space for documenting and transferring Indigenous botanical knowledge.

Conclusion

A perceived benefit of eToro is that it covers the knowledge gap between the young and old generations of the Long Lamai community. The youth have ICT skills: thus they feel more confident helping in the documentation process; ultimately, they are part of the learning cycle. The collected data, pictures and videos can also be used in tourism promotion activities. eToro helps in the training of young people, so the community will have more trained human resources as guardians of the rainforest.

In addition, the project produced a database of 50 plants, a data collection instrument and software, a secured content management system and written cultural protocols for researchers to follow and practice. The concept of developing cultural protocols is being adapted and replicated in the Telecenter project for Indigenous communities in Pahang and Kelantan, two states in West Malaysia.

The Indigenous knowledge that was carefully crafted around observing nature, adapting and surviving in the changing world, obtaining sustenance from the plant and animal world, is now made available using ICT technology. Community elders pay special attention to preserving and sharing knowledge based on cultural heritage and traditions in the process of mentoring youth on IBK. Efforts are particularly taken to emphasize the harnessing of community team spirit, developing creative expressions and living in harmony with the environment while nurturing life-long learning processes. Some common themes surrounding the knowledge acquisition process include care for the environment, responsibility toward nature, deeper inquiry into the connectedness of things and a need for a balanced lifestyle.

The experiences with eToro have given rise to the concept of ICTs for Indigenized Development in which the core characteristics of community-owned ICTs lend themselves well to solutions for many of the problems the world's Indigenous people share. By empowering communities to use appropriate ICTs, they put these technologies to use within development activities of their choice, as opposed to those that are imposed from outside and often assume absorption of minority cultures into mainstream society. The close engagement between the research team and the community has resulted in: i) a project that is firmly embedded in the problems, aspirations and opportunities the community itself identifies; ii) community capacity in the appropriating of mobile and other technologies; and iii) a methodology for community-based research applicable in Indigenous contexts.

Acknowledgement

The authors are grateful for the institutional support and Dana Principal Investigator research funding from the University of Malaysia Sarawak, and they acknowledge and are thankful for the continuous support from the community of Long Lamai.

References

Ara Irititja Team. "Ara Irititja: Cultural Knowledge in the Digital Age." *Our Place* 40, no. 3 (2011): 14–16. Accessed March 28, 2015. http://www.icat.org.au/wp-content/uploads/2012/12/ourplace40/.

Duncker, Elke. "Cross-Cultural Usability of the Library Metaphor." In Proceedings of the 2nd ACM/IEEE-CS Joint Conference on Digital Libraries, 2002, 223–230.

Hunter, Jane. The Role of Information Technologies in Indigenous Knowledge Management. *Australian Academic and Research Libraries* 36, no. 2 (2005): 113–127.

Juan, Sarah Flora, Suriati K. Jali, Edwin Mit and John Phoa. "Measuring the Performance of 'Train a Trainer' Programme: A Case Study in Long Lamai, Sarawak, Malaysia." *Proceedings of the ICERI2010 Conference*, 2010, 3118–3123.

Kargbo, John Abdul. "Indigenous Knowledge and Library Work in Sierra Leone." *Journal of Librarianship and Information Science* 38, no. 2 (2006): 71–78.

Kwache, Peter Zakawa. "The Imperatives of Information and Communication Technology for Teachers in Nigeria Higher Education." *MERLOT Journal of Online Learning and Teaching* 3, no. 4 (2007): 395–399.

Lyndon, Novel, A. C. Er, S. Sivapalan, Hasnah Ali, A. C. R. Rosniza, A. M. Azima, A. B. Junaidi, M. J. Fuad, Mohd Yusoff Hussein and A. R. Mohd Helmi. "The World-View of Penan Community on Quality of Life." *Asian Social Science* 9, no. 14 (2013): 98–105.

Open Data Kit (ODK). Accessed March 28, 2015. opendatakit.org.

Rodil, Kasper, Søren Eskildsen and Matthias Rehm. "Demonstration of a Cultural Indigenous Knowledge Transfer Prototype." Paper presented at Indigenous Knowledge Technology Conference 2011 (IKTC2011): Embracing Indigenous Knowledge Systems in a New Technology Design Paradigm, Windhoek, Namibia, 2011.

Sercombe, Peter G. "Small Worlds: The Language Ecology of the Penan in Borneo." In *Encyclopedia of Language and Education*, edited by Nancy Hornberger, 3068–3078. New York: Springer, 2008. DOI=10.1007/978-0-387-30424-3_230.

Siew, Siang-Ting, Alvin W. Yeo and Tariq Zaman. "Participatory Action Research in Software Development: Indigenous Knowledge Management Systems Case Study." In *Human-Computer Interaction. Human-Centred Design Approaches, Methods, Tools, and Environments,* edited by Masaaki Kurosu, 470–479. Berlin and Heidelberg: Springer, 2013.

Velden, Maja Van Der. "Knowledge Facts, Knowledge Fiction: The Role of ICTs in Knowledge Management for Development." *Journal of International Development* 14, no. 1 (2002): 25–37.

Winschiers-Theophilus, Heike, Nicola J. Bidwell and Edwin Blake. "Community Consensus: Design beyond Participation." *Design Issues* 28, no. 3 (2012b): 89–100.

Winschiers-Theophilus, Heike, Kasper Jensen and Kasper Rodil. "Locally Situated Digital Representation of Indigenous Knowledge." In *Proceedings of the Cultural Attitudes Towards Technology and Communication,* Murdoch University, Australia, 2012a, 454–468.

Yeo, Alvin Wee, Faisal S. Hazis, Tariq Zaman, Peter Songan and Khairuddin A. Hamid. "Telecentre Replication Initiative in Borneo, Malaysia: The CoERI Experience." *The Electronic Journal of Information Systems in Developing Countries* 50 (2011): 1–15.

Zaman, Tariq, and Alvin Wee Yeo. "Ensuring Participatory Design through Free, Prior and Informed Consent: A Tale of Indigenous Knowledge Management System." In *User-Centric Technology Design for Nonprofit and Civic Engagements,* edited by S. Saeed, 41–54. Cham, Switzerland: Springer, 2014.

17 Language Vitalization through Mobile and Online Technologies in British Columbia

Peter Brand, Tracey Herbert and Shaylene Boechler

British Columbia's Language Context

British Columbia, Canada, is home to 203 First Nations communities, 34 languages and 61 dialects. Approximately 60% of the First Nations languages of Canada are spoken in BC. According to the *2014 Report on the Status of B.C. First Nations Languages*, 4.08% of BC First Nations speak an Indigenous language, 59% of speakers are 65 years or older and 65% of communities have access to recordings of their language. Research has shown a link between a strong linguistic and cultural identity and an increase in social, mental and physical wellbeing (First Peoples' Cultural Council 2014). Communities engaged in language vitalization have observed increased rates of high school graduation, which include mature students returning to complete basic education and heightened motivation to attend University and other post-secondary programs. Currently, language apprentices from across BC are completing Master's Degrees to prepare them to support language vitalization in their communities.

The First Peoples' Cultural Council (FPCC) is a First Nations-run Crown Corporation established in 1990 with a mandate to support the vitalization of Aboriginal language, arts and culture in BC. The organization works with a community-based committee of cultural experts representing the 34 BC languages to develop strategies that meet the unique needs of communities at various stages in their language vitalization efforts. FPCC supports communities to maintain, preserve and restore their languages by providing funding, training, capacity building and advocacy for language immersion. These goals are achieved through collaborating, planning, and archiving programs that support language and culture (First Peoples' Cultural Council). The organization also monitors the status of BC First Nations languages and publishes these findings to provide communities with baseline information and advocate for additional investments in language activities.

By the year 2000, the beginning of its second decade of service to BC language groups, FPCC was receiving numerous funding applications from community language teams seeking to record elements of their language on compact disc, the preferred language documentation technology at the time. Recognizing the challenges associated with distribution and updating of language CDs for their community members, the FPCC Advisory Council decided to seek a more cost-effective investment that would better serve the diverse needs of First Nations language documentation activists.

The Beginnings of FirstVoices.com

FirstVoices is the innovation of elementary school teaching colleagues John Elliott and Peter Brand. John is the son of the late Dave Elliott, a language pioneer who in the 1970s single-handedly developed the unique orthography used by the WSÁNEĆ First Nation to write their SENĆOŦEN language. Following in his father's footsteps, John committed his career to his SENĆOŦEN language and culture, leading the language program at ȽÁU,WELṈEW Tribal School, a First Nations elementary school located on Vancouver Island, BC, from 1978 to the present. Peter's earliest exposure to Indigenous languages occurred in 1971–1972 while teaching at an outback school for Warrumungu children in Australia's Northern Territory. Both men ascribe these early influences as critical elements of their passionate commitment to Indigenous language documentation and vitalization.

The duo first collaborated with Angus Gratton, an Australian high school student, on the development of the multimedia authoring software, Vocab LanguageLab in 2000. Vocab LanguageLab was designed to enable young users to create media-rich presentations by incorporating text in their Indigenous language, images, audio recordings and video. The enthusiasm of their students and fellow teachers for their initial innovation inspired John and Peter to envision similar language resources for a wider audience via the Internet.

A Collaboration with First Peoples' Cultural Foundation

The office of FPCC is only a couple of kilometers from ȽÁU,WELṈEW Tribal School. Hearing about the work of John and Peter, then-Executive Director Simon Robinson stopped by the school unannounced on February 28, 2001, to meet with the two teachers and discuss their recent language technology experiments. When the pair described their vision for a multilingual web-based language repository, Simon recognized the ideal solution to the pressing needs of FPCC. As Simon drove away from that first meeting at 10.54 AM, the 6.8 Nisqually Earthquake shook the Pacific North West—an interesting foreshadowing of a groundbreaking new innovation for Indigenous languages!

When presented with the two teachers' vision for a web-based language archiving tool, the FPCC board committed seed funding to the development of a prototype, and Peter was granted a one-year leave of absence from his teaching post to lead the project for the First Peoples' Cultural Foundation (FPCF), the sister organization of the Council. From its roots as a simple table of words and phrases for the SENĆOŦEN language, FirstVoices would grow to become a sophisticated database-driven online archive hosting more than 70 Indigenous languages in Canada, Australia and the USA. These humble roots have contributed to the project's success, and particularly to the trust placed in the project by so many First Nations.

The Launch of FirstVoices.com

FirstVoices.com officially launched in June 2003 following two years of development, including design input and focus group testing by Aboriginal consultants (Figure 17.1). In 2005, a companion set of interactive online games was designed to present the archived FirstVoices language data in creative learning activities.

Figure 17.1 FirstVoices.com website.

At its core, FirstVoices is an interactive multimedia dictionary and phrase collection containing thousands of text entries in many diverse Aboriginal writing systems, enhanced with sounds, pictures and videos. FirstVoices also offers tools for the recording of media-rich song and story collections. Some language archives at FirstVoices are publicly accessible, whereas others are password protected at the request of the language community. Using

FirstVoices, Indigenous communities can document their languages, and manage their own language resources. Once the data is recorded and stored, Indigenous communities may choose to make their language archive available via the Internet for study and cultural transmission. This makes the archives useful to many audiences, particularly teachers, who can incorporate the data in lesson plans and other language teaching applications delivered to Indigenous students.

In 2005, the BC Head Start program administered by Health Canada approached the FPCC seeking better access to FirstVoices language data for their young pre-reader clients. The request resulted in the development of FirstVoicesKids, which provided a new interface that is high on visual cues and low on text. Watching pre-readers skillfully navigate the site to learn and practice their language skills instantly confirmed the success of yet another community-driven innovation at FirstVoices.com.

FirstVoices provides an excellent opportunity for a two-way mentorship between youth and Elders. Recording language and tradition requires an intergenerational partnership. Elders speak the language and have knowledge of the traditions and culture, whereas young people are comfortable using technology. In this partnership, the Elders provide words, phrases, songs, stories and grammatical input while the youth use the FirstVoices application to upload and organize the data. This data can then be expanded upon and used by the community in the vitalization of their language for generations to come.

FirstVoices – Early Challenges

Most of the 34 distinct BC First Nations languages and multiple dialects use a variety of special characters not available on a standard English keyboard to represent the unique phonemes of each language. The adoption of these special characters occurred over several decades as various linguists, priests, missionaries, teachers, anthropologists and community members either borrowed from the International Phonetic Alphabet (IPA) and the American Phonetic Alphabet (APA) or created their own custom characters. By the year 2000, most BC languages had settled on orthographies for documentation and teaching. Some had already established large bodies of documentation whereas others were just in the beginning stages.

Prior to the introduction of Unicode technology, computer fonts were limited to 256 characters, including uppercase, lowercase, numbers and symbols. This meant that a typical font was already heavily loaded with Roman-based characters before the addition of Indigenous language characters. A document created in a particular font and shared with someone else could only be opened and read in that font if the recipient of the document had the font installed on his or her computer. The sharing and installation of special fonts added yet another burden to the often non-technical Elders attempting to adopt computers into their language programs. Also, early custom fonts used by First Nations language activists could not be employed successfully in web browsers. Displaying First Nations language orthographies at FirstVoices and in related language documentation applications was one of the early challenges for the development team.

Font technology at the time was on the cusp of the 'Unicode revolution'. Wikipedia (2014) describes Unicode thus:

> *Unicode is a computing industry standard for the consistent encoding, representation and handling of text expressed in most of the world's writing systems. Developed in conjunction with the Universal Character Set standard and published as* The Unicode Standard, *the latest version of Unicode contains a repertoire of more than 110,000 characters covering 100 scripts and multiple symbol sets.*

Although taken completely for granted today, the simultaneous arrival of the Internet and Unicode technology in the lives of First Nations language activists is worth noting because of the profound impact and demands it placed on pioneering enterprises like FirstVoices.

Early in the FirstVoices prototype development, the team discovered Keyman, a keyboard remapping software developed by Tavultesoft, a small company based in Tasmania, Australia. The son of a missionary working in Laos, founder and Keyman developer Marc Durdin began exploring keyboarding technologies at age fourteen. By the time the FirstVoices team discovered his work in 2001, Marc's Keyman software was already in use by hundreds of minority languages around the world. FPCC bought a global license on Keyman to provide free custom keyboards for all BC First Nations users of Windows-based computers. Matching keyboards with identical keystrokes were simultaneously developed for Mac users.

Another early challenge confronting the FirstVoices team was the alphabetization of the online dictionaries. Unicode characters do not alphabetize in the same way Roman alphabets do. The problem was overcome with a simple but ingenious solution devised by technical analyst Alex Wadsworth. As part of the preparation of each new archive, community language teams are required to submit their orthography ordered in the way the community has agreed upon. Alex devised a tabular system to assign a number to each character. The FirstVoices database then utilizes the numbers in the table to artificially 'alphabetize' the orthography.

The other major challenge facing the FirstVoices team was the fact that every one of the 34 BC languages and multiple dialects uses a different character set and font. Fortunately the team discovered Chris Harvey, the self-titled 'language geek' at LanguageGeek.com. Chris had already devoted hundreds of hours to the development of font and keyboard technologies for Canadian Indigenous languages when the FirstVoices team discovered his exceptional expertise and commitment. LanguageGeek provides free fonts and free keyboard layouts that try to cover all of the characters necessary for writing Native languages. Following the international Unicode standard, languages can be typed, read, printed, emailed and put on the web using these fonts (LanguageGeek 2014).

With the help of Chris Harvey, FirstVoices promoted, pioneered and lobbied for Unicode technology for all BC First Nations languages, and today more than 70 keyboards are available as a free download from FirstVoices. com. Keyboards can be activated to type in various word processing applications and a web version can be activated directly within the FirstVoices. com interface. From their pioneering work during the 1990s to the present time, Marc Durdin and Chris Harvey have been critically important members of the FirstVoices family. Indigenous language speakers around the world have much to thank them for in the technically challenging road to language literacy.

FirstVoices Training Model

The FirstVoices training model encourages engagement of as much of the local community as possible. If community leaders are interested in using FirstVoices to archive their language, they must first host a meeting to discuss the project and gain support from the community. Once the community agrees to archive with FirstVoices, they submit a letter from their policymakers and administration in support of the project. A contract is then drafted to state that the language data is owned by the community, whereas rights to the technology are owned by the First Peoples' Cultural Foundation.

The FirstVoices training model continues to evolve with the advent of new technologies. The first training session was delivered as a summer course at the University of Victoria, with trainees attending from 15 First Nations across BC and several from San Diego County in California, USA. It soon became evident that sending one or two trainers to a remote First Nation was far more cost effective than bringing trainees to a central location. In addition to being able to work with multiple trainees in the community, the FirstVoices trainers were also in a position to assess and resolve the variety of technical challenges that inevitably arose in the communities.

Because the FirstVoices development trajectory closely tracked the advent of the Internet in many of the small remote client communities, FirstVoices trainers often found themselves on the front line of community technical support. Professional technical support services were sporadic and expensive, with contractors billing for hours of travel to resolve relatively simple issues. The one- and two-week FirstVoices training sessions required the trainers to live in the community for the duration of the session, placing them in an ideal position to detect hardware and software issues and effect or recommend replacements and repairs. These services strengthened the bond between trainers and community members, paving the way for long-term relationships and ongoing technical support long after the trainers had returned to headquarters. Often, community language teams would be found to be making do with older hand-me-down computers, some of which were incapable of handling even the most basic networking and Unicode requirements. This recurring problem soon prompted the allocation of a brand new computer and peripherals to each FirstVoices team as an essential requirement of each new project. This of course required recognition of the need and benefits by the relevant national and provincial funding agencies.

Currently, the advent of web-based 'remote access' enables the FirstVoices technical team to offer both training and technical support from headquarters. Some training and technical support is now delivered via video conference, but many client communities still prefer training on site, when funding allows.

FirstVoices and the Development of Mobile Technologies

Ever since its first tentative experiments developing tools for Indigenous language revitalization, the FirstVoices team has relied on First Nations community language champions and the First Peoples' Cultural Council Advisory Committee for inspiration, direction and feedback. In this way FirstVoices remains dynamic and responsive to its core user base. With the advent of mobile devices, the FirstVoices toolset expanded to include mobile dictionary and language tutor apps.

In 2007, the FirstVoices team partnered with Udutu Online Learning Solutions, a Victoria-based company specializing in e-learning software. In 2010, the FirstVoices Language Tutor was developed in response to requests from communities for an online language-learning tool.

The FirstVoices Language Tutor provides First Nations students with access to graduated interactive web-based vocabulary and conversation building exercises. Using this app, users are able to listen to a word or phrase, record themselves speaking, and compare the result with a recording of a fluent speaker. The online program also includes a student tracking system so parents and teachers can monitor progress through the lessons.

In 2007, Internet access was still not a reality for many remote First Nations communities in BC. In order to support these communities, the FirstVoices team needed to find a way to provide Language Tutor access, independent of the Internet. After multiple working meetings with Udutu, the concept for the FirstVoices Language Lab was conceived.

The FirstVoices Language Lab is an iPad-based language-teaching laboratory designed to deliver Language Tutor lesson content via an app. The lab contains its own small laptop server, Wi-Fi antenna, power bars, portable case and a set of iPads. No Internet access is required, thus enabling remote First Nations communities access to lesson content.

By 2009, the increase in the use of mobile devices prompted First Nations youth to request access to mobile language technologies. To keep up to date with the latest technologies, the FirstVoices team developed dictionary apps for use on the iPod, iPhone and more recently, the iPad. The dictionary apps use a subset of content archived at FirstVoices.com and contain text, audio, images and video. As of September 2014, FirstVoices has developed 12 interactive dictionary/phrase apps, which are available as a free download from the iTunes store.

Whereas the mobile dictionary apps allow users to access language content from their mobile devices, the apps do not allow users to text in their Indigenous languages. In addition, the regular keypads on mobile devices were not capable of generating many of the special characters of Indigenous languages, making texting in these languages impossible for most First Nations people.

A special grant from the Department of Canadian Heritage in March 2010 funded research into the keyboarding requirements for mobile messaging in First Nations languages. In December 2011, FirstVoices approached the First Nations Technology Council (FNTC), a not-for-profit enterprise created to support the technology needs of First Nations in BC, to request support for the development of a keypad and chat app to allow users to compose and text using the unique characters of their Indigenous language. The FNTC agreed to fund the development, and in the spring of 2012, FirstVoices Chat was launched, allowing users to text using Facebook Chat and Google Talk.

FirstVoices Chat had a high-profile world premiere when BC's Lieutenant Governor, Steven Point, and his wife, Gwen Point, exchanged text messages in their Halq'eméylem language before several hundred delegates at the annual FNTC Technology Summit on February 24, 2012. At present, FirstVoices Chat provides custom keypads capable of texting in more than 100 Indigenous languages in Canada, Australia, New Zealand and the United States.

Conclusion

The common thread running through FirstVoices' 14-year success story is the extraordinary level of community engagement. The project was initially envisioned by two elementary school teachers with limited technical expertise working in a tiny tribal school. Their early experiments, intended to engage their increasingly tech-curious young students, demonstrated a valuable place for technology in the school language program. The lucky convergence of a need recognized by the leadership of the First Peoples' Cultural Council and the vision of the two teachers created a perfect opportunity for an innovative new use for the Internet.

An important reality of innovations like FirstVoices is the ever-evolving technical landscape. The underlying hardware and software employed by each element of FirstVoices require regular updating to keep abreast of new technological advances. Ongoing funding, innovative thinking and strong leadership are critically important to ensure that the priceless linguistic data collected at FirstVoices.com remains secure and accessible to language teachers and learners. Numerous First Nations have entrusted First Peoples' Cultural Council with their language resources for safe-keeping and free, open access online and via mobile apps. The organization takes this responsibility very seriously.

Since colonial times, First Nations people have witnessed successive waves of innovation and experimentation, often at their own expense. The FirstVoices story is different. From the moment of the original concept to the present, the project has been led by First Nations people, for First Nations people. Each FirstVoices innovation has been carefully weighed by First Nations community representatives to ensure that it meets the needs and wishes of their constituents, particularly youth. One of the most rewarding and exciting aspects of the project is the forum it provides for enduring partnerships between First Nations elders and technically savvy youth. Each member of these partnerships brings important skills, knowledge and experience to the project, and invariably their lives are enriched by newfound mutual respect. The program has never been about profit, which in turn has drawn exceptional hard work, dedication and commitment from the many innovators and language champions who have contributed to its success.

References

FirstVoices. Accessed August 30, 2014. http://www.firstvoices.com.

First Peoples' Cultural Council. Accessed September 24, 2014. http://www.fpcc.ca/.

First Peoples' Cultural Council. *Report on the Status of B.C. First Nations Languages 2014.* 2nd ed. Brentwood Bay, B.C.: First Peoples' Cultural Council, 2014.

LanguageGeek. Accessed September 24, 2014. http://www.languagegeek.com.

Wikipedia. Accessed September 29, 2014. http://en.wikipedia.org/wiki/Unicode.

18 The Influence of Mobile Phones on the Languages and Cultures of Papua New Guinea

Olga Temple

Introduction

"Society is purely and solely a continual series of exchanges" made possible by the "admirable and wise invention of language", claimed Destutt de Tracy, the French philosopher and political economist two centuries ago (de Tracy 1817). This continual succession of exchanges gives us, he argued, three very remarkable advantages in terms of:

- **Productivity:** "the labour of several men united is more productive, than that of the same men acting separately";
- **Acquisition and utilization of knowledge:** "No man is in a situation to see everything, and it is much more easy to learn than to invent";
- **Division of labour:** "when several men labour reciprocally for one another, everyone can devote himself exclusively to the occupation for which he is fittest" (Tracy 1817).

It stands to reason, of course, that these great benefits of society will become even greater if social exchanges are conducted more effectively. This is why the invention of writing marked a pivotal stage in human development: it enabled us to accumulate and share information across geographical distances and through time to the benefit of all. Writing gave us control over our internal biological memory by externalizing it.

> *Everything that civilized humanity remembers and knows at present, all the accumulated experience in books, monuments and manuscripts—all this colossal expansion of the human memory, without which there could be no historical and cultural development, is due precisely to external human memorization based on symbols.*
>
> *(Vygotsky 1930)*

Because "Memory is enhanced to the extent that systems of writing and of symbols, together with the methods for using those symbols, are enhanced" (Vygotsky 1930), the Internet and the new mobile communication technologies have marked another pivotal point in the development of our civilization by enabling instant series of exchanges globally and in real time. Digital technologies have connected our world, erasing geographical and political barriers that had separated societies and cultures. By giving individuals and organizations access to all human knowledge ever accumulated, they have expanded our collective RAM exponentially.

Yet, nothing is ever simple; new technologies have also created a new kind of language barrier—a generational divide ("text-speak", also known as "netlingo", " tech-speak", "Youthspeak", etc.), which now separates users from non-users of digital communication tools. With new concepts/words being created almost every minute, according to Global Language Monitor (2015), the number of English words has already exceeded one million—ten times more than in Shakespeare's time!

"The World Is Flat", declared Thomas Friedman (2005), writing his *Brief History of the 21st Century*. He identified the unprecedented diffusion of information (one of those remarkable advantages of social exchanges), made possible by new technologies, as the main cause of the world becoming one global "information society". The digital tools we now use to conduct our social exchanges have set in motion powerful socio-economic forces that are rapidly homogenizing humanity. As all tools can be (often are) used both for "the good" and for "the bad" (relative as these concepts are), there is a lot of divided opinion and debate about the ways in which digital technologies have affected our lives. Anthropologists, sociologists, economists, political scientists and communication professionals have been analyzing from every angle the dramatic "flattening" of our world and all the associated issues and cultural changes (Handman 2013; Hannerz 2000; Watson 2011, etc.).

In this chapter, I want not only to describe some technologically induced changes in the languages of Papua New Guinea (PNG), but also to discuss their social implications, such as shifts in people's attitudes and perceptions of traditional cultural values and "national" identity. Important issues all around our newly flattened world, they are of particular significance in PNG's uniquely multicultural society, where government policies in the area of education and cultural development are bound to have a huge impact on the entire future of the country. Amidst the clamor for cultural "preservation", which generally dominates public discourse, there are also caveats against possible manipulation of people's insecurities at a time of change:

> *There may be a preoccupation with cultural autonomy and the defense of a cultural heritage for its own sake, yet frequently this rhetoric of culture is closely linked to power and material resources as well.*
>
> *(Hannerz 2000)*

Courtney Handman (2013) of Reed College, Portland, Oregon, gave a noteworthy historical account of the politics and ideologies of cultural development and language planning that have been playing out in PNG since before Independence. To add depth to our subsequent discussion, let us first examine some factual evidence of ongoing changes, obtained from a series of sociolinguistic studies conducted at the Linguistics Department, School of Humanities and Social Sciences, University of Papua New Guinea since 2008.

The Internet and mobile phones have transformed traditional "series of exchanges" in PNG. The story of this transformation can be best told through the words people use in conducting their exchanges. Having arrived in Port Moresby in 1998, I witnessed the technological revolution in this country first hand and attempted to take the pulse of the linguistic change occurring. My students and I recorded more than a thousand words of the emerging students' text speak on the two campuses of the University of Papua New Guinea (UNPG) in 2008 (Temple et al. 2009). Two years later, we researched the impact of mobile technologies on some PNG vernacular (Tok Ples) languages (Temple et al. 2011). I will now summarize the findings of these studies and compare them with the findings of a follow-up survey we completed in September 2014. On the basis of this evidence, I will then share with you my thoughts about how mobile phones are impacting the PNG ideological landscape.

Linguistic Overview of Papua New Guinea

For millennia, the rugged mountains of this breathtakingly beautiful island had kept hundreds of its ethnic communities in relative isolation from each other and from the rest of the world, thus "culturing" a diversity of distinct languages. The Summer Institute of Linguistics (SIL) website lists 850 individual languages for Papua New Guinea (838 as living, and 12 as extinct). Of the living languages, 43 are institutional, 303 are developing, 348 are vigorous, 108 are in trouble and 36 are dying. To put these figures into perspective, we must remember that there are only about 6,732,000 Papua New Guineans speaking all those 838 languages! (SIL 2015) Most of the languages (78%) are Papuan (spoken on the island of New Guinea and several neighboring islands in the West Pacific), and about 20% are Austronesian (they belong to one of the world's major and geographi-cally most widely dispersed language families, on par with Indo-European, encompassing maritime Southeast Asia, Oceania, Madagascar, Taiwan, Micronesia, Hawaii and New Zealand). Papua New Guineans fondly refer to all of their Indigenous languages as *Tok Ples* ("Talk of the Place"/"Talk of the Home Village").

At the 2014 Linguistic Society of Papua New Guinea confer-ence in Madang, Professor Aikhenvald of James Cook University, Australia (an authority on Papuan languages), gave a breakdown of the approximate numbers of Papuan/Austronesian languages and their speakers (Table 18.1):

Table 18.1 The languages of Papua New Guinea and approximate numbers of speakers. (Adapted from: Aikhenvald 2014)

Number of speakers	Austronesian	Non-Austronesian
over 10,000 – 100,000	about 12	about 60
1,000 – 10,000	about 100	about 230
over 500 – under 1000	about 30	100 and more
between 100 – 500	under 50	170 and more
fewer than 100	about 10	about 50

The only Tok Ples spoken by more than 100,000 people, according to Aikhenvald, is Enga (Eastern Highlands Province of PNG). By all accounts, these figures bring into question the future survival of at least 410 languages (90 Austronesian and 320 Papuan).

Colonial expansion and the military conflicts of the 20th century thrust the over 600 islands of New Guinea, big and small, into the wider world. In 1949, the Territory of Papua and New Guinea was established, eventually culminating in the birth of the independent state of Papua New Guinea (September 16, 1975). Considering that, as recently as in June 1954, an aerial survey revealed several previously undiscovered highland valleys, inhabited by up to 100,000 people, it is truly amazing that Papua New Guinea is now an integral part of our interconnected Flat World, buzzing 24/7 with "continual series of exchanges" (Tracy 1817). Powerful socio-economic and political forces have shaped its composite national identity, held together by three lingua francas— English, Tok Pisin and Hiri Motu. Tok Pisin ("Talk Pidgin") is a creole language, whose words derive mostly from English (with an admixture of vernacular, Portuguese and German) but whose syntax is reminiscent of simplified vernacular structures. Hiri Motu (also known as Police Motu, Pidgin Motu or just Hiri) is a simplified version of Motu (an Indigenous Austronesian language); Hiri Motu evolved as a language of communication between the British/Australian administrators and the Papuan Motu-speaking population.

Words—Reflections of the Changing World by Human Minds

This complex linguistic mosaic has changed dramatically in the past few years. Since 2007 mobile phones have become ubiquitous in urban centers and even in remote villages of PNG, spawning with their spread numerous varieties of text-speak whose word-meanings are unintelligible to the uninitiated (Figure 18.1).

Figure 18.1 Man texting in Goroka, Eastern Highlands Province. (Photo: Irene Gashu, July 2014).

This fascinating "live" display of linguistic change in action prompted many studies into the impact of the Internet and mobile technologies on the languages of PNG (Handman 2013; King 2014, etc.).

The studies we conducted at UPNG in 2008 and 2010, apart from recording the new words, used the dialectical method of linguistic analysis to explain how and why they had emerged in the first place (Temple et al. 2009, 2011). Here is a brief summary of the theoretical framework of our research projects, as well as their aims, objectives, findings and general conclusions.

Theoretical Framework: Dialectical Analysis

Modern linguistic theory is the product of a deeply rooted tradition of scientific analysis of observable facts, which has dominated linguistic inquiry since the Enlightenment, particularly in the past century. Yet, there is more to language than just its observable physical structures. The atomistic, descriptive method of linguistic analysis, while providing a wealth of observable detail, "misses the forest for the trees". Once broken into its smallest bits, the Humpty-Dumpty of language cannot be put together again for, as Aristotle said in Book I of *Metaphysics*, "The Whole is more than the sum of its parts".

To understand the workings of a complex whole, we must identify its smallest functional unit and study its properties, which determine how the units interact within the whole, thus ultimately shaping the behavior of the complete system. Inexplicably, this method of analysis, axiomatic in the physical and natural sciences, has not yet been generally accepted in linguistics, despite numerous historic enquiries into the language/thought relationship—from Plato's *Cratylus* to the Speculative Grammar of the Scholastics (von Erfurt 1310/1972), the Rational Grammar of Port Royal (Arnauld and Lancelot 1975), Herder's (1772) *Treatise on the Origins of Language*, Wilhelm von Humboldt's *Heterogeneity of Language*, the linguistic relativism of Sapir and Whorf and the poststructuralist theories of Michel Foucault and Jacques Derrida, etc. More recent work emphasizes the very mechanism of human thought as the key to understanding language (Christiansen and Chater 2007; Sakai 2008; Temple 2009, 2011, 2012, 2013). Christiansen and Chater (2008, 489) suggest that:

> *apparently arbitrary aspects of linguistic structure may result from general learning and processing biases deriving from the structure of thought processes, perceptuo-motor factors, cognitive limitations, and pragmatics.*

Combining the advantages of both synthesis and analysis in its approach, the dialectical method of analysis views language as verbal thought—every word is already a generalization and, therefore, an act of social thought; it reflects reality as it is understood by each society in concrete time-space. Word-meaning, therefore, is the smallest functional unit of language, with all of its psychological, physical, social and historical properties. A person's behavior can only be understood in the context of all the interrelated aspects of his or her being ("body and mind" in concrete time-space). Likewise, word behavior (syntax) can only be understood in the context of all these interrelated aspects of word-meanings.

The dialectical perspective and method of analysis provide a valid and verifiable explanation of how abstract thought (also known as generalization), governed by the universal principles of human understanding, generates all words and phrases. Therefore, we view all meaning, all verbal thought (conceptualization) as *generalization* (i.e., synthesis and analysis of ideas by resemblance, contiguity in space-time and cause/effect). In the context of our research, this means that the analysis of all the new words and phrases recorded has shown that mental associations by resemblance, contiguity in space-time and/or cause/effect had caused their emergence.

Spread of Mobile Technology in PNG

Until mid-2007, mobile phones were prohibitively expensive, which put them out of reach of ordinary people in PNG. Commercial competition, introduced into the mobile telecommunication sector that year, flooded the market with cheap mobile phones and expanded network coverage across most of the country (Watson 2013).

Indeed, in the cities, as well as in the villages of Papua New Guinea, most people nowadays use cell phones to communicate with family, friends and *wantoks* ("one talk" people/tribesmen, speaking one language). Smartphones are now becoming more common (and more popular!) in urban centers, where there is better network coverage. Over four-and-a-half million Papua New Guineans out of about seven million are mobile phone users, with about 300,000 of them on Facebook (FB) (*The National* August 6, 2014, 49). Telikom (and their subsidiary Citifon), Bmobile/Vodafone and Digicel have significantly improved and expanded their network coverage and wireless services in PNG (Figure 18.2).

Figure 18.2 A Telikom/Citifon outlet in the Vision City Mall, Port Moresby. (Photo: Irene Gashu).

Today, "Even the little village kids know what Google or Facebook is," wrote an Enga student of mine in his essay on the impact of the Internet on Tok Ples languages of Papua New Guinea (Language, Culture, and History exam, October 2014); "When you travel around the city, you will notice that kids will sit in pairs, discussing what they saw on FB or Google in their vernacular languages. FB and Google are really becoming part of everyday life of the people. When they go to FB or Google and find anything new or interesting, they will … use these words with their fellow students or mates. And also they will go around telling everybody that they have discovered something on FB or Google, and telling the names of these things, and it becomes a new word … And this goes on and on."

   With increasing numbers of smartphone users, newer and cheaper ways of instant messaging, such as WhatsApp, Skype, Facebook Messenger, Google Chat and Hangouts seem to have started taking over from SMS. If only a couple of years ago relatively few students were seen using personal computers, notepads and smartphones on UPNG campuses, the situation has changed dramatically in 2014. Now, all UPNG students have personal e-mails and Internet access in the University Library, and many also have their own laptops, notepads and smartphones (See Figure 18.3):

Figure 18.3 My students "taking down notes" in class, UPNG. (Photo: Irene Gashu).

   It is now common to see students, faculty and staff, even security officers and cleaners totally focused on their mobile devices, messaging, playing games, surfing the net, downloading or sharing music and photos, taking selfies, videos—you name it (Figure 18.4)! Despite the higher cost, smartphones, with their multiple functions, are now displacing the old mobile phones in Port Moresby and other cities of PNG.

Figure 18.4 UPNG student using her smartphone. (Photo: Irene Gashu).

Research Methods

All of our three studies (in 2009, 2010 and 2014) aimed to (1) record changes in the languages spoken in PNG and particularly in the vernacular Tok Pisin, and (2) publish a baseline lexicon of neologisms in these languages, both with the objective of explaining the ongoing language change using the dialectical method of linguistic analysis.

Our methods of data collection included observation of language use in the student community and *wantok* groups; harvesting data from popular social networking sites, such as Facebook and Twitter and a cross-sectional survey based on convenience sampling. The survey consisted of anonymous pre-tested questionnaires, designed to record Tok Pisin and Tok Ples texting lingo used amongst the UPNG student community and their *wantok* groups, as well as in the wider *wantok* communities, including the researchers' villages.

The focus of the 2009 study was on student texting lingo, then emerging on the Waigani and Taurama campuses of the University; English and Tok Pisin words and phrases constituted most of our "harvest", with basically no feedback for the Tok Ples languages. The 2010 study attempted to fill in the gap: nine researchers collected lists of text-speak in Boiken, Dobu, Foi, Hiri Motu, Iduna, Iatmul, Keapara, Kuanua, Loniu, Solos, Taulil, Weri and Yil languages.

Our 2014 survey, conducted by eighty-four UPNG students under my supervision, was a follow-up on the earlier studies, designed to identify and describe any new developments and trends in language use that may have occurred since 2010. We recorded more than 2,300 words and phrases of Tok Pisin text-speak, as well as a considerable number of words and phrases from 31 Indigenous languages of Papua New Guinea and two languages of the nearby Solomon Islands. These languages are: Abadi, Agarabi, Ajera, Avaipa (Bougainville), Boiken, Darno (Eastern Highlands Province), Enga, Foi, Gabadi (Central Province), Gela (Solomon Islands), Hiri Motu, Hoatana (Solomon Islands), Huli, Imbongu, Kame'a (Gulf Province), Kanite, Kewabi, Kuanua, Kuman, Kurti (Manus), Mamusi, Melpa, Mengen, Mokorua (Tufi), Motuna, Nakanai, Nikanik Pau, Nunga Naga, Telei (Bougainville), Tiang Tigak, Tungak, Wedau and Wiru. Numbers of words collected vary from language to language, due to the researchers' productivity or competence, among other factors. However, we have obtained, on average, up to a hundred new words and phrases in these languages.

The list of all 2,312 Tok Pisin words, acronyms and phrases recorded in 2014 is available on my website <http://www.templeok.com/research.htm>; the data obtained for the Indigenous Tok Ples languages we have examined are available upon request.

Findings of the Three UPNG Studies Compared

Findings: 2009

Our 2009 survey recorded more than 2,000 words and acronyms of the UPNG students' texting lingo (1,113 for English and 1,026 for Tok Pisin, to be exact). It was the first description of the emerging text-speak amongst the UPNG students. As stated before, no data was obtained for Tok Ples languages.

Findings: 2010

The 2010 study suggested that the Internet, mobile phones and, in particular, SMS technology had impacted Tok Ples languages differently from the way they did English and Tok Pisin:

- The average amount of data collected in each of the 13 Tok Ples languages we worked on was much scantier than the well over 1,000 items in the English and Tok Pisin lists that we had "harvested" in 2009. Respondents were often at a loss, when asked to write down the Tok Ples words they commonly used in texting, saying that they "just use the ordinary words" they would normally use when speaking (suggesting that they do not use any texting lingo).
- Most of the words in the lists we obtained were of common everyday use, sometimes (but not always) abbreviated—this was in sharp contrast with the thousands of acronyms and shorthand of Tok Pisin and English SMS lingo we had obtained in 2009.

- All of the word lists contained a large proportion of Tok Pisin words and acronyms, suggesting a widespread Tok Pisin influence on the Indigenous languages spoken in urban centers, such as Port Moresby.
- English/Tok Pisin borrowings in the Tok Ples lists of computer terms exhibited a marked influence of Tok Ples phonology (i.e., in the Foi language of North-Eastern PNG "bulete" for "bullet", "borosa" for "browser", "bolde" for "bold", etc.).

We concluded that the reasons for our relatively poor 2010 "harvest" of Tok Ples SMS were that:

- Residents of Port Moresby tend to use Tok Pisin when texting, because of their wider, cross-cultural communication needs.
- SMS is generally not very popular in *wantok* communities (particularly in the villages), due to low Tok Ples literacy levels, lack of language development, or a combination of both.
- Some tonal languages (like Weri, which has two phonemic tones) are difficult to transcribe due to lack of appropriate symbols on cell phone keypads; in order to avoid misunderstanding, speakers resort to using Tok Pisin SMS.
- Tok Pisin SMS adequately satisfied the communication needs of all Tok Ples speakers in terms of referencing new concepts that had no Indigenous names.
- Internet use had not yet become commonplace in village communities, due to lack of infrastructural development, power supply and education; therefore, rural dwellers' use of social networking sites, such as Facebook or MySpace, was almost non-existent.

The characteristics of Tok Ples SMS, described in 2010 included:

- Tok Ples SMS, and particularly the new technical/computer terms, are typically borrowings from Tok Pisin which, in turn, feeds on English words/ jargon.
- The Tok Ples borrowings from Tok Pisin/English were heavily influenced by Tok Ples phonology, with insertion of epenthetic vowels even more common than in Tok Pisin; for example, Foi *kibode* for "keyboard," *kibede* for "keypad," etc. (Epenthic vowels are inserted to break up consonant clusters.).
- Tok Ples SMS samples collected exhibited the same morphological changes that have been observed in other texting lingos, i.e.:
  - Words tend to be shortened through vowel deletion or clipping in easily recognizable words (i.e. "npra" for Solos "napora" (basket), "vavi" for Iduna "vavine" (girlfriend, wife), etc.;
  - People commonly used acronyms, i.e., CU for "see you," IOU for "I owe you" or IOT for "in order to," etc.
  - Creativity often trumps time and cost constraints: i.e., use of fancy symbols, such as "@" or "2x," etc. as expressions of individual flair.

- Code switching from Tok Ples to TokPisin and English was common, especially when communicating on less traditional subjects; for example, Solos "N'meh ere" for Tok Pisin *Km lo hia/* English *Cme h're*, etc.
- As in Tok Pisin, in Tok Ples SMS words of English origin, people generally ignore conventional English spelling, e.g. call → kol, feel → fil, etc.

Quite apart from borrowing and code-switching, Tok Ples SMS exhibited common structural features with other SMS lingos (use of acronyms, vowel deletion, clipping, etc.).This was attributed to the universality of a) mechanisms of human thought, and b) incentives and constraints of the SMS communication medium.

Our results indicated that new technologies, by introducing novel concepts and by providing new channels for both *wantok* and cross-cultural communication, had accelerated the processes of linguistic change in PNG. Most of the Tok Ples languages examined showed a strong and growing influence of English and, particularly, Tok Pisin. We concluded that the observed strong influence of English on Tok Pisin and the overwhelming influence of Tok Pisin on Tok Ples communication in multicultural environments indicated long term Tok Ples vulnerability.

Findings: 2014

The 2014 study yielded very interesting results for messaging in Tok Pisin. Data collected (more than 2,300 words and phrases) showed a tremendous increase in the influence of English on Tok Pisin. This was evident in:

- Frequent code switching (search, love, device, balance, voucher, recharge, date, etc.);
- Numerous English borrowings: lots of new concepts have been incorporated into the language. Having undergone phonological and morphological adaptation, they qualify as new Tok Pisin words by regularity and frequency of use (i.e., *usa aidi* = user ID; *ringim vois mail* = to call a girlfriend; *Intanet*; *pon/fon* = phone; *yunit* = unit; *miskol* = missed call; *risivim/painim netwok* = to receive/ find network; *rechagim Digicel flex akaunt* = to recharge Digicel/flex account; *blututim piksa/music* = to Bluetooth photo/music, *daialim* = to dial; *daunloudim* = to download; *teksim* = to text; *konfemim/ confemim* = to confirm; *koneksen* = connection, etc.);
- Increasingly frequent omission of the Tok Pisin transitive verb marker, i.e., *saikup* = to psych up; *skrol* = scroll; *konekt* = to connect; *fesbuk* = to facebook; *akses Intanet* = to access Internet; *andu* = undo; *topap* = top up; *ringaut* = to ring out, etc.;
- Frequent use of English functional morphemes (*bat* = but; *and*; *so*; *tru* = through, etc.) and inflections (the plural – s: *frens* = friends; *kontaks* = contacts; *yunits* = units; *fons/pons* = phones, etc.;

- Increasingly less frequent epenthesis : this may also be due to the general tendency to leave out vowels when texting: OBT – *operesen burukim tulait* → *operesen brukim tulait* (studying until dawn); *onst* – *turu* → *trutru* (honest); *poket buruk* → *poket bruk* (I'm broke); *sigirapim* → *sigrapm* (scratch it), etc.

Similar findings were reported for Tok Pisin by Courtney Handman (2013) and Phil King (2014).

With regard to the Tok Ples data collected in 2014: Many neologisms, particularly the technical concepts, are borrowings from English, often via Tok Pisin: i.e., Melpa: *combita/computa/komputa onim* = start a computer; *pawa onim et men mel* = power switch, *e-mail adres* = email address, *ip adres* = IP address, etc. However, most of them are derived from Indigenous words, mostly by clipping: i.e., Hiri Motu: *lebux* < *lebulebu* = show off; *lalo* < lalokau = love; *gagaru* < *garugaru* = infant; Melpa: *pa* < *panga* = go, etc.

Most of our 2010 generalizations about Tok Ples languages were confirmed in 2014; i.e., the use of vernacular languages in texting is restricted by very practical concerns: people often feel that Tok Ples words are "too long" to be typed out, and there are no symbols on phone keypads for the sounds and tones of many Indigenous languages, such as Weri, Kame'a, Melpa, etc. However, compared to our 2010 survey, we noticed a sharp increase in the numbers of new words and phrases (both Indigenous and borrowed) in many of the 31 Indigenous PNG languages examined. This may be attributed to the explosive growth in the numbers of mobile device users, better network coverage and purposely targeted messaging (coded *wantok*). Our data suggest that the main reason for this increase is that people conceptualize new reality mainly through adopting ready-made English or Tok Pisin words and phrases—hence, the extensive borrowing. Acronyms and clipping of Indigenous words, as well as sound change and use of trendy suffixes, such as *–ex* and *–x*, are also common: Mamusi: *pri kol* = free call; *putu* = photo; *tatex* = brother; saltex = daydream; Tok Pisin: *gelix* = "meri man," womanizer; *hunx* = hungry; *jux* = junior, etc.

Despite these developments and government encouragement of Tok Ples use, i.e., the policy on vernacular education (1990s-2012), our 2014 data still points to a general Tok Ples vulnerability, caused, in our view, by the decreasing *utility* (social function) of the smaller Tok Ples languages: to conduct social exchanges in modern Papua New Guinea, people need one indispensable tool—a common language.

Research Significance

This research provides evidence of rapid linguistic change as a direct consequence of mobile phone use in PNG. The smaller languages will inevitably lose their usefulness in social exchanges, which will probably lead to their eventual demise. As noted above, there are more than 400 PNG languages

with fewer than 1,000 speakers; they can no longer perform the social function of language—that is, to satisfy their speakers' communication needs
outside of their villages. It is hoped that our research, by providing evidence
of long-term Tok Ples vulnerability, may motivate action, particularly in the
field of language documentation and education.

Our hypothesis regarding the mechanism of conceptualization (which
causes new words) was convincingly validated: every word we recorded was,
indeed, a *contiguity* of concept, *caused* by *resemblance* (i.e., generalization).
For example, Tok Pisin: *kaix* = food, eat; *laix* = likes—resemblance to *kaikai*
and *laiks*, respectively; *lalu ful bar*—resemblance to "love you" and analogy
with a strong Wi-Fi signal, "full bar," etc. Generalization, of course, works
hand-in-glove with phonological (sound) change—our anatomy and our collective generalizing minds together cause all languages and language change.

Mobiles and Cultural Change in Papua New Guinea

Words reflect people's understanding of their world; mobile phones have
spawned hundreds of new concepts embodied in trillions of new words
around the world. Despite the obvious enthusiasm with which the new
technologies have been embraced by all sectors of PNG society, there is a
seemingly contradictory, but also an unmistakable negativity toward the
changes they have brought. What is peculiar about it is that it emanates
from the intellectual circles of PNG—academics and educators—people
who use (and benefit from) the new tools the most. This general undercur-
rent regularly surfaces in public discourse and is particularly manifest during cultural events. The 5th Melanesian Festival of Arts and Culture, which
took place in June-July 2014, celebrated cultural diversity in PNG and in
the other independent states of the South Pacific region. It was, indeed, a
magnificent display of traditional song and dance, as well as a venue for
group discussions and seminars on culture-related issues. Artists and academics (including scholars from Western universities), government agencies and NGOs, as well as representatives of the business community, all
stressed the importance of cultural preservation, extolling the value of cultural diversity. National papers, such as the *Post Courier* and *The National,*
reported widely on all the happenings and carried numerous articles on cultural issues. Dr. Steven Winduo, a UPNG academic and poet, expressed the
quintessence of this popular sentiment in his "Steven's Window" column in
the *Weekender*:

> *The event has brought to light the question of leaders and their respon-*
> *sibility in making sure our cultures are celebrated, promoted, and pro-*
> *tected. Where such responsibility is absent, we are likely to go through*
> *a process of cultural death that does no good to our people. In many of*
> *our Melanesian states, there are unwarranted acts of cultural sabotage*
> *that threaten our own declarations to uphold our cultures and sing out*

aloud the praises of our common heritage. Each Melanesian country uses culture to build and strengthen its nationhood. ... Nations are vindicated by their cultures. Without culture, a nation cannot claim a political identity. Nationalism is a defense of a nation's cultural inventions. Nationalism vindicates its own inventions by politicizing its culture. A nation is denied of its identity, once culture is separated from it.
(Weekender, July 11, 2014)

From the dialectical perspective, culture is not a fixed entity that can be separated from a nation—it is a people's way of life, and life does not stand still. The "culture" concept encompasses all the systems of meaning created by societies in the course of their interaction with their constantly changing environments, as well as social behaviors shaped by people's understanding of the world they live in. In other words, culture is what people think, know and believe, what they do, and what they produce at any given place and time. Yet, in Papua New Guinea, the term is generally used to refer to the traditions of revered ancestors whose beliefs and behaviors must be "preserved" in order to create and maintain a distinct national identity which will, purportedly, ensure national development. Why is this ideology so widespread in Papua New Guinea and other Melanesian states?

This is a complex phenomenon, and the dialectical method of analysis demands that we view it in all its interconnectedness, development and change. It may be that the trauma of the colonial experience caused nostalgia for the certainty of the past familiar ways, reinforcing people's natural desire to justify ancient social values, thus redeeming a fractured sense of self-esteem. Difficult-to-rationalize upheavals always cause a sense of insecurity, which often gives rise to nationalist ideologies. In Papua New Guinea, since even before Independence, the nationalist movement has been receiving a lot of support and encouragement from the Australian government (Simet and Iamo 1992), as well as from Christian missions and Western linguists and anthropologists (Handman 2013).

Jacob Simet and Wari Iamo's (1992) well documented account of the crystallization of *The Melanesian Way* ideology effectively transmits the political atmosphere of the time through numerous quotes from the writings of nationalist leaders, such as Bernard Narokobi:

> *Without the wheel, gunpowder and kings, queens, emperors or high chiefs, PNG maintained a fine sociological balance based not on large nation states, but on small communal autonomy. Some people call this disunity anarchy, even chaos. I say it is an aspect of unity, because no one group can claim on the basis of its technology that it is better or superior to another. All of us had a common belief in the living reality of our ancestors. We believed even the trees, the rocks and the natural life had souls or entities to which life can be attributed.*
> (Narokobi 1980, 24–25)

Thus, PNG's motto "Unity in Diversity" became a key component in the development of Papua New Guineans' national identity:

> *Despite the acknowledgement of the cultural diversity, the educated Papua New Guineans do have a "national culture." Diversity in cultures and languages is talked about as having a uniting effect. ... Communities are centered around tracing their kinship roots back to a village and land, having faith in beliefs in magic and sorcery, death after life, and customs and traditions. ...*
>
> *(Simet and Iamo 1992, 12–13)*

The work of missionaries and prominent linguists and anthropologists also helped popularize this "Unity in Diversity" concept in PNG. In a meticulously referenced historical analysis of the changing social attitudes toward Tok Pisin, the linguistic anthropologist Courtney Handman (2013, 268) documents the role of missionaries and Australian National University linguists in promoting nationalism in Papua New Guinea:

> *While missionaries and linguists were instrumental in developing and promulgating Tok Pisin as a language of the nation-state that was separable and autonomous from English (just as they hoped PNG could become separable and autonomous from Australia), urban and higher-class speakers have rarely been as troubled about maintaining the boundary between these two languages.*

This clearly highlights the futility of promoting (or legislating) language use in society, once it loses its *utility* in communicating meaning. Despite the emotional appeal of trying to save the endangered languages from dying, few attempts at resuscitation "are likely to lead to communities raising children in the language, which is the only way a language exists as its full self" (McWhorter 2015).

Human creativity in devising thousands of different ways to express their thoughts is, indeed, amazing and deserves study and appreciation, just as does the study of various architectural styles. Yet, practicality is a major factor determining their use in different environments, social or physical:

> *the existence of so many languages can also create problems: It isn't an accident that the Bible's tale of the Tower of Babel presents multilingualism as a divine curse meant to hinder our understanding.*
>
> *(McWhorter 2015)*

Despite the impracticalities of vernacular education and a strong popular opposition to it, which has recently caused a reversal in the government education policy, most linguists and anthropologists are totally committed to the cause of preserving PNG languages and cultures. This commitment was evident in the many papers presented at the LSPNG 2014 conference "*Celebrating Tok Pisin and Tok Ples.*"

Yet, despite all the psychological, material and political pressures to preserve ancestral cultures, our research findings show major changes in the way people speak, think and live, indicative of an ongoing cultural shift toward the "Global Village" ways. Almost 90% of the 84 students who participated in our 2014 study, claim that mobile technologies have changed people's lives for the better, viewing the ongoing linguistic and cultural changes as necessary and irrevocable (UPNG Students' Voices 2014).

Conclusion

Social exchanges (and their benefits) are inconceivable without language, whose social function is to communicate meaning. In a country as linguistically diverse as Papua New Guinea, false appeals to emotion must give way to pragmatism. In the PNG context, English is the master key that can power the engine of national development. Boosted by digital technologies, it is becoming the language of the young. Mobile phones have awoken people to the reality of our Global Village, abuzz with new, profitable exchanges. Have these new "series of exchanges" yielded the benefits de Tracy wrote about? Multinational energy companies have flocked to this resource-rich country, investing in numerous development projects but also causing environmental damage and associated problems. And yet, people are now benefiting through tapping into the reservoir of accumulated human knowledge, through improved communications and new forms of social networking; this is evidenced by high levels of public awareness about national and world events. Government departments, NGOs, businesses, banks, universities, the country's first free-to-air commercial television service (EMTV) and national dailies have all gone online, and dozens of excellent social and news media websites have sprung up (i.e., www.pngloop.com, www.pngperspective.com, www.garamut.wordpress.com, etc., etc.). Mobile phones have penetrated all the nooks and corners of PNG, catalyzing the development of a common language of communication (English/ Tok Pisin) and, through it, raised the series of social exchanges here to a new level. Thanks to mobile technologies, an awareness of our common humanity and shared concerns is growing in PNG, to the benefit of all.

Acknowledgements

I am grateful to Dr. Laurel Dyson, Max Hendriks and Stephen Grant of the University of Technology, Sydney (Australia) for welcoming my thoughts on this fascinating subject, as well as for all the support they have given me in the past few months. Editorial comments I received were very helpful—my sincere thanks to the Editors whom I, unfortunately, cannot name.

I also want to thank Professor P. J. Yearwood of the University of Papua New Guinea for his invaluable advice and patience in reading twice through the manuscript. Finally, I am grateful to my husband, Prof. V. J. Temple, for his constant support and encouragement.

References

Aikhenvald, Alexandra Y. "Living in Many Languages: Linguistic Diversity and Multilingualism in Papua New Guinea." Plenary address at the LSPNG 2014 Conference September 17–19, 2014. *Language & Linguistics in Melanesia* 32, no. 2 (2014):. I–XVII. Accessed March 28, 2015. www.langlxmelanesia.com.

Arnauld, Antoine, and Claude Lancelot. *General and Rational Grammar: The Port-Royal Grammar*. Janua Linguarum, Series Minor, No. 208. Edited and translated by Jacques Rieux and Bernard E. Rollin. The Hague: Mouton, 1975.

Christiansen, Morton H., and Nick Chater. "Language as Shaped by the Brain." SFI Working Paper 2007–01–001. Santa Fe, CA: Santa Fe Institute, 2007. Accessed January 27, 2015. http://www.santafe.edu/media/workingpapers/07–01–001.pdf.

Christiansen, Morton H., and Nick Chater. "Language as Shaped by the Brain." *Behavioral and Brain Sciences* 31 (2008): 489–558. doi:10.1017/S0140525X08004998.

de Tracy, Antoine Destutt. *A Treatise on Political Economy*. Part I, Chapter 1: "On Society." Georgetown, DC: Joseph Mulligan, 1817. Accessed October 5, 2014. https://mises.org/books/tracy.pdf.

Friedman, Thomas L. *The World is Flat: A Brief History of the Twenty-First Century*. New York: Farrar, Straus and Giroux, 2005.

Global Language Monitor. "Number of Words in the English Language." http://www.languagemonitor.com/number-of-words/number-of-words-in-the-english-language-1008879/. Accessed January 9, 2015.

Handman, Courtney. "Text Messaging in Tok Pisin: Etymologies and Orthographies in Cosmopolitan Papua New Guinea." *Culture, Theory and Critique* 54, no. 3 (2013): 265–84. DOI: 10.1080/14735784.2013.818288.

Hannerz, Ulf. "Flows, Boundaries and Hybrids: Keywords in Transnational Anthropology." Stockholm: Stockholm University, 2000. Accessed October 10, 2014. www.transcomm.ox.ac.uk/working%20papers/hannerz.pdf.

Herder, Johann Gottfried. *Treatise on the Origin of Language*. 1772. Accessed January 26, 2015. https://www.marxists.org/archive/herder/1772/origins-language.htm.

Humboldt, Wilhelm von. *The Heterogeneity of Language and Its Influence on the Intellectual Development of Mankind*. 1836. Accessed January 27, 2015. http://www.scribd.com/doc/90736832/Humboldt-Uber-die-Verschiedenheit-des-menschlichen-Sprachbaus-und-seinen-Einflu%C3%9F-auf-die-geistige-Entwicklung-des-Menschengeschlechts#scribd.

King, Phil. "Tok Pisin and Mobail Teknoloji." *Language & Linguistics in Melanesia* 32, no. 2 (2014): 118–52. Accessed January 27, 2015. www.langlxmelanesia.com.

LSPNG. *Proceedings from LSPNG 2014 Conference in Madang*. In *Language and Linguistics in Melanesia* 32, no. 2 (2014). Accessed March 28, 2015. www.langlxmelanesia.com/issues.htm.

McWhorter, John. "What the World Will Speak in 2115." *The Wall Street Journal, The Saturday* Essay, January 2, 2015. Accessed January 25, 2015. http://www.wsj.com/articles/what-the-world-will-speak-in-2115–1420234648.

Narokobi, Bernard. *The Melanesian Way*. Boroko, PNG: Institute of Papua New Guinea Studies, 1980.

Plato. *Cratylus*. Translated Benjamin Jowell. 360 B.C.E. Accessed January 26, 2015. http://classics.mit.edu/Plato/cratylus.html.

Sakai, Yuuko. *Universal Sentence Structure: A Realization in Spanish*. 2008. Accessed October 10, 2014. www2.teu.ac.jp/media/~sakai/xuss/ust.pdf.

Sapir, Edward. *Language: An Introduction to the Study of Speech*. New York: Harcourt, Brace, 1921.

Simet, Jacob, and Wari Iamo. "Cultural Diversity and the United Papua New Guinea." NRI Discussion Paper No. 64, presented in a Melanesian Studies Occasional Seminar Series at the National Research Institute, February 27, 1992. ISBN 9980750383.

Summer Institute of Linguistics (SIL). "Papua New Guinea." Accessed March 28, 2015. http://www.ethnologue.com/country/pg/default/***EDITION***.

Temple, Olga. "Language: Captured 'Live' through the Lens of Dialectics." *Language & Linguistics in Melanesia* 29 (2011): 31–53. Accessed March 28, 2015. www.langlx melanesia.com.

Temple, Olga. "Limitations of Arbitrariness." *The South Pacific Journal of Philosophy & Culture* 10 (2008–2009): 107–25.

Temple, Olga. "Society—the Foundry of Human Minds." *Pacific Journal of Medical Sciences* 7, no.2, Dec. (2010): 23–38.

Temple, Olga. "Syntax through the Wide-Angle Lens of Dialectics." *Language & Linguistics in Melanesia* 30, no. 2 (2012), 31–46. Accessed March 28, 2015. www.langlxmelanesia.com.

Temple, Olga. "The Rational Language Mechanism: Key to Understanding Syntax." *Journal of English Studies* 1 (2009): 62–81.

Temple, Olga. "The Syntax of Semantics: The Basics of Building Complex Structures of Meaning." *Language & Linguistics in Melanesia* 31, no. 2 (2013): 43–60. Accessed March 28, 2015. www.langlxmelanesia.com.

Temple, Olga. "Tok Ples in Texting & Social Networking: PNG 2010." *Language & Linguistics in Melanesia* 29 (2011): 54–64. Accessed March 28, 2015. www.langlxmelanesia.com.

Temple, Olga, Alopi Apakali, Dorothy Bai, Lilly John, Gabriel Matiwat and Malinda Ginmauli. *PNG SMS Serendipity, or sms@upng.ac.pg*. Port Moresby: UPNG University Press, 2009.

Temple, Olga, Lin Berry, Jack Bobby, Charlotte Laudiwana, Nadia Lawes, Emmanuel Maipe, Deborah Salle, Filomina Sion and Xavier Winnia. *Tok Ples in Texting and Social Networking*. Port Moresby: UPNG University Press: University Bookshop, 2011.

UPNG Students' Voices. 2014. Accessed March 2015. http://www.templeok.com/research.htm.

von Erfurt, Thomas. *De modis significandi, seu, Grammatica speculativa*. Upper Saddle River, NJ: Prentice Hall Press, 1310/1972.

Vygotsky, Lev S. "Chapter 2: Three Planes of Psychological Development." In *Primitive Man and His Behaviour* by Alexander R. Luria and Lev. S. Vygotsky. Translated by Evelyn Rossiter. Orlando, FL: Paul M. Deutsch Press, 1930. Accessed October 5, 2014. https://www.marxists.org/archive/vygotsky/works/1930/man/.

Watson, Amanda H. A. "Early Experience of Mobile Telephony: A Comparison of Two Villages in Papua New Guinea." *Media Asia* 38, no. 3 (2011): 170–80.

Watson, Amanda. "Mobile Phones and Media Use in Madang Province of Papua New Guinea." *Pacific Journalism Review* 19, no. 2(2013): 156–75.

19 An Example of Excellence
Chickasaw Language Revitalization through Technology

Traci L. Morris

Introduction

There are 566 federally recognized American Indian tribes in the United States, and currently little if any data exists on connectivity or uses of technology by either tribal governments or tribal citizens. There is a persistent and pervasive digital divide on tribal lands and very little data about Internet use or broadband connectivity in tribal areas. Further, there are few studies on how digital technologies are used for language and cultural preservation or on how Tribal governments use these technologies to officially enhance cultural or linguistic programming. Likewise, no data exists on how Indian Country[1] residents use social media or the Internet and on how technology aids individuals in perpetuating their learning and use of tribal languages, although anecdotally we know it is happening. Technology holds great promise in stabilizing tribal languages, which are in danger of extinction, allowing connections between remote peoples and between young and old.

Access to telecommunications and the Internet in Indian Country communities is often limited and sometimes non-existent. Native North American communities are affected by the digital divide, not only between their communities and the rest of the United States, but between the boundaries of their respective tribal reservations or tribal regions, where some areas are connected (at lower speeds than the rest of the nation) and other areas of the tribal homeland are not connected—there are multiple divides. Despite the digital divide, recent research indicates that Native Americans utilize digital communications technologies at rates much higher than national norms (Morris and Meinrath 2009). This does not mean that broadband access is widely available on Tribal lands, but rather that individuals manage to find ways to access broadband resources and that there is a great demand for these resources among Native American community members. Tribal members are extremely resourceful and very mobile, often traveling to urban areas surrounding their tribal lands giving them digital access via their mobile devices. In fact, like other minority populations, tribal populations are mobile adopters. Indeed, Native people have a history of adopting new technologies and then adapting them to cultural uses.

What little research there is on connectivity on Tribal lands demonstrates that the 5.2 million Native Americans in this country are tech savvy, no doubt due in part to the fact that this is a young population with a median age of 31 as compared with the national median of 37 (US Bureau 2014). Interestingly, 39 percent have a bachelor's degree in a science or engineering field (US Bureau 2014). It seems that these demographics are indicative of a connected population, but the reality is there is little data on how many Native Americans are connected, how they connect, or how they use their time on the Internet. One recent study, *Digital Inclusion in Native Communities: The Role of Tribal Libraries*, released by the Association of Tribal Archives, Libraries and Museums, included interview data from tribal librarians, and many spoke of how patrons used their library and connectivity available at their library for the perpetuation of their respective tribal languages. This study is the only one of its kind with regard to tribal libraries.

Native North American languages are in extreme danger of disappearing. According to the *American Community Survey Brief on Native North American Languages* released in December of 2011, of American Indians and Alaska Natives (AI/AN) just 5.4 percent spoke a Native language, identified as solely AI/AN and lived in their community; additionally 1 in 5 of these people were over 65 (US Bureau 2011). These numbers are even more staggering for those Native Americans living off reservations and/or in urban settings. According to the Census, just 0.7 percent of AI/AN alone or in combinations speaks a Tribal language (US Bureau 2011).

One American Indian Nation—the Chickasaw Nation—is working hard to grow the number of speakers of their language, and they are using technology in innovative ways to bolster their efforts. The Chickasaw Nation is a federally recognized American Indian Tribe located in south-central Oklahoma. The jurisdictional area includes 7,648 square miles and encompasses all or parts of 13 counties in Oklahoma (Official Site of the Chickasaw Nation 2015). However, the Chickasaw Nation population is not limited to these boundaries; the Nation has a significant "at-large" population, meaning that large numbers of enrolled tribal members do not live in the Tribal jurisdictional boundaries. The Chickasaw Nation has 59,919 tribal citizens; 17,486 of those live in the jurisdictional boundaries of the nation and 24,742 live in other states, with predominant population centers in Texas and California.[2] The Nation currently has less than 65 native speakers, the youngest born in the late 1940s; these individuals acquired Chickasaw as their first language.[3] Early in 2014, the Nation lost Emily Johnson, the last monolingual speaker of the Chickasaw language, making the mission to grow speakers of paramount importance (Alan 2014). The data tells us that without young people newly equipped to communicate in Chickasaw, people who are able to really *live* in the language, Chickasaw as a spoken language will likely become only a memory in the next 20 to 30 years—this author's lifetime.

This chapter examines uses of technology in Native American language revitalization. Specifically the author will look at how Chickasaw Nation tribal citizens—independent of the Chickasaw Nation—are using technology to create and access cultural and linguistic programming and how the Chickasaw Nation Language Revitalization Program is using technology in an effort to expand their educational outreach. Chickasaw citizens have developed a very active Facebook Group and a less active Google+ Page. The Chickasaw Nation has developed both mobile and desktop technologies in order to perpetuate the Chickasaw Language and to serve those within and outside the jurisdictional boundaries in other ways, including Internet language websites; a language app for phones, tablets and desktops; language videos disseminated on Internet sites and tribal language interfaces for browsers, devices and computers.

Background

American Indian Tribes and Native American individuals are using digital technologies in cultural language preservation. This chapter seeks to document one specific example of use of technology in cultural and linguistic preservation and perpetuation, the Chickasaw Nation of Oklahoma and its citizens. While there is much published on Native American language revitalization, there are few academic articles and only a handful of newspaper articles regarding technology and Native American languages. Academically, particular attention has been paid to technology and Indigenous language revitalization with regard to the Ojibwe language demonstrating the use of technology for communications, material production, documentation and archival efforts (Hermes and King 2013); the discussion of technology-driven multimedia language software project for language revitalization (Hermes, Ban, Marin 2012); the Hawaiian language (Warshauer 1998; Galla 2009) and, increasingly the Chickasaw Nation via newspaper articles (Richmond 2014; Russon 2014). Additionally, two recent dissertations have been written; one examines the Chickasaw Language Revitalization Program (Ozbolt 2014, Davis 2013), and the other discusses uses of technology and in particular mobile apps for language learning (Begay 2013).

The number of Tribal language apps in the iTunes and Android stores is constantly growing. Additionally, there are now a number of companies creating Tribal language apps, including Thornton Media and Ogoki Learning Systems. In addition to language apps, a number of Native Nations have developed or are developing officially sanctioned tribal programs and programming, which may make use of Internet language websites; language apps for phones, tablets and desktops; language videos disseminated via Internet sites; tribal language interfaces for browsers, devices and computers; and now gaming and virtual worlds with cultural and linguistic programming (Byrd 2014).

These digital tools are being created by a number of American Indian Nations and governments in an effort to promote and perpetuate language preservation. Meanwhile, Native American individuals are using social media, especially Facebook groups, Google+ Groups, Twitter and YouTube in increasingly high numbers to interact in and learn tribal languages, in many cases bridging the gap between elders and youth via technology (Alan 2014).

History of Technology and Innovation in Native American Communities

Native communities have always embraced new technologies; thus the venture of American Indian Tribes into new media, mobile and Internet technologies is a natural area for expansion. Tribes embraced newspapers as early as the 1820s and radio as early as the 1920s. In fact, the Southeastern tribes including the Chickasaw, Cherokee, Choctaw and Creek were the primary producers of pre-civil war Native journalism, printing bilingual and in some cases, trilingual newspapers (Murphy and Murphy 2014). Tribal broadcasting has its roots in tribal papers. The first known radio show was developed by a writer for a tribal newspaper in Wyoming. A member of the Eddleman family, Ora Eddleman Reed, who wrote for The Twin Territories, started the first talk radio show in 1924 on KDFN, Wyoming's first radio station (Trahant 1995). She had an hour-long show and was billed as the Sunshine Lady, as she was known for pointing out the positive. In the 1930s, John Collier, the Commissioner of Indian Affairs, carved out a budget for radio communications and sponsored national programming about tribal affairs that was aired on 170 stations nationwide. This show started in 1937 (Trahant 1995). By the 1950s, radio stations in the southwest were hiring Navajos for shows, announcements and commercials, which were broadcast in Navajo (Trahant 1995).

The history of the first Native-owned station to go on the air is contested. However, Alaskan tribes were among the first to create radio stations, in part a response to legislation, the Alaska Native claims Settlement Act of 1971 (Wilkins 2007). Two Native stations were started in part to keep Alaska's Natives informed about the issues affecting their lands and rights after the legislation was passed (Wilkins 2007). In 1971, the first noncommercial Native station KYUK-AM in Alaska went on the air. Other rural Alaskan stations followed in 1973 and 1975. Since 1971 about 10 stations a decade have launched (Keith 2008). Currently, there are 53 tribally owned radio stations on air (Native Public Media 2014).

Community-Centric Language Learning via Social Media

Chickasaw citizens—unaffiliated with the Chickasaw Nation Government— are very active and have created their own language learning tools via such social media outlets as Google+ Groups, Google Hangouts, YouTube and Facebook. Given that a large portion of the members of the Chickasaw Nation do not live in close enough proximity to the Nation to attend language classes, citizens have created their own means of facilitating language use. Internet technology has created a new virtual community within a widely geographically disbursed tribal population.

In 2013, several Chickasaw citizens—independent of the Tribal Government—worked with the Google suite of products including Google+, Google Groups, Google Hangouts and YouTube to set up a closed language group enabling other Chickasaw Tribal members to interact in real time. The *We Speak Chickasaw* Google+ Page was created by several enrolled members of the Chickasaw Nation in conjunction with the Google American Indian Network or GAIN team who informally helped set up the group and train the group creators in the use of the technology. The intent of this online community is to create a more formalized class-like learning environment using Google Hangouts, online and live, making it accessible to multiple people all over the country. Each Google Hangout live learning session is then archived on YouTube, making the lesson available to others who may not have been able to attend the live Google Hangout. The *We Speak Chickasaw* Google+ Group was created in 2013. However, it is not particularly active as the technology of the Google groups proved restrictive, cumbersome, and difficult for some Tribal members to use. Some of the group founders had difficulties archiving the Google Hangouts on YouTube and group members did not like the requirement of having a G-Mail address and a Google+ page, which is at this time a requirement of using the Google suite of products.

The most active community-learning group is the Facebook Group *We Speak Chickasaw*. A Chickasaw citizen living in Arizona created this members-only closed group. The group is very popular and used daily; as of the writing of this article there are 500 plus members. The purpose of the group is for "Chickasaw language questions and study discussions" and to "Interact for practice and improvement in Chickasaw language mastery." Interactions on *We Speak Chickasaw* are mostly limited to day-to-day conversational questions about how to convert a word, concept or phrase from English into Chickasaw, although there are intermittent uses of the group for actual language lessons. Some members of the group are very active using the print workbook *Chickashashanompa' Kilanompoli'* or *Let's Speak Chickasaw*, and posts from more literate speakers often include active drills encouraging other members to conjugate verbs and tenses. Members will find a printable list of resources, including some group user produced flash cards for language drills. Vocabulary words are often posted in this group as well.

Texting

A fascinating, applied use of the technology and the language in the real world is the growing use of Chickasaw textisms by citizens on their mobile devices and phones. In the same manner as English, users of Chickasaw have created shorthand versions of common phrases for incorporation in their text communications. Examples include the Chickasaw phrase *Ayalimak illa* and texting slang AYN in lieu of the English texting phrase "I've got to go" or texting slang GTG. Another Chickasaw example is *Chipisala'cho* shortened to CPL; the English equivalent phrase is "I'll see you later" and the shortened text is "cul8tr." The Chickasaw language phrase *Ollalili*, is shortened to OLL for texting; this translates to "I'm laughing" commonly texted as "LOL." This mirrors the use of texting in other tribal communities for language perpetuation, but as of yet there have been no studies of this phenomenon.

Chickasaw Nation Language Program: Language Engagement with Technology Since the 1990s[4]

In a very different but complimentary, the Chickasaw Nation Language Revitalization Program, the official program of the Chickasaw Nation Tribal Government under the direction of Joshua Hinson, has embraced technology in a very robust manner in an effort to reach all Chickasaw citizens regardless of where they live and to provide them quality language learning materials in various new platforms. Starting with language on cassettes then moving to CD-ROMs the Nation developed both mobile and online technologies in order to perpetuate the Chickasaw Language, including Internet language websites; a language app for phones, tablets and desktops; language videos disseminated on Internet sites and tribal language interfaces for browsers, devices and computers.

A Chickasaw Dictionary CD-ROM

A Chickasaw Dictionary, co-authored by Reverend Jesse J. Humes and Mrs. Vinnie May James Humes, was first published in hardback by the Chickasaw Nation in 1973 and later reissued in paperback. Some years following the publication of the book, Mrs. Humes began recording the entries on reel-to-reel and later audiocassette formats. In the mid-1990s the Chickasaw Historical Society in collaboration with Various Indian Peoples (VIP) Publishing released a CD-ROM version of Mrs. Humes' audio. The formatting is identical to the dictionary, with an English entry followed by a Chickasaw entry and a phonetic pronunciation, with linked audio files.

The CD-ROM is still available from various online sources, but this is increasingly dated technology. In order to ensure that this material remains accessible, Joshua Hinson, Director of the Chickasaw Nation Language Department, began an expanded and updated edition of the dictionary in the spring of 2014. This version will be an update to the original text with entries in the original orthography and the modern Munro-Willmond orthography.[5] The updated edition will include new entries that Mrs. Humes committed to audio prior to her death in 1996, edits to the original text that she herself made in the audio and copyediting of errors in the original text that were not necessarily noted by Mrs. Humes in the audio. In each case, audio is carefully transcribed to accurately reflect the spoken Chickasaw.

Phraselator

In 2008, the Chickasaw Language Revitalization Program began working with Thornton Media Inc. (TMI), to create a Chickasaw translation program for use on the Phraselator, a mobile technology device used by the United States Armed Forces and law enforcement worldwide for the purposes of mobile translation of critical languages (Buckley 2008).[6]

TMI, following the introduction of a child's language learning toy, adopted the Phraselator for use in language revitalization environments. Its three-way translation mode and incredibly rugged construction made it ideal for a wide range of applications. Perhaps the most significant feature was the ability of communities to create their own custom content for on-demand use. The Phraselator allowed access to real-world, communicative language, recorded by native speakers, without the need to ask speakers repeatedly about a particular word or phrase.

The process for creating the Phraselator content included the development of a script, which incorporated basic vocabulary, cultural content and communicative daily language, recording that content, importing and transcribing the content using Module Builder Pro software and exporting to the device.

According to Hinson, the Phraselators were an integral component in a program called Community Council Language Outreach. Chickasaws living at large (outside the service area in south-central Oklahoma) are organized in groups called Chickasaw Community Councils, comprised of citizens living in a particular geographic area, and each interested community council was given a Phraselator and a two-day training to assist them in implementing their own language-learning program. The Phraselator was the first concerted effort made by the Chickasaw Language Revitalization Program to address the issue of language learners at large or citizens who desire access to their language of heritage but have no opportunity to interact with a native speaker.

Chickasaw.tv

Chickasaw Nation TV[7] is a high-definition online video network owned by the Chickasaw Nation. The language pages found at *Chickasaw Nation TV* are one piece of a larger effort to utilize modern media and the Internet to extend the educational reach of the Chickasaw Nation into the larger world. The language resources on Chickasaw.tv include profiles of speakers, program employees, informational videos on aspects of the language including history, linguistic varieties and neologisms, as well as a portal to the language application and interactive language learning games and videos.

The Nation's *Chickasaw Nation TV* maintains a video library of over 1600 videos and is still growing. The site includes several sections, each of which are links, including Discover Chickasaw, Arts and Humanities, History and Culture, Language, Health and Wellness and Oklavision. Each section includes subsections. Discover Chickasaw includes News and Events, Profiles of a Nation, Our People, Citizen Benefits and Community Partners. Arts and Humanities include the following subsections: Visual Arts, Fashion, Performing Arts, Humanities, Education and Events. The Health and Wellness section includes subsections on Fitness, Food and Recipes, Wellness and Diabetes Prevention. Oklavision links to a new page presented by the Chickasaw Nation but generally about Oklahoma as a state.

Finally, the Language section of *Chickasaw Nation TV* has four subsections: Lessons, Programs, Language Keepers and the Language Web App (Anompa). The Lessons section has 23 short video lessons. The Programs section has short videos discussing two distinct programs, the Master-Apprentice Program and Camps and Clubs. The Language Keepers section has three segments with videos: one about the Chickasaw Nation Language Department, one on Language Education and a section of videos on Speaking and Sharing the Chickasaw Language. Finally the last subsection of the Chickasaw Language section of *Chickasaw Nation TV* is a link to the *Anompa Language Web App*.

Chickasaw Nation TV has had a designated language channel since February of 2014, before which material was embedded in other subsections. Website analytics tell of the strong performance of the language channel: people are using it. A total of 9,855 people watched a language video between February 1, 2014 and December 28, 2014. Additionally, they spent an average of 4:44 hours on the site.

Anompa: iPod/iPad/iPhone application/ANOMPA website

Following the successful implementation of the Community Council Language Outreach, using Phraselator, a new technology burst upon the scene—Apple Computer introduced the iPhone and the first generation of the iPod Touch. With similar functionality to the Phraselator, a simple application for Apple iTouch devices could surmount the main issues with the Phraselator devices—their cost. The Chickasaw Nation Language Revitalization Program leadership determined that the development of an app would move closer to having quality language learning materials at the fingertips of any Chickasaw citizen worldwide.

Again the Chickasaw Language Revitalization Program teamed with TMI to develop the application. Beginning with the content originally programmed for the Phraselator, a full-function application was developed for iPhone, iPod and iPad. The topical language sections were augmented with Chickasaw-Choctaw hymns and two short videos, one an animated *shikonno'pa'* (traditional animal story) and the other a video of *Chipota Chikashshanompoli* (Children Speaking Chickasaw) Language Club.

The *ANOMPA: Chickasaw Basic* application was released in 2009. It was the first Indigenous language application for mobile devices created by a tribal government for its people. The Chickasaw Language App *Anompa* is both an online web app and a mobile app available for the iPhone and Android platform.

Roughly a year following its release, Hinson indicated that Chickasaw citizens began to ask for an Android version of the application. At that time the Department of Chickasaw Language (founded in 2009) chose to create a website with hyperlinked audio that would function on all platforms, rather than develop another device-specific application. The flexibility of a website also ensures that additional content can be added as needed without constantly having to update the application architecture and content. The web version of *Anompa* is located at www.chickasaw.net/anompa and has been visited on average by 15,000 unique users per year, since it went live (Lokosh, e-mail messages to author, March 2015).

Conclusion

This chapter demonstrates how Chickasaw citizens, independent of Chickasaw Nation governmental programs, are using technology to create and access cultural and linguistic programming and how the Chickasaw Nation Language Revitalization Program is using technology in an effort to expand their educational outreach and create a balance between creating new speakers and meeting the general enrichment needs for a large and dispersed population. Chickasaw citizens have developed a very active Facebook Group and a less active Google+ Page. The Chickasaw Nation has developed both mobile and desktop technologies in order to perpetuate the Chickasaw Language and to serve those within and outside the jurisdictional boundaries in other areas, including Internet language websites; a language app for phones, tablets and desktops; language videos disseminated on Internet sites and tribal language interfaces for browsers, devices and computers.

According to Joshua Hinson, the Director of the Department of Chickasaw Language, more people access resources for language learning online (as the numbers have demonstrated), but it is clear that in-person learning is much more effective in terms of actually learning the language. However, the use of the technologies employed here—whether by Chickasaw citizens for citizen-directed learning or by the Chickasaw Nation to engage a dispersed population—goes far beyond creating a linguistically literate population; the technology is creating a culturally literate Chickasaw population. Moreover, the example of innovation and excellence this Nation sets for other American Indian Nations and Indigenous peoples is farther reaching.

The scholar reading this chapter will undoubtedly see an example of excellence in uses of technology by the Chickasaw Nation in encouraging the growth of their language. This is what the author has endeavored to demonstrate. Indeed, through this article, in the Western academic sense, we see technology, adaption and adoption, synthesis and innovation by the Chickasaw Nation. But those from Native communities, including this author, see self-determination in praxis. Language is a matter of cultural survival. It is the ability to tell our own story of our own tribal identity and nationhood—it is cultural sovereignty.

Notes

1. While the term Indian Country is a legal, philosophical and geographical designation for Native peoples who are members of a Native Nation, for the context of this research, the term refers to people who are enrolled members of Native American Nations as a whole comprised of many different tribes in the United States.
2. Lokosh (Joshua D. Hinson). Director, Department of Chickasaw Language, Chickasaw Language Revitalization Program, Chickasaw Nation Division of History and Culture. The entire section on the Chickasaw Nation Language Program is based on e-mail correspondence with Lokosh March 2015.
3. Ibid.
4. Ibid.
5. Linguist Pamela Munro of the University of California Los Angeles developed the standard Chickasaw orthography with native speaker Catherine Willmond.
6. Product information about the Phraselator P2 can be obtained from the Voxtec website: http://www.voxtec.com/phraselator/ (accessed December 30, 2014). Information about features and building modules can be accessed on: <http://www.voxtec.com/phraselator/features/> and <http://www.voxtec.com/modulebuilder>.
7. https://www.chickasaw.tv/language.

References

Alan, Silas. "Outlook 2014: Chickasaw Nation Uses Technology to Broaden Tribe's Reach | News OK." *The Oklahoman*, April 27, 2014. Accessed March 28, 2015. http://newsok.com/article/3942556.

Begay, Winoka Rose. "Mobile Apps and Indigenous Language Learning: New Developments in the Field of Indigenous Language Revitalization." Master's thesis, University of Arizona, 2013.

Buckley, Carrie. "Holding Words and Phrases in the Palm of Your Hand: Military Technology Advances Chickasaw Language." *The Chickasaw Times*, August 2008: 8. Accessed March 28, 2015. http://digital.turn-page.com/i/24086/7.

Byrd, Christopher. "In 'Never Alone' Native Alaskans Explore the Future of Oral Tradition." Video Game Review, *The Washington Post*, December 29, 2014. Accessed March 28, 2015. http://www.washingtonpost.com/news/comic-riffs/wp/2014/12/29/never-alone-review-native-alaskans-explore-the-future-of-oral-tradition/.

Chow, Kat. "What Happens When a Language's Last Monolingual Speaker Dies? Code Switch: NPR." *NPR Code Switch: Frontiers of Race, Culture & Ethnicity*. 2014. Accessed March 28, 2015. http://www.npr.org/blogs/codeswitch/

2014/01/07/260555554/what-happens-when-a-languages-last-monolingual-speaker-dies.

Davis, Jennifer Lynn. "Learning to 'Talk Indian': Ethnolinguistic Identity and Language Revitalization in the Chickasaw Renaissance." PhD diss., University of Colorado, 2013.

Eaton, Kristi. "Last Monolingual Chickasaw Speaker Dies in Oklahoma." *Associated Press*, January 5, 2014. Accessed March 28, 2015. http://www.nativetimes.com/index.php/life/people/9423-last-monolingual-chickasaw-speaker-dies-in-okla.

Galla, Candace K. "Indigenous Language Revitalization and Technology from Traditional to Contemporary Domains." In *Indigenous Language Revitalization: Encouragement, Guidance & Lessons Learned*, edited by J. Reyhner and L. Lockard, 167–82, 2009. Flagstaff, AZ: Northern Arizona University.

Haag, Marcia. "Chickasaws Are on the Move" | Linguistic Society of America, 2014. Accessed December 30, 2014. http://www.linguisticsociety.org/chickasaw.

Hermes, Mary, and Kendall King. "Ojibwe Language Revitalization, Multimedia Technology, and Family Language Learning." *Learning, Language, & Technology* 17, no. 1(2013): 124–44.

Hermes, Mary, Megan Bang, and Ananda Marin. "Designing Indigenous Language Revitalization." *Harvard Educational Review* 82, no. 3 (2012): 381–402.

Hinton, Leanne, and Ken Hale. *The Green Book of Language Revitalization in Practice*. Brill, 2013. Accessed March 28, 2015. http://mesamedia.org/uploads/Hinton_article.pdf.

Jorgensen, Miriam, Traci L. Morris and Susan Feller. *Digital Inclusion in Native Communities: The Role of Tribal Libraries: Report for Printing.pdf*. 2015. Accessed January 2, 2015. http://www.atalm.org/sites/default/files/Report%20for%20Printing.pdf.

Keith, Michael C. (Ed.). *Radio Cultures: The Sound Medium in American Life*. Peter Lang, 2008.

McKnight, Chelsey, Becky King and Gina Brown. "Chickasaw Nation Stats," December 30, 2014.

Morris, Traci L., and Sascha D. Meinrath. *New_Media_Technology_and_Internet_Use_in._Indian_Country.pdf*. Native Public Media & New America Foundation, 2009. http://oti.newamerica.net/sites/newamerica.net/files/policydocs/New_Media_Technology_and_Internet_Use_in_Indian_Country.pdf.

Murphy, James E, and Sharon M Murphy (Eds.). *Let My People Know: American Indian Journalism, 1828–1978*. Norman, OK: University of Oklahoma Press, 1981.

Native North American Languages Spoken at Home in the United States: 2006–2010 - acsbr10–10.pdf. U.S. Census Bureau, 2011. http://www.census.gov/prod/2011pubs/acsbr10–10.pdf.

"Native Public Media." Accessed December 30, 2014. http://www.nativepublicmedia.org/radio.

Office of Head Start Tribal Language Report. Arlington, VA: US Department of Health and Human Services, Administration for Children and Families, 2012.

"Official Site of the Chickasaw Nation | Geographic Information." Accessed January 2, 2015. https://www.chickasaw.net/Our-Nation/Government/Geographic-Information.aspx.

Ozbolt, Ivan Camille. "Community Perspectives, Language Ideologies, and Learner Motivation in Chickasaw Language Programs." PhD diss., University of Oklahoma, 2014.

"Progress Report 2013.pdf." Accessed January 2, 2015. https://www.chickasaw. net/chickasaw.net/media/Progress-Reports/2013/2013-Progress-Report.pdf? ext=.pdf.

Ramswell, Prebble. "Ayali: Is it Time to Say Good-Bye to American Indian Languages?" *Indigenous Policy Journal* 23, no. 1 (2012.): 1–21.

Rindels, Michelle. "Tribes Turn to Technology to Save Endangered Languages." *Associated Press*, 2013. Accessed March 28, 2015. http://www.nativetimes.com/ index.php/culture/language/8638-tribes-turn-to-techn.

Russon, Mary-Ann. "Chickasaw Nation: The Fight to Save a Dying Native American Language." *International Business Times*, May 8, 2014. Accessed March 28, 2015. http://www.ibtimes.co.uk/chickasaw-nation-fight-save-dying-native-american-language-1447670.

Trahant, Mark N. *Pictures of Our Nobler Selves: A History of Native American Contributions to the News Media.* Nashville, TN: The Freedom Forum First Amendment Center, 1995.

Turin, Mark. "Orality and Technology, or the Bit and the Byte: The Work of the World Oral Literature Project." *Oral Tradition* 28, no. 2 (2013): 173–86.

US Census Bureau. "Profile Facts for Features: American Indian and Alaska Heritage Month." *US Census Bureau News*, November, 2014. Accessed March 28, 2015. http://www.census.gov/newsroom/releases/pdf/cb13ff-26_aian.pdf.

Warschauer, Mark. "Technology and Indigenous Language Revitalization: Analyzing the Experience of Hawai'i." *Canadian Modern Language Review* 55, no. 1 (1998): 139–59.

Wilkins, David E. *American Indian Politics and the American Political System.* Lanham: Rowman and Littlefield Publishers, 2007.

Wilson, Samantha, and Leighton Peterson. "The Anthropology of Online Communities." *Annual Review of Anthropology* 31 (2002): 449–67.

Young, Amber, and Shaila Miranda. "Cultural Identity Restoration and Purposive Website Design: A Hermeneutic Study of the Chickasaw and Klamath Tribes." In *Proceedings of the Hawaii International Conference on System Sciences (HICSS)*, 2014, 3358–67. IEEE Computer Society.

Epilogue

Laurel Evelyn Dyson, Stephen Grant and Max Hendriks

We are always astounded by the extent of innovation found when investigating Indigenous use of mobile technologies. Whether it is tracking reindeer using GPS or climbing coconut palms to obtain a signal, the ideas and usage patterns covered in this book leave the reader with a feeling of amazement.

Most Indigenous communities until recently had no exposure to mobile technologies. The mobile revolution, created by low-cost connectivity allied with cheap, smart mobile devices, has allowed underserved people to become connected. The introduction of communication towers, linked to the world's backbone infrastructure, has given Indigenous communities the means to participate in the modern world. And this has all happened in the last decade.

Because Indigenous people are generally not the wealthiest, their rapid uptake of this latest technology has surprised many. It has been achieved by individuals taking ownership of their personal mobile devices by choosing pre-paid options over monthly plans, sharing devices and a whole range of other cost management strategies. Their usage patterns, when little or no credit is available, reveal a deep-seated commitment to this technology, even when they have zero credit balance.

Using mobiles requires choices to be made. Keeping the connection with family and friends is the primary function as it keeps community and cultural links active and is vital for retaining Indigenous identity. Phones allow the transfer of funds and payment of bills, completion of government forms and communication for work. And the listening to music and playing of games becomes a regular part of the daily life of a phone owner.

Communities are being transformed by applying these technologies to social and economic development in areas such as health, education and business. They are the means for reaching young people and giving them access to important health information. They provide mobile learning, including distance educational solutions for people in remote regions. Moreover, they empower people in both rural and urban communities to improve their economic well-being and employment opportunities. Mobiles empower!

Self-esteem is being enhanced through cultural and linguistic mobile applications both informally and formally. In addition, everyday practice includes the use of text messaging that communicates and keeps language alive and multimedia recordings that allow the sharing of local cultural activities. There is now a wide range of often highly sophisticated systems that have been developed to retain culture and language. These are important for keeping culture and language alive for the future generations.

Overall the chapters of this book reveal a dynamic embracing of mobile technologies by Indigenous people. Through their use they are creating an environment where self-determination becomes achievable through the revitalization of identity, which in turn will shake off the effects of colonization. It will be interesting to observe how future wireless embedded systems that enhance communication and multimedia content creation will be used to empower communities and individuals to achieve this end.

Contributors

About the Editors

Laurel Evelyn Dyson is a Senior Lecturer in Information Technology at the University of Technology, Sydney, Australia, and President of anzMLearn, the Australian and New Zealand Mobile Learning Group. She is one of the founders of her university's Indigenous Participation in Information Technology Program and provides support for Aboriginal IT students in her faculty. Dr Dyson's research interests center on Indigenous people's adoption of ICT, in particular mobile technologies, as well as the use of mobile technologies in education. She has worked in Indigenous communities in Cape York on a number of mobile technology projects, led an evaluation for UNESCO of one of their Indigenous ICT projects and visited, run workshops and given keynote addresses at the University of Malaysia in Sarawak. For many years she has been an active member of a local organization working for reconciliation between Aboriginal and non-Aboriginal people in Australia. She loves teaching and has won five teaching awards. In her spare time she researches and writes about the history of Australian cookery, including Aboriginal and Torres Strait Islander foodways, reads, gardens, walks, swims and enjoys the company of friends in beautiful Sydney.

Stephen Grant is a Lecturer in Information Technology at the University of Technology, Sydney, Australia. Since 2002 he has taken a key position in the Indigenous Participation in IT Program at UTS, and the success of his work has been recognized with a UTS Equity, Social Justice and Human Rights Award. In addition, he has taken a lead role in the new Cisco Networking Academy at Redfern, Sydney, where he has lectured Aboriginal students in Internetworking and has trained Redfern Aboriginal Academy instructors at UTS. He was presented with the Cisco Achievement Award 2012 for his work with Aboriginal students within the Cisco Academy Program. He is one of a small number of qualified Indigenous IT professionals working in Australia with qualifications and industry experience in engineering and IT as well as experience in Indigenous affairs. His research interests are mobile networks and autonomous systems. He lives in an inner city terrace house with his partner Hazel and cat Bella and follows the Sydney Swans football team.

Max Hendriks lectures in Internetworking Science at the University of Technology, Sydney, Australia. He has been an educator for over 40 years and has taught all grades from Pre-School through to University postgraduate students, as well as holding senior executive positions in education. He has been a strong advocate for the rights of all people and seeks to increase the numbers of Indigenous Australians studying and working in ICT. His research interests are in Internetworking and Indigenous people and their innovative use of technology. Of particular interest to him are security technologies within wireless networks. He lives in the inner city and enjoys riding a bicycle.

About the Authors

Karla Alfaro completed her Master's degree in Public Management at the Instituto Tecnológico y de Estudios Superiores de Monterrey, Mexico. She is the co-founder of Asamblea Interactiva, a non-profit dedicated to promote citizen participation in the municipality of Pachuca. She received a diploma in documentary film from the Universidad Nacional Autónoma de México and a Bachelor's degree in Political Science (honors) from the Instituto Tecnológico y de Studios Superiores de Monterrey Campus Ciudad de México. Her research and development interests include participatory governance, e-governance, communication and education technologies for rural communities in Latin America.

Maria Augusti is an instructional designer at the Open University of Tanzania. Her work focuses on providing opportunities for making more efficient use of the existing expertise and resources, and she facilities e-learning tools development. She works with academic staff and leads professional development workshops. Maria heads the section of ELearning Development and Multimedia (EDMS) at the Institute of Educational and Management Technologies (IEMT). She is also a coordinator for the OUT OER (Open Education Resources) course, which is expected to be an African Council for Distance Education MOOC. She implemented a mobile application at the Open University of Tanzania (OUT) through a change project sponsored by SIDA. She has been a project team leader and a consultant in various projects at the institution. She also participated as one of the evaluation team for the ICT in Teachers Training Colleges Project in Tanzania.

Gyanendra Bajracharya is the director of the price statistics section at the Central Bureau of Nepal, Kathmandu. He has been working there since 1996. He was one of the members of the core team of the Population and Housing Census 2011. He has completed a post-graduate degree in Population and Human Resources from Flinders University, South Australia, in 2004. He also holds a Master's degree in Statistics from Tribhuvan University, Nepal, from which he graduated in 1995.

Brian Beaton has been developing and working on innovative communication technology projects with First Nations since 1983. His work began with IT training projects in Northern Ontario and in 1987 expanded to distance education programs with remote First Nations. From 1994 to 2013, he was the Coordinator of KO-KNET, the telecommunications division of the Keewaytinook Okimakanak (Northern Chiefs) Tribal Council based in Sioux Lookout, Ontario, Canada. With the KO-KNET team and collaborating First Nations, he worked to support the development of local First Nation broadband infrastructure, regional backbone networks, a First Nations social media service and email service, the Northern Indigenous Community Satellite Network and the innovative Keewaytinook Mobile (KMOBILE) cellular service. KO-KNET supports the Keewaytinook Internet High School (KiHS) and KO Telemedicine (KOTM). Since 2004, Brian has also been a partner on several national research initiatives in Canada, including the First Nations Innovation (http://fn-innovation-pn.com) and First Mile (http://firstmile.ca) projects. Brian has a BMath (Mathematics/Computer Science) from the University of Waterloo and a M.Ed. (Critical Studies) from the University of New Brunswick where he is currently pursuing his PhD in Education research—a critical analysis of skills training and entrepreneurship in remote Indigenous communities. Contact: brian.beaton@unb.ca.

Kirsten Black, PhD, MPH, RD is an Instructor at the Centers for American Indian and Alaska Native Health at the University of Colorado-Anschutz Medical Campus. Currently, she directs all day-to-day aspects of the Teen Pregnancy Prevention among Native Youth of the Northern Plains project. Previous research experience includes directing translational studies on practice improvement strategies for diabetes and depression and evaluation of an Internet-based behavior change program. Her areas of research interest include health behavior change, health promotion, and maternal and child nutrition. She has also taught in a family medicine residency program, worked as a state WIC consultant and been the evaluator for the Nurse Family Partnership program.

Shaylene Boechler, as FirstVoices Coordinator for the First Peoples' Cultural Council, conducted training workshops and managed much of the day-to-day community-based FirstVoices programming. Shay currently manages outreach efforts for the Endangered Languages Project, an online collaborative network that aims to protect global linguistic diversity. Shay holds a Bachelor of Arts degree in Applied Linguistics from the University of Victoria.

Fiona Brady works with government and small isolated communities in the Cape York Peninsula and the Torres Strait Islands, Australia. She holds a Master of Learning Management degree and has been involved in community development throughout the region. She has a passionate interest in realizing the promise of technology to improve the lives of Indigenous people in the bush. Fiona has been actively researching this theme from the time of the introduction of the first computers into workplaces in remote communities through to the advent of game-changing social media and mobile phone technologies. She is currently working on applications of mobile technology to support Indigenous languages.

Peter Brand is the co-visionary of FirstVoices.com, a suite of online Indigenous language revitalization resources launched in 2003 and administered by the First Peoples' Cultural Council in British Columbia, Canada. Peter spearheaded the development of the suite of FirstVoices Web and mobile applications, including multiple dictionary, tutor and chat apps. Peter retired from his role as FirstVoices Manager in 2013.

Terence Burnard is the Network Administrator of KO-KNET, the telecommunications division of the Keewaytinook Okimakanak (Northern Chiefs) Tribal Council based in Sioux Lookout, Ontario, Canada. He worked on the team to develop the online management software supporting the KMOBILE cellular service. Beginning work with KO-KNET Services in 2005, Terence developed and supported numerous network management tools for KO-KNET and the First Nation partners. Terence is responsible for performing analytical, technical and support work in the planning, implementation, documentation and administration of all Cisco network infrastructures. This includes the day-to-day operational tasks such as proactive maintenance, management, monitoring performance, incident and problem management, security and backup and recovery across the network infrastructure. Contact: terenceburnard@knet.ca.

Ivo Burum is a journalist and an award-winning writer, director and television executive producer. He has more than 30 years' experience working across genres including frontline international current affairs. A pioneer in user-generated story (UGS) creation, Dr. Burum is an author and lectures in multimedia storytelling and convergent journalism. He runs Burum Media, a mojo and web TV consultancy that provides training for journalists, educators and remote communities internationally.

Coppélie Cocq, PhD in Sámi Studies, is Research Fellow at HUMlab, Umeå University, Sweden. Her research interests include storytelling, place-making and folklore in digital environments. She is currently involved in a research project about the production and transmission of Indigenous knowledge that examines means and strategies developed by Sámi communities in order to ensure the transmission of culturally specific knowledge. Dr. Cocq's research focuses more particularly on how digital environments are shaped and used for the production and transmission of knowledge in outreach and inreach initiatives.

Lorenzo Dalvit is the MTN Chair of Media and Mobile Communication in the School of Journalism and Media Studies at Rhodes University in Grahamstown, South Africa. He previously headed the ICT Education

Unit in the Education Department. He also worked as Research and ICT Coordinator in African Language Studies in the School of Languages and as Researcher in Multilingualism and ICT in the Computer Science Department. His areas of academic interest include Mobile and ICT for rural development, hyperlocal media, mobile services and localization in African languages. He is currently developing an additional research interest in the use of mobile devices by disabled people. He is involved in various ICT for development initiatives and international collaborations with partners in Europe and Southern Africa. Professor Dalvit has coauthored over 100 publications and supervised more than 30 students across various disciplines, e.g., Media Studies, Education, African Languages and Computer Science. His work has appeared in both English and Italian. He is a rated NRF researcher and has attracted local and international funding.

Lee R. Duffield is a Senior Lecturer and Postgraduate coordinator in Journalism at Queensland University of Technology. He is a staff representative on the governing body, the QUT Council, and member of the University Academic Board. During more than 20 years in journalism, mostly with Australian ABC, he became the first news editor on the JJJ youth network, and European Correspondent at the fall of the Berlin Wall. Dr Duffield conducts field trips for students to work as overseas correspondents for campus-based media. His research interests include media in Europe, new media practice and development journalism notably in the Pacific Region.

Pedro Ferreira is a PhD student at the Royal Institute of Technology in Stockholm, Sweden, both at the Interaction Design department as well as the Mobile Life Centre. Ferreira is interested in technology use and development, having published on ethnographic fieldwork in Rah Island at a time of technological change, studying the introduction of mobile communications with a focus on technological play. Recently he has been an intern at the Technology for Emerging Markets group in Microsoft Research India, Bangalore, where he designed and developed a social networking system for low-literate farmers, allowing him to explore further the connection among technology, play and development work. Ferreira has received awards for two of his publications and is due to complete his doctoral dissertation in the summer of 2015.

Amanda Gaston (Zuni Pueblo) is the Project Manager for *Native It's Your Game* (NIYG), administered by Project Red Talon at the Northwest Portland Area Indian Health Board, in Portland, Oregon. NIYG is a multimedia sexual health and STD/HIV prevention curricula for American Indian/Alaska Native (AI/AN) middle school students aged 12 to 14 years. Amanda is working with a multisite research team to adapt and evaluate the original *It's Your Game (IYG)* program to better meet the needs of AI/AN youth. Amanda also works on a project developing a national multimedia health resource for Native teens and young adults (www. WeRNative.org). Amanda has worked as an *International Baccalaureate* (IB) teacher in Thailand for the three years before coming to the Board. She received her Master of Arts for Teaching at Oregon State University in 2007.

Gwenda Gorman, BS (Navajo) serves as the Health and Human Services Director for the Inter Tribal Council of Arizona, Inc. (ITCA). She has worked with ITCA for the past 12 years. Her primary role is developing grants and programs for tribes in Arizona in the areas of human services and tobacco, teen pregnancy, and STI prevention. She received a Bachelor of Science Degree in Family Studies and Human Development from Arizona State University. Some of her recent projects include collaborating with other tribes and organizations in adapting evidence-based programs for AI/AN teens and AI women around STI prevention and promoting healthy relationships. She enjoys working with youth of all ages, emphasizing the importance of living a healthy lifestyle.

Tracey Herbert is a member of the St'uxwtews First Nation. Tracey is the Executive Director of the First Peoples' Cultural Council, an organization with a mandate to vitalize and support First Nations languages, arts and cultures in B.C. Under her leadership the FPCC has developed into a globally recognized organization guiding innovations in Indigenous arts and language programming.

Kristina Höök is a professor in Interaction Design at the Royal Institute of Technology and also works part-time at Swedish Institute of Computer Science. She is the director of the Mobile Life Centre. Höök has published numerous journal papers, books and book chapters, and conference papers in highly renowned venues. A frequent keynote speaker, she is known for her work on social navigation, seamfulness, mobile services, affective interaction and lately, designing for bodily engagement in interaction through somaesthetics. Her competence lies mainly in interaction design and user studies helping to form design. She has obtained numerous national and international grants, awards, and fellowships including the Cor Baayen Fellowship by European Research Consortium for Informatics and Mathematics, the INGVAR award and she is an ACM Distinguished Scientist. She has been listed as one of the 50 most influential IT women in Sweden every year since 2008. She is an elected member of Royal Swedish Academy of Engineering Sciences.

Cornelia M. Jessen, MA, is a Senior Program Manager for the Alaska Native Tribal Health Consortium (ANTHC), a tribal health organization that provides specialty medical care, community health services, construction of clean water and sanitation facilities and a broad range of health system support programs to 130,000 Alaska Native people from 231 federally recognized tribes in Alaska. She manages the HIV/STD

Prevention Program with a focus on adolescent sexual health and STD/HIV prevention through resources like the youth wellness website iknowmine.org and community-based participatory research projects to develop, adapt and evaluate sexual health interventions for AI/AN youth. She has collaborated with state and other non-profit partners on projects aimed at preventing violence and promoting healthy and safe relationships. Cornelia received a BA in Anthropology and an MA in Applied Anthropology concentrating on Medical Anthropology and Public Health at the University of Alaska, Anchorage.

Carol Kaufman, PhD, is an Associate Professor at the Centers for American Indian and Alaska Native Health at the University of Colorado-Anschutz Medical Campus. She has over 13 years of research experience working with Native American communities on reproductive health issues, with a special emphasis on the development and evaluation of culturally appropriate and theoretically based interventions. She has worked steadily to develop community partnerships that have served to support and facilitate scientifically rigorous projects in Native American community settings.

Kevin R. Kemper is an assistant professor at the College of Social and Behavioral Sciences, University of Arizona, Tucson, USA, where he researches issues about Indigenous peoples. Kemper, who is mixed-race Choctaw and Cherokee Indian, has a PhD in journalism and a juris doctorate from the University of Missouri-Columbia. Kemper, a former news reporter and publisher, also earned a Master's of Law in Indigenous people's law and policy from the University of Arizona's College of Law.

Paul Kim is the Assistant Dean and Chief Technology Officer for Stanford University School of Education. He conducts research on sustainable educational technology for the developing regions, including Latin America, South East Asia, and East Africa. He is also a senior researcher for a National Science Foundation-funded project POMI (Programmable Open Mobile Internet) at Stanford. He currently serves on a National Academies Committee on Grand Challenges in International Development and Board Member for WestEd.

Narayanan Kulathuramaiyer is a Professor of Computer Science at the Faculty of Computer Science and Information Technology, Universiti Malaysia Sarawak (UNIMAS). He received his PhD in Computer Science from Graz University of Technology, Austria. He has served as the Dean of Faculty for over 10 years, and also previously headed the Centre for Applied Learning and Multimedia. He is now a visiting Professor at Graz University of Technology, Austria, and Braunschweig University of Technology, Germany. He is the Director of the Web Intelligence Consortium (WIC), Malaysia Research Centre, Editor-in-Chief for the Journal of Universal Computer Science and the Senior Fellow of the Information Society Institute. He has been a Senior Fellow of the

Institute of Social Informatics and Technological Innovations (ISITI-CRI), UNIMAS. He has won a number of National and International awards for his role in the eBario project and the e-Toro Indigenous Knowledge Management System project, Semantic Clustering Toolkit and the e-Co Outcome-based profiling system. His research interests include Semantics-Aware Systems, Technology Assimilated Learning and Future Web developments.

Travis L. Lane, BA, (Navajo/Southern Ute) currently serves as a Health Program Specialist for the Inter Tribal Council of Arizona, Inc. In this capacity, he administers, evaluates and provides training and technical assistance for teen pregnancy prevention programs in various tribal communities across Arizona. He also coordinates a health research project to determine the effectiveness of a computer-based teen pregnancy prevention curriculum for American Indian/Alaska Natives middle school youth. He previously served as the Outreach Coordinator for the Indians into Medicine (INMED) program, a health career recruitment program. Mr. Lane serves on numerous education-related boards and committees including the Arizona Tri-Universities for Indian Education, the Arizona Indian Education Association, and serves as an Ambassador to the Gates Millennium Scholars Program. Travis received his Bachelor of Arts in Political Science with a minor in American Indian Studies at the University of Arizona.

Adi Linden is the Manager of Technical Services of KO-KNET, the telecommunications division of the Keewaytinook Okimakanak (Northern Chiefs) Tribal Council based in Sioux Lookout, Ontario, Canada. He led the technical development of the KMOBILE cellular working closely with other members of the KNET team. He began working with KO-KNET in 1998 and helped to develop the service to become the largest Indigenous-owned and managed converged network company specializing in rural and remote First Nations and communities. Adi's work continues to support KO-KNET's growth and includes operating a vast, complex and diverse network along with its partners spanning several Canadian provinces. His technical interests and skills span ham radio, internet, VoIP, cellular, video operations and his family of three young children. Contact: adilinden@knet.ca.

Christine Markham is Associate Professor of Health Promotion and Behavioral Sciences at the University of Texas School of Public Health and Associate Director of the Center for Health Promotion and Prevention Research. She has over 20 years' experience in child and adolescent health promotion programming, including family- and school-based programs, with a primary focus on sexual and reproductive health, and chronic disease management. She has expertise in the development of theory- and evidence-based approaches to program development, implementation, evaluation and dissemination, cultural adaptation, the application of technology in adolescent health promotion programming and qualitative inquiry. Christine received her doctoral degree in Behavioral Sciences from the University of Texas and her Master's degree in Anthropology from the University of Pennsylvania.

Leigh Anne Miller completed her Master's degree at Stanford University's School of Education in 2012. Her work includes building a text message utility to improve the agency of public servants in fragile states by offering them Twitter-like streams to deliver vocational training. Prior to Stanford, she conducted special projects for the International Security and Foreign Policy program of the Smith Richardson Foundation with a focus on fragile states (Sudan, Afghanistan, Nepal). In 2005–2008, Leigh Anne moved to India where she founded and ran the Rainbow School, a Telugu-medium government school in a slum in Hyderabad. She works for Pratham, an education NGO helping 5 million children in Asia and Africa.

Traci L Morris, PhD (Chickasaw Nation of Oklahoma), Director of the American Indian Policy Institute at Arizona State University, has a diverse professional background, including academia; university teaching; book, article, and white paper researcher and author; and public speaker. She has worked with Native American tribes; Tribal businesses; Native American non-profits; Native media makers, artists and galleries; written a college accredited curriculum in Native American new media; and has advocated for digital inclusion at the Federal Communications Commission and on Capitol Hill. Dr Morris's research and publications on Native American media and the digital divide are focused on Internet use, digital inclusion, network neutrality, digital and new media curriculums and development of broadband networks in Indian Country. As an entrepreneur prior to her current appointment, she founded *Homahota Consulting LLC*, a national Native American woman-owned professional services firm working in policy analysis, telecommunications, education, and research assisting tribes in their nation-building efforts and working with Native Nations, tribal businesses and those businesses working with tribes. Traci has a MA and PhD from the University of Arizona's American Indian Studies, in addition to a BA from Colorado State University.

Doreen Richard Mushi is an instructional designer and an assistant lecturer at the Open University of Tanzania (OUT) within its Institute of Educational and Management Technologies (IEMT). Her main responsibility is to support and facilitate effective and efficient delivery of education through the use of educational technologies. She has been engaged in various projects which involve design, creation and evaluation of online courses; customization of learning management systems based on the targeted learning environments; and integrating learning materials with multimedia elements in order to enhance teaching and learning experiences. Doreen is also an academic at the Faculty of Science, Technology and Environmental Sciences at OUT. She works with the Department of Information and Communication Technologies (ICT) to design and teach academic courses. Furthermore, she is an aspiring researcher and her key areas of interest include software engineering, e-learning and multimedia development. She is currently the research coordinator at IEMT with focus on educational technologies and Open Educational Resources (OER). She has also done consultancy work in areas of ICT, e-learning and instructional design.

Susan O'Donnell has been researching the social, community and political aspects of digital technologies and communications since 1995. Her research with First Nation partners and communities began in 2005. She is the lead investigator of the First Nations Innovation project (http://fn-innovation-pn.com) and co-investigator on the First Mile project (http://firstmile.ca). Susan's research has included studies and evaluations of ICT and marginalized communities in Canada, Ireland, Britain, Finland, Italy, Germany and Denmark with the European Commission amongst others. Prior to her research career Susan was a senior editorial consultant in Ottawa specializing in Indigenous issues, including work with the Royal Commission on Aboriginal Peoples and Assembly of First Nations. Susan has a PhD (Communications) from Dublin City University, an MA from Cardiff University and a BA from the University of Ottawa. Since 2004 Susan has been a Researcher and Adjunct Professor at the University of New Brunswick and Senior Researcher at the National Research Council of Canada in Fredericton, New Brunswick. Contact: susanodo@unb.ca.

Gino Orticio is a Sessional Academic at the School of Cultural and Professional Learning, Queensland University of Technology (QUT). His present research interests are actor-network-theory (ANT), digital/Indigenous studies, and social theory. His PhD research focused on the complexities of digital technologies and Indigenous communities in the Philippines.

Sojen Pradhan is a lecturer at the University of Technology, Sydney. He has been teaching at this university since 2001 and is currently co-ordinating Information Systems subjects. He holds a Master's degree in Statistics from Tribhuvan University, Nepal (1995) and an MBA in Finance and IT. He is a PhD candidate and working in the area of recommendation systems and social networks.

Taija Koogei Revels, BS, is an enrolled member of the Central Council of Tlingit and Haida Indians of Alaska. Taija belongs to the Kaagwaataan Tlingit clan. Her hometown is Juneau, Alaska, and her ancestral roots are from Hoonah. She received her Bachelor of Science in Women's Studies and Pre-Medical Studies from Portland State University in 2011. Taija has worked in the fields of sexual health, domestic violence and suicide prevention, with a special focus on Native youth and LGBT youth. In 2012 she joined the Alaska Native Tribal Health Consortium HIV/STD Program Services as a Research Associate, where she works on cultural and technological adaptations of evidence-based sexual health interventions for the Alaska Native and American Indian populations across the US. She is the editor of iknowmine.org, a comprehensive Alaskan youth wellness website and affiliated social media outlets. She also serves on the National Native AIDS Prevention Center Community Advisory Committee, the State of Alaska HIV Planning Group and the National Minority AIDS Council Youth Scholars Initiative.

Stephanie Craig Rushing, PhD MPH, is a Project Director at the Northwest Tribal Epidemiology Center, affiliated with the Northwest Portland Area Indian Health Board. Dr Rushing directs Project Red Talon, a sexual health and STD/HIV prevention project, and several other adolescent health promotion projects. Stephanie contributes to mixed methods, community-based participatory research activities at the regional and national level. Current activities include developing and adapting technology-based sexual health interventions targeting AI/AN youth. She has 20 years of experience working in tribal communities and organizations. Having worked at the Northwest Portland Area Indian Health Board for twelve years, Stephanie has focused her efforts on developing culturally-appropriate educational materials and social marketing campaigns, facilitating intertribal coalitions and action plans, focusing on community health and social change. She completed her Masters of Public Health concentrating on International Health Development at Boston University, and completed her PhD in Public Administration and Policy at the Hatfield School of Government at Portland State University.

Ross Shegog, PhD, is an Associate Professor of Health Promotion and Behavioral Sciences at the University of Texas School of Public Health and holds an adjunct appointment with the University of Texas School of Health Information Sciences. His research interest is in the application of instructional technology in health promotion and disease prevention to find creative solutions to the challenges of optimally impacting adolescent health behavior. His recent projects have focused on using computer-based education and decision-support programs to enhance the management of pediatric asthma by children, families, and community physicians; computer- and Internet-based applications for smoking cessation and prevention in adolescent populations; computer-based HIV/STI/pregnancy prevention in middle school children; Internet-based violence prevention in high school and college populations; and clinic-based decision-support for epilepsy management in adult patient populations. He has a doctoral degree in Behavioral Sciences, a Masters in Health Promotion, and a diploma in Biomedical Communications from the University of Texas and a post-graduate diploma in Nutrition and Dietetics from the University of Sydney.

Lisa J. Switalla-Byers is from New Zealand. *Ko Hikiroroa te mauka* (Hikiroroa is my Mountain). *Ko Waikouaiti te awa* (Waikouaiti is my river). *Ko Kai Tahu te iwi* (I am a Kai Tahu descendent). *Ko Huirapa te hapu* (Hui Rapa is my subtribe). *Ko Puke-te-Raki te Marae* (Puke-te-Raki is my marae). In *Te Reo* (Māori Language) when we greet or introduce ourselves we recite our *pepeha*. I have learned *te reo* as a second language learner and continue to seek opportunities to practice and immerse my family within the Māori language and culture. It is really important to us that our children can walk with equal confidence within the Māori and European Cultures. As a family, we love to spend time outdoors making the most of the natural environment. I am a mainstream teacher in a New Zealand who has a passion for learning and teaching *Te Reo*. I believe in integration and using technology to promote conversational language. Having recently completed my diploma in Māori Language I am proud to support the Kai Tahu Language in Homes Project.

Olga Temple, born in Latvia, studied English Philology at the Latvian State University (1971–1974) and the Moscow State University (1974–1977), graduating with a Masters in Romanic and Germanic Philology in June 1977. She has lived and taught in Europe, Africa, America and the South Pacific. Her main academic interests have been as wide ranging as philosophy of language, syntax and semantics. Since March 2000, Olga has been teaching a wide variety of linguistics courses at the University of Papua New Guinea. This exposure helped her develop a dialectical view of language, which she has advocated since 2007. Her research and publications examine language from the dialectical perspective, which provides a convincing, albeit unorthodox, explanation of the nature, history and workings of human language. Most of her books, articles, and seminar presentations are available on her personal website: www.templeok.com.

Jennifer Torres, MPH, is a Research Coordinator with the University of Texas Center for Health Promotion and Prevention Research (CHPPR). Since she joined the CHPPR in 2007, she has worked on various projects focused on the evaluation, adaptation and dissemination of HIV, STI and teen pregnancy prevention curricula for middle school youth. Currently, she serves as a trainer for the It's Your Game (IYG) curriculum and program director for a project to adapt IYG for American Indian and Alaska Native youth. She received a BA in Anthropology from the University of Notre Dame and her MPH from the University of Texas School of Public Health. Her research interests include adolescent sexual health, underserved populations and the dissemination of evidence-based programs.

Amanda H. A. Watson, PhD, is a mobile phone researcher with the Economic and Public Sector Program in Papua New Guinea and a Visiting Fellow with the School of International, Political and Strategic Studies at Australian National University. Her research interests lie in mobile telephony, its use by communities and potential for its application in development efforts. She has published in *Pacific Journalism Review*, *Media Asia*, *Australian Journalism Review* and *The Australian Journal of Emergency Management*. www.ahawatson.com.

Jennifer Williamson is a Special Projects Coordinator for the Alaska Native Tribal Health Consortium. Jennifer joined the HIV/STD Prevention Program in the spring of 2013 to coordinate research activities and site coordinator training for a teen pregnancy prevention and healthy life skills curriculum research project. She has worked in Community Health Services at the Alaska Native Tribal Health Consortium for 6 years where she managed a worksite wellness research project with a focus on chronic disease prevention. Jennifer Williamson began her non-profit work 13 years ago at the YWCA of Anchorage Women's Health Programs, then the Alaska Health Fair, Inc. focusing at both on early detection and health promotion. She has enjoyed working collaboratively towards creating supportive environments for people to make healthy choices. She also enjoys traveling around her home state.

Alvin W. Yeo is the Director of the Institute of Social Informatics and Technological Innovations (ISITI-CRI), and a Professor at the Faculty of Computer Science and Information Technology, University of Malaysia Sarawak. He has expertise in the area of Information and Communications Technology for Rural Development (ICT4RD). Alvin has evaluated Malaysian federal-funded ICT4RD initiatives, and worked with United Nations Economic and Social Commission for the Asia Pacific (UNESCAP). He has been involved in the e-Bario Project which garnered numerous awards including the Commonwealth CAPAM Innovation award. In addition to ICT4RD research, Alvin is also active in Human Computer Interaction research, specifically in software internationalization, multimodal interaction, gaze-based systems, and the use of ICTs for the preservation of Indigenous languages. He currently heads the Sarawak Language Technology (SaLT) Research Group, is involved in numerous national projects, and two European Union projects. Alvin earned his PhD from the Computer Science Department, University of Waikato, New Zealand.

Tariq Zaman earned his PhD from the Faculty of Computer Science and Information Technology, University of Malaysia Sarawak (UNIMAS). Tariq's PhD project *eToro* garnered 6 international, 1 national and 1 university level award. His research for his PhD was in the formulation of cultural protocols with and for Penan Indigenous communities in Malaysia. Working as a Postdoctoral fellow in the Institute of Social Informatics and Technological Innovation at UNIMAS his research is funded by 2014 ISIF Asia grants. He has recently been selected as research associate for the IPinCH Project Simon Fraser University, Canada. His interests include Indigenous Knowledge Management (Governance), Indigenous Communities, Rural ICT, Community Informatics and ICT4D. His projects and publications equally reflect the multiple voices of Indigenous wisdom and cultural understanding by converging local, scientific, traditional and cultural knowledge.

Index

For Product Safety Concerns and Information please contact our EU
representative GPSR@taylorandfrancis.com
Taylor & Francis Verlag GmbH, Kaufingerstraße 24, 80331 München, Germany